I WANT TO BE A MATHEMATICIAN

Paul R. Halmos

I WANT TO BE A MATHEMATICIAN

An Automathography

With 43 Photographs

Springer-Verlag
New York Berlin Heidelberg Tokyo

Paul R. Halmos
Department of Mathematics
University of Santa Clara
Santa Clara, CA 95053
U.S.A.

AMS Subject Classifications: 01A65, 01A70, 01A80

Library of Congress Cataloging in Publication Data
Halmos, Paul R. (Paul Richard)
  I want to be a mathematician.
  Includes index.
  1. Halmos, Paul R. (Paul Richard),
  2. Mathematicians—United States—Biography.
  I. Title.
  QA29.H19A35   1985   510′.92′4 [B]   84-26688

Typeset by Composition House Ltd., Salisbury, England.
Printed and bound by R. R. Donnelley & Sons, Harrisonburg, Virginia.
Printed in the United States of America.

9 8 7 6 5 4 3 2 1

ISBN 0-387-96078-3 Springer-Verlag New York Berlin Heidelberg Tokyo
ISBN 3-540-96078-3 Springer-Verlag Berlin Heidelberg New York Tokyo

Dedication

To Ambrose, Doob, and von Neumann, who, without knowing it, made me what I am today.

Overture

This is an automathography, a mathematical biography written by its subject. It is definitely not a mathematics book, and, even more definitely, it is not the story of my origins and my life. Sure, I had parents (two), and wives (two, one at a time, the present one for forty years), and cats (eight, two at a time, the present two for three years). I had and have many faults, and, I'm pretty sure, a few virtues. I like Haydn, long walks, Nero Wolfe, and dark beer, and for a few years I tried TM. All that is true, but it's none of your business—that's not what this book is about.

The book is about the career of a professional mathematician from the 1930's to the 1980's. It is presented, more or less, in chronological order, from high school to retirement, but its sections are organized by substance rather than time. It tells about the University of Illinois, the Institute for Advanced Study, and the universities commonly identified as Chicago, Michigan, and Indiana. It expresses prejudices, it tells anecdotes, it gossips about people, and it preaches sermons. It tells about taking prelims, looking for a job, writing a book, travelling, teaching, and editing.

Fifty years ago I was cocky, iconoclastic, eager, ambitious, in a hurry, ignorant, insecure. I have slowed down, mellowed (?), and learned a few things. To some extent the book is from the me of today to the me of yore, revealing some of the secrets that I desperately wanted to know then.

P.R.H.

Thanks

Six people read every word of a typescript version of this book. Their comments (cut this out—who was he?—not so bad—tone it down—are you sure?) cheered me up, or made me mad, but, in either case, spurred me on. They are John Ewing, Elena Fraboschi (who typed it all, much of it several times), Leonard Gillman, Virginia Halmos, Walter Kaufmann-Bühler, and Peter Rosenthal. Everything they said helped, and I am truly grateful for their help. Thanks a lot.

Table of Contents

Part I. Student

CHAPTER 3
Graduate school

CHAPTER 4
Learning to study

CHAPTER 5
Learning to think

CHAPTER 6
The Institute

Part II. Scholar

CHAPTER 11

The fabulous fifties

Part III. Senior

CHAPTER 12

How to teach

CHAPTER 13

To Sydney, to Moscow, and back

CHAPTER 14
How to do almost everything

CHAPTER 15
Service, one way or another

Coda

My expository style relies heavily on the exemplary singular, and the construction "everybody . . . his" therefore comes up frequently. This "his" is generic, not gendered. "His or her" becomes clumsy with repetition and suggests that "his" alone elsewhere is masculine, which it isn't. "Her" alone draws attention to itself and distracts from the topic at hand. "Their" solves the problem neatly but substitutes another. "Ter" is bolder than I am ready for. "One's" defeats the purpose of the construction, which is meant to be vivid and particular. "Its" is too harsh a joke. Rather than play hob with the language, we feminists might adopt the position of pitying men for being forced to share their pronouns around.

PART I
STUDENT

CHAPTER 1

Reading and writing and 'rithmetic

Words

I like words more than numbers, and I always did.

Then why, you might well ask, am I a mathematician? I don't know. I can see some of the reasons in the story of my life, and I can see that chance played a role at least as big as choice; I'll try to tell about all that as we go along. I do know that I wasn't always sure what I wanted to be.

The sentence I began with explains the way I feel about a lot of things, and how I got that way. It implies, for instance, or in any event I mean for it to imply, that in mathematics I like the conceptual more than the computational. To me the definition of a group is far clearer and more important and more beautiful than the Cauchy integral formula. Is it unfair to compare a concept with a fact? Very well, to me the infinite differentiability of a once differentiable complex function is far superior in beauty and depth to the celebrated Campbell–Baker–Hausdorff formula about non-commutative exponentiation.

The beginning sentence includes also the statement that I like to understand mathematics, and to clarify it for myself and for the world, more even than to discover it. The joy of suddenly learning a former secret and the joy of suddenly discovering a hitherto unknown truth are the same to me—both have the flash of enlightenment, the almost incredibly enhanced vision, and the ecstacy and euphoria of released tension. At the same time, discovering a new truth, similar in subjective pleasure to understanding an old one, is in one way quite different. The difference is the pride, the feeling of victory, the almost malicious satisfaction that comes from being first. "First" implies that someone is second; to want to be first is asking to be "graded on the curve". I seem to be saying, almost, that clarifying old mathematics is more moral than finding new, and that's obviously silly—but let me say instead that insight is better without an accompanying gloat than with. Or am I just saying that I am better at polishing than at hunting and I like more what I can do better?

I advocate the use of "words more than numbers" in exposition also, and in the exposition of mathematics in particular. The invention of subtle symbolism (for products, for exponents, for series, for integrals— for every computational concept) is often a great step forward, but its use can obscure almost as much as it can abbreviate.

Consider, as an example of the point I am trying to make, a well-known elementary theorem with a well-known elementary proof. The theorem is Bessel's inequality in inner-product spaces; it says that if $\{x_i\}$ is an orthonormal set, x is a vector, and $\alpha_i = (x, x_i)$, then $\sum_i |\alpha_i|^2 \leq \|x\|^2$. A standard way of presenting the proof is to write $x' = x - \sum_i \alpha_i x_i$ and then compute:

$$
\begin{aligned}
0 \leq \|x'\|^2 = (x', x') &= (x - \textstyle\sum_i \alpha_i x_i, x - \sum_j \alpha_j x_j) \\
&= (x, x) - \textstyle\sum_i \alpha_i(x_i, x) - \sum_j \bar\alpha_j(x, x_j) + \sum_i \sum_j \alpha_i \bar\alpha_j(x_i, x_j) \\
&= \|x\|^2 - \textstyle\sum_i |\alpha_i|^2 - \sum_i |\alpha_i|^2 + \sum_i |\alpha_i|^2 \\
&= \|x\|^2 - \textstyle\sum_i |\alpha_i|^2.
\end{aligned}
$$

Rigorous and crisp, but not illuminating. To my mind the ideal proof is the one sentence: "Form the inner product of $x - \sum_i \alpha_i x_i$ with itself and multiply out." That should be enough to get the active reader to pick up his pencil and reproduce the chain of equations displayed above, or, if he prefers, to lean back in his chair, close his eyes, and look at those equations inside his head. If he is too lazy for that and wants more of his work done for him, the sentence can be padded. "The result is positive and it consists of four terms: the first is $\|x\|^2$, orthonormality implies that the last is $\sum_i |\alpha_i|^2$, and both of the cross-product terms are equal to $-\sum_i |\alpha_i|^2$. Cancel one of them against the positive sum; what remains is the asserted inequality."

As far as length is concerned, the words use about the same amount of paper as the symbols. The words are clearer, and they have, moreover, the advantage that they can be communicated during the walk back from class to the office, with no blackboard or chalk handy. They are also more likely to deepen insight by leading to the discovery of the appropriate general concepts and context (such as the pertinence of projections).

I do not advocate but I observe in myself a liking for almost everything about words. I like to play with words—anagrams, James Thurber's super-ghosts, Jotto, and Scrabble—and I enjoy their etymologies—where they came from, what they used to mean, how they have changed, and precisely what they mean now. What I like used to be called philology—the love of words. The modern word is "linguistics", and I don't like it. It is too much like mathematics, and, what's worse, it is like the "numbers" kind of mathematics. Instead of tackling the difficult abstract semantic concepts of natural languages (such as "meaning" and "interpretation"), it is preoccupied almost exclusively with the symbolic calculations of formal syntax.

Not all mathematicians like words. I suspect but I have no evidence for a correlation: algebraists, such as Artin, do; analysts, such as Courant, don't; and geometers split according to whether they are of the algebraic (Zariski) or analytic (Lefschetz) variety. Another correlation I wish I could prove is that the ones who do like words become famous and beloved for their clear explanations, whereas the ghosts of the others are muttered and sworn at by exasperated students every day. Whether I can prove the general theorem or not, the correlation, the analogy, is emotionally convincing to me. Set theory and algebra seem to stand to arithmetic and differential equations in the same ratio as words stand to numbers.

Do I dare generalize? I want to say that as an indication of mathematical ability liking words is better than being good at calculus. Many graduate-school advisors have noted that an applicant for a mathematics fellowship with a high score on the verbal part of the Graduate Record Examination is a better bet as a Ph.D. candidate than one who did well on the quantitative part but badly on the verbal.

Books

How does a preference for "words more than numbers", or the other way around, come about? I think it's either congenital, or, in any event, acquired very early in life, and once we have acquired it, in either direction, we're stuck with it. In my own case the tendency was beginning to be visible when I was 4 and was solidified by the time I was 8. I learned to read early and easily, and then to write; arithmetic was not hard, but it came later and slower, and was neither conspicuously easy nor a source of great pleasure.

My family lived in a third floor apartment, in Budapest, that faced out on a busy street (now called Lenin Boulevard). It was an exciting street—colorful, crowded, noisy. There were many shops—a glamorous hardware store displaying shiny knives behind its huge plate glass front, several bookstores with books of many colors piled and strewn around, coffee houses with grouchily servile waiters carrying white napkins on their black left sleeves, and stores full of toys and candy and crutches and clothes and shoes and watches. The sidewalk was broad, and milling crowds of people separated the shop windows from the curb-side trees and scales (your weight for a penny) and newspaper kiosks and taxi stands. The crowds seemed always to be there—they were there when I went to school early in the morning and they were there on the rare occasions when I was brought home late at night from an excursion or from a movie. Later, when I grew up, went to Hungary as an American tourist, and was out *real* late at night, the crowds were still there. The lights were bright and gypsy music could be heard from the coffee houses.

The boulevard was broad—it had room for two trolley lines and two more lanes of traffic—and the electric trolleys (what fun to put a penny on the track and pick it up squashed flat after the metal wheels ran over it) and the taxicabs (horsedrawn when I was very young and motorized later) were never quiet. The corner at Lenin Boulevard and Rákoczi Road is not exactly like Broadway and Seventh Avenue, but it's a lot more like that than it is like downtown Scarsdale. When at the age of 13 I was moved to a residential neighborhood in Chicago, it took me a while to get used to streets that were quiet in the day and almost deserted at night.

Long before I started school I entertained myself by gaping out the front windows, enjoying the action, and puzzling about the large letters that spelled the names of the shops. "What does that say?", I demanded to know; "what is the name of that letter there?". The answers to my questions didn't exactly constitute a systematic course in reading. The procedure would probably horrify the proponents of both the look-say and the phonic schools, but it worked. The result of my insistent curiosity was that by my fourth birthday I not only knew the alphabet but I could read, and I started working through my grandfather's large private library; before long the lending library a few doors down the street became useful too. Grandfather's books were in glass-covered bookcases, under lock and key, and the keys were kept in a drawer, which was locked by a key that the old man kept on his person. He let me have any book I pointed to, no questions, no censorship, but he had to be there to open things up.

I read a lot. When I started in school, at the usual age of 6, I had homework to do—yes, homework, every day, from the age of 6 on—but it left ample time for reading, during breakfast, before supper, in the bathtub, and, best of all, in the evening. Before lights out I would read in bed, listening with earphones to radio music from Vienna and Paris on a small crystal set. I read and enjoyed the Grimm stories, and Hans Christian Andersen, and Aesop, and La Fontaine—and, not much later, *Scaramouche* and *The Prisoner of Zenda* (all in high literary Hungarian). There was Hungarian literature too, of course, Jókai and Karinthy, and both good stuff and exciting trash from all over the world. Captain Nemo and Nick Carter, Sherlock Holmes and Arsène Lupin were all among my friends. American literature was represented by James Fenimore Cooper and Mark Twain; I probably knew as much about Chingachgook and Tom Sawyer as most boys in Kansas City.

Looking back on those days now, 50 to 60 years later, I appreciate and admire the atmosphere. Culture was not ridiculous, unusual, or sissy— it was taken for granted. Books and music were regarded as a part of everyone's common heritage. At school we discussed the latest Nick Carter, sure, but we talked about d'Artagnan too, and we could tell each other that we had just read something without being considered odd.

Writing

I knew how to read when I started grade school, but I still had to learn to write. Freud would perhaps deduce that it must have been a traumatic experience for me: I remember almost nothing of it. By concentrating I can bring forth a dim picture of a room, not very large, no larger than my living room now, with some faceless children seated in it, and a faceless woman, the teacher. We were told to use specially ruled paper. The small letters, like e and o, had to fit between the two middle lines, close together; the up and down strokes, as in t and p, reached to the top and bottom lines farther away. The result looked like this:

It was hard for me, and I felt insecure about it. I was absent the day capital M was taught, and I had to invent my own way of doing it— that bothered me for a long time.

When I came to America I quickly noticed that my handwriting was "different"—it was legible, but it had, in an odd way, a foreign accent. Later, when I started teaching, that became a small but inconvenient communication handicap. I re-taught myself so that, at least as far as handwriting was concerned, I could pass for someone born in Omaha.

So much for writing in the mechanical sense. The other sense, writing as a way of saying something, is much more important, much harder, and took much longer to learn. In the first such writing that I remember I was trying to express my patriotism.

The main focus of patriotic fervor in those days was the treaty of Trianon, which carved up Hungary after the first World War and left it a third as large as it was before. "No, no, never!" was the national motto. In school we recited the Hungarian equivalent of the pledge of allegiance to the flag every day. I loved it; I felt indignant about Trianon, and I wished for the return of the empire. I described my politics by saying that I was a royalist. When we, in my class, were told to write a patriotic essay, I won a prize with mine. I believed in the cause, and I already felt stirrings of my subsequent love for words. The result was purple prose, and I was proud.

How did it come about that I love writing and do it easily? I can only guess. The three main influences I am aware of are that I read a lot, write a lot, and love languages—and I suppose at bottom it all comes down to liking words.

I am sure that reading Voltaire and Cervantes, or for that matter Dumas and Schnitzler, influence and improve my mathematical style and help make me a better teacher than I would be otherwise. Reading enlarges my diction and my world view and my understanding of the audience I want to reach. I am horrified at my friends who proudly announce that they never read fiction. Some of them call all fiction lies, or, in a more charitable phrase, made-up, artificial stories instead of true, real ones. The true stories are about oppression and depressions and wars, and about campaigns and elections and corruption at high levels; they are much more "significant" and "relevant" than Jane Austen and Dickens and Galsworthy and Somerset Maugham. No, no, no! I am emotionally opposed to that point of view and operationally opposed. I am not advocating withdrawal from the world and retirement to a hermitage—but I insist that I couldn't be a writer of any sort without having read the Old and the New Testaments (King James version), and *Alice in Wonderland*, as well as *Macbeth*, *The Declaration of Independence*, and *Man and Superman*.

As for writing—I write all the time, and I have done so as far back as I can remember. I write letters, from time to time I have kept diaries, and I write notes to myself, explicit ones with sentences, not just "try power series expansion" or "see Dunford–Schwartz for $p > 1$". I think by writing. In college I wrote notes—that is, I transcribed abbreviations scribbled in class to legible and grammatical sentences. Later, when I started trying to prove theorems (the acceptably low-key phrase for the more stuffy-sounding "doing research"), I would keep writing, as if I were conducting a conversation between me and myself. "What happens if I restrict to the ergodic case? Well, let me see. I have already looked at what happens when S is ergodic, but the useful case is when both S and T are... ."

I am tempted to preach, and to say: this is the right way, do it this way, do it my way or else you'll fail—but all I can be sure of is that this is the right way for me, there is no other way that I can do things.

Languages

Reading and writing and languages are the three factors that contribute to my writing, and I have explained about reading and writing. As for languages, my feelings might be due, at least in part, to having learned a foreign language before I learned my mother tongue. I was taught German by the Tyrolean nurse who joined our family when I was less than two years old, and I have been wishing all my life that it had stayed with me. I can hear German just fine (especially Austrian), but I cannot produce it: except for the technical language I learned later, my vocabulary is that of a four-year old, and verbs and the der-die-das are beyond me.

Latin entered my bloodstream when I was about 6. One of my brothers, five years older than I, mumbled his memorization lessons at the communal table where we frequently played and studied, and some of them rubbed off on me. I still remember a list of Latin prepositions (ante, apud, ad, adversis, circum, circa, citra, cis, ...): are they the ones that go with the accusative?

Hebrew came a little later. Religious instruction in school (separate for Catholics, Protestants, and Jews) had two parts: one of them was fun and the other dreadful. The fun part was the ineffectual old cantor with small chin whiskers who sat at a harmonium and taught us to sing the Hebrew words to some hymns. He was scared of us, I think, but the peppy tunes carried the day, and we almost looked forward to our weekly lesson. I still remember some of the tunes and the first couple of lines of one of the hymns. The dreadful part was the Hebrew lesson. It was taught by a severe, smooth-shaven, round-faced man, who took himself seriously. He didn't teach us any Hebrew, he didn't even try; that wasn't what the job description called for. What he did was convert us into machines that could vocalize printed Hebrew text. I could look at a page of the Old Testament, printed in punctuated Hebrew, and without knowing the meaning of a single word, I could convert the symbols that entered my eye into sound that issued from my lips.

Both in school, and at home too, there was a cosmopolitan emphasis. French (which somehow I never got around to learning) was not just a subject, but something alive—it was what my father spoke on the telephone when he didn't want his nosy kids to understand.

I feel that even baby German, second-hand Latin, mechanical Hebrew, and secret French enlarged my literary field of view—if nothing else, they taught me to sense similarities and differences among various ways of expressing the same thought.

Some graduate students nowadays object to being made to learn to read two languages as a Ph.D. requirement. "Why should we learn about flowers and families and genitives and past participles?—all we want is to read last month's Paris seminar report." Some go further: "Who needs German?—for me Fortran is much more relevant."

Horrors! I am upset, and I predict that the result of such anti-linguistic, anti-cultural, anti-intellectual attitudes will lead to a deterioration of international scientific information exchange, and to a lot of bad writing. Every little bit I ever learned about any language was later of help to me as a writer. That is true of the Danish and Portuguese and Russian and Rumanian that I learned for specific mathematical reasons, but it is true also of the hint or two of Greek and of Sanskrit that I managed to be exposed to. I have always rued that I was never taught Greek; every ounce of it would have paid off with a pound of linguistic insight. In the course of the years I managed to pick up quite a few Greek root words; my source for them was my shelf of English dictionaries, especially the American Heritage and the second edition of Webster. I feel that I need to look up the etymologies of words before I can use them precisely, and I know (a small matter, but here is where it belongs) that the reason I have no trouble spelling in English is that even a nodding familiarity with other languages makes me aware of where most of the difficult words come from.

To give the devil his due, I admit that substituting Fortran for German is only 90% bad, not 100. What it loses in the understanding of culture and mastering the *art* of communication, it gains in meticulous attention to detail and moving closer to mastering the *science* of communication. A knowledge of the theory and practice of formal languages might be a help for writing with precision, especially to students whose talents are not mathematical, but it is of no help at all for writing with clarity. The distinction is sometimes ignored or even argued away, but that is a sad error—there is all the difference in the world between an exposition that cannot be misunderstood and one that is in fact understood.

Numbers

I learned to count at home, meaning that I learned the names of the members of a small initial segment in the set of positive integers; I could recite them in order, and I could tell the difference between five cookies and seven. One of my earliest arithmetic experiences put me in contact with my glamorous but shadowy father; it had to do with the clock. He cut out a cardboard circle, marked it with the roman numerals I to XII, pinned hands to the center, and taught me to tell time. The lessons were in the large room that was his domain and where it was a rare privilege to be allowed to enter. The room served as his office (he was a doctor) and his study (during rare moments of leisure he would work chess problems at his desk) as well as his bedroom. The bed was a wide and luxurious couch in the daytime, covered by a gaudy many-colored Persian rug. The room was sunny and cheerful with three large windows on one wall and glass-covered bookcases facing them. The clock lessons lasted over a week. I finally got it right and I was thrilled and proud to be able to say "three quarter nine" when the clock showed 8:45.

I heard about trigonometry too in those days, but it was a great and scary mystery. My brother, the one who mumbled Latin prepositions, memorized opposite versus hypotenuse and various identities; I could never remember any of that then. I don't even know the Hungarian words for the concepts anymore. Many years later, when I learned the subject, I had to conquer the fear before I had a chance to see that really it was all very easy.

In elementary school I learned arithmetic, meaning such things as the multiplication table through 9×9. I wasn't especially good at it, and certain products, such as 6×7 and 7×8, never became automatic for me—I keep having to stop and think them through. (Were they taught the same day as capital M?) The ones I do know (I am quite sure that $3 \times 7 = 21$) seem to be in my auditory memory; I do not so much "see" them or "understand" them as I *hear* them.

I hear them, moreover, in my mother tongue, in Hungarian, and sometimes that disturbs me. Although I have been speaking, reading, writing, and dreaming English for well over a half century, the arithmetic I hear is always in Hungarian, and when I balance my checkbook and mumble the numbers, I mumble them in Hungarian. Not "nine and eight is seventeen, carry one", but "kilenc és nyolc az tizenhét, marad egy". Why? Perhaps because arithmetic is something all of us mumble every day— most other subjects can often be put on the shelf for a week or a month at a time.

My Hungarian education in arithmetic was a small handicap in my early years of teaching in America. To see what the trouble is, look at these two layouts:

$$
\begin{array}{r}
1932 \times 473 \\
\hline
7728 \\
13524 \\
5796 \\
\hline
913836
\end{array}
$$

and

$$
\begin{array}{r}
1963 \div 29 = 67.68 \\
-174 \\
\hline
223 \\
-203 \\
\hline
200 \\
-174 \\
\hline
260 \\
-232 \\
\hline
28
\end{array}
$$

It doesn't take long to figure out that they are a long multiplication and
a long division, but they are "different"; they don't look the same
as the layouts in this country. That's how I was taught them, and my
freshman algebra students, troubled enough at having to divide 1963 by
29, couldn't cope with having to do it in a way that seemed backward to
them. Here too, just as with handwriting, I had to re-learn something.
I did it, and I became that much more of a Nebraskan. I confess, however, that even now, when I do long multiplication or division in public,
I must consciously keep myself from reverting to type.

Still another arithmetic handicap, but not one that ever caused me
a moment's worry, is that at school I was never taught some of the more
sophisticated algorithms, such as the one for square roots. Long after
school, a friend once showed me how to take square roots, and I remembered the technique for a couple of weeks.

When my cousin and I were 8 and 9 (he's a year older than I),
our grandfather used to race us against one another in the multiplication
of three-digit numbers (in our heads of course—no paper and pencil).
I neither loved nor hated the game. I liked to win, and I liked getting the
coin that was awarded to the winner. (Is that when I turned pro?)
We were about equally good.

Study or worry

Are all children scared of school? I sure was—not so much in grade
school as in the next step, the secondary school called Gymnasium (with
the G pronounced as in GIVE). "Scared" is too soft a word for what I
felt; a better description is a continual nagging worry, apprehension, and,
yes, fear. I was never beaten, but the overpowering moral force of
sarcasm and public disapproval was enough to motivate being "good".
To be "bad" meant not so much to be dirty, or noisy, or destructive, or
cruel—mainly it meant not to be prepared to recite.

The Gymnasium curriculum was an eight-year one, including eight
years of Latin, four years of Greek, five years of mathematics, and generous amounts of natural science, geography, history, and literature. I
entered on schedule, when I was 10 years old, but I didn't stay for
the gruelling comprehensive examination at 18, called the "matura",
because I was transplanted at the age of 13. The emphasis in Gymnasium
was on what in the U.S. used to be called a liberal arts education—culture, broad and deep, with absolutely no vocational or professional training.

There were textbooks, heavy ones, to be lugged back and forth between school and home, and many of us had what I now view as a

deplorable habit: on Tuesday morning we would tear out the 4 or 5 pages pertinent to Tuesday's class work, and leave the rest of the book at home. So much less to carry—and it had to be carried, because we could be called upon, in class, to translate from p. 87 of the Latin book, or to solve the problems on p. 43 of the algebra book. The chance, the threat, of being among the ones who had to recite on Tuesday was always with us. We gambled of course. We knew that the probability of being picked again on Tuesday after having recited on Monday was small, and we tried to guess the habits of the teachers—one would go alphabetically, where another would keep harassing the boys seated in the back, the dummies.

The recitation could sometimes be short and done from your seat ("Who was the successor of King Matthias?", "Which is the longest river in Italy?"), or long, up front (write your Latin translation of an assigned passage on the board, solve an assigned quadratic equation by completing the square). Unacceptable work frequently met with public ridicule, or being kept after school, or extra homework, and, of course, a bad grade for the day, which then led to a bad report card to be taken home at the end of the year.

The worst part of a bad recitation was feeling like a fool, standing there with everybody looking at you. The students were no better and no worse than students usually are. They weren't geniuses sneering at a dope, and their attitude outside class was in great part the customary callow anti-intellectualism of students everywhere—candy bars and soccer and movies and girls were far more important than lessons and homework and recitations and grades—but when you were being grilled and doing badly, silent or stuttering, you felt that everyone held you in contempt, and you'd never be able to hold your head up again.

I was not very good at Latin and even worse at geography. The teachers in those subjects frightened me more than the ones in literature and mathematics; I don't know which was the cause and which the effect. All the more I recollect with a thrill of warm pleasure, almost as alive after the passage of 55 years as it was then, a time when, for once, I did the right thing. I was writing a Latin sentence on the board, and at the end of a line I came to a word that I had to divide between syllables. The rules for the syllabication of Hungarian words are not the same as the ones our text said held for Latin, and I remembered to do it the Latin way. Some of the class—who never knew, or who forgot—snickered at my "mistake", and I wasn't so secure that I could keep from worrying. The teacher, Mr. Esztergomi (I swear that was his name—I'll never forget him), glanced at what I'd written and growled at the snickerers: "He's doing all right"—and he made my day. My century.

In my memory (nostalgia?) I see my teachers of those days, the ones I was afraid of as well as the ones I liked, as serious people who really knew and cared about their subjects. Even in my subteens I began to get the idea that learning was good—better than just studying and getting good grades.

Physics was a spectator sport; we were shown some prepared experiments, such as the Magdeburg hemispheres, and we were told how pumps work. I was about 12 when on Christmas Eve, in a scheme of my own invention, designed to calm myself and to fill the agonizing time till present opening, I made a very careful and beautiful engineering drawing of a water pump. I had just learned about it in school.

The mathematics I was being taught was elementary algebra (parentheses, quadratic equations, two linear equations in two unknowns)—it was neither hard nor conspicuously easy for me. I was good enough to be above average and to be noticed for it. One summer the father of a schoolboy of my own age found me at a resort and paid me (!) to tutor his son. The boy's problem was that he flunked mathematics the preceding year, and he needed to pass a special exam before he could be allowed to register next fall. The pay was not high (an hour's wage bought a slice of cream pie), but it was the first I ever earned (as opposed to won). The boy passed.

Learning English

My father, a widower, emigrated to America when I, his youngest son, was 8 years old. When he got established, he remarried, presented us with two step sisters, and began to import us: first my two brothers, and later, almost immediately after he became a naturalized citizen, myself. In view of my father's citizenship I became an instant American the moment I arrived, at the age of 13.

In anticipation of the move, arrangements were made for us to learn English. They didn't work very well. John Fegan, an elderly Irishman, came once a week to teach us. I have no idea why he was in Budapest and how he made contact with us in the first place. I suspect he was a remittance man of some sort, and I know for sure that he had no talent for teaching languages, or, for that matter, learning them. In all the years he lived in Hungary he learned only three words of Hungarian. One of them was "transfer" (for the streetcar) and the other two were "scrambled eggs".

Fegan's principal teaching technique was to give us set pieces that he cut out of English magazines, and have us copy them. They were the kind of articles that nowadays you find in the Reader's Digest. One of them, for instance, was about the sinking of the Titanic; I learned much more about the subject than I wanted to know. Having copied the piece, I was then expected to read it out loud, have my pronunciation corrected as I went, and answer questions. When did the disaster occur? (April 15,

1912); how many passengers died? (1500). As far as I was concerned the procedure was like school without the redeeming feature of sociability with my friends. I did as little as possible, and I learned very little indeed.

Five minutes after I arrived in New York (accompanied by a friend of my father's who was emigrating at the same time), a kind customs official thought I was at a loss and asked if he could help. I could just barely understand the word "help" and I proudly answered: "We are waiting for a..."—the word "porter" got away from me. My pronunciation was atrocious, but he got it, smiled, and supplied my missing word. (The sentence I was trying to say had come up in one of Fegan's lessons.)

Less than a month later I was attending my first high school class at Waller High, on Chicago's near northside. Somebody showed me to a classroom in which, I still remember, a very nice man was talking about physics. I listened dutifully, but I didn't understand a single word of what was being said. At the end of the hour all the students got up and went to some other room. I felt stupid and awkward—no one had explained the procedure to me—what was I to do?—where was I to go? I just sat there, bewildered. The teacher, Mr. Payne, came over to my seat and asked me something. I shrugged my shoulders helplessly. We tried various languages. A few Latin words, a few French words, and some sign language—he finally succeeded in telling me that I had to go to Room 252. I went to room 252, and, somehow, I survived my first day in an American high school.

Learning a new language is easy when you're 13, even if you are self-conscious and proud and shy—easy, but not always pleasant. I remember one incident from a couple of months later—I remember it with a shudder of shame and anger. We were taking an exam, and the boy next to me asked the one in front for the loan of an eraser. When he was finished using it, I too needed it, and with an apologetic grin at the owner I borrowed it from the borrower. The owner glared at me and muttered: "Goddam foreigner". I'll never forget that—I feel bad and bitter and angry each time it comes to mind. I see him and I hear him, and he probably explains one of my irrational prejudices—he must be the major cause of my instant dislike of round-faced, square-shouldered "football player" types.

Three months later I spoke rapid, incorrect, heavily accented, colloquial English. Reading was easier and quicker. The big thrill came on a day during the Christmas vacation of 1929, when I first read a book in English all the way through (it was a Sherlock Holmes story), not for study but for joy, not with strain, as for a lesson, and without having to stop to look up words. Joy!

Then there was the accent. I was a foreigner, with or without pejorative adjectives, I felt like one, and I sounded like one. I was studying in a temporarily deserted classroom one day, when a woman stuck her head in and asked me where the teachers' room was. Hearing my answer, she didn't go there, but came in instead. She was working for the Board of

Education; her job was to find foreigners, like myself, and help them. She had linguistic training, and she was going to use it to make me sound American. She didn't succeed, but she helped a lot, and I never stopped being grateful to her.

I used to say VAHT (for WHAT); the distinction between v's and w's was too subtle for me. My accent de-programmer cured most of that. Say "HOO", she said. Say "AH", she said. Say the letter "T", she said. Now say them one after another, spaced out, nice and slowly: HOO-AH-T. Faster. Still faster. Quickly now—and, sure enough, I said WHAT, just as if I'd been born on Fullerton Avenue, four blocks away.

The family language at home was English almost all the time, and my knowledge of Hungarian froze at the 13-year-old level. I might have been a clever 13-year-old and a well-read 13-year-old, but I was 13 years old just the same. Most of the words of philosophy and politics (such as insight, morality, to vote, to govern) were not in my active daily vocabulary, and neither were the adult words of practical life (such as salary raise and building contractor). Although I can still speak Hungarian fast and with a nearly perfect accent, my control of words and their grammar is feeble and causes me to feel awkward and uncomfortable.

That I have not completely lost my foreign accent in pronouncing English is the bane of my existence. I tried. I tried very hard. It would be comforting sometimes to be inconspicuous, to be unnoticed, to meld into the crowd. I would like to be just one of the boys, and not to be asked "and where are *you* from?", and "how long have you been *here*?". I never again want to be told, "Oh, you have such an *interesting* accent". I may be too sensitive about it, but I always hear "and where are you from?" as "you are a stranger and we don't much want you around here". I have spent almost my entire life in the U.S.A., and in every way—in behavior, in language, in taste, and in culture—I feel like an American. I don't like being described as Hungarian. I am not one of *them*, I am one of *us*.

High school

The eight-year Gymnasium curriculum came after four years of elementary school—just the other way around from the American way, in which four years of high school followed eight grades. The difference made some chicanery possible and, in a sense, saved four years of my life. My stepmother was always convinced that everything in her life was the solution of an extremum problem: the poor people she knew were poorer than all others and the rich richer; her university offered the best law degree; her butcher had the best meat; her children (and that came to include me) were the cleverest in Chicago. With the help of an amiable acquaintance at the Hungarian consulate she had my Gymnasium records translated (course for course, grade for grade), and the official translation certified that I had completed three years of secondary school. Result: at 13 I was admitted to high school, standing somewhere between the juniors and the seniors. She was gambling that I was clever enough to skip four grades.

The chicanery made my education curiously spotty (I never had a course in biology of any kind, but I learned to solve quadratic equations over and over again); other than that it did no harm. It put me in a class where everybody else was older than I, but that didn't matter. I was tall for my age, and cocky (according to my dictionary that means "cheerfully self-assertive or self-confident"), and as soon as the language problem was solved I had none of the social problems that the age difference might have caused. For the next ten or fifteen years I was usually the youngest person at most parties; it has been quite a wrench, during the past ten or fifteen years, to get used to being the oldest.

The language problem was greatest at the beginning, of course, and unfortunately at about the same time I had to take the course, namely civics, where language was the most important. To make matters worse, Mr. Throckmorton, the teacher, had an accent I found difficult (Welsh?), and we presently came to a friendly understanding. I would just have to repeat civics another semester; for this one, I was to use the class as a study hour, and neither of us would bother the other. (The routine of "drop-and-add" didn't exist at Waller High.)

History, with Miss Talbot, came next semester, and I had no trouble with it—in fact I loved it. The course was about American history, and constitutional law was a part of it. I couldn't get enough. I organized a petition that an additional course about the constitution be offered the semester after that, but it came to naught.

The mathematics courses I was exposed to were called algebra and geometry. Algebra began with "high school algebra", which I liked (and most of which I already knew). I continued with "advanced algebra" and since I did well in that too, I went on (still in high school) to "college algebra". In college, by the way, I had to begin with "freshman algebra" in order to become prepared for "higher algebra" (matrices), after which came "modern algebra" (permutation groups and polynomials). The culmination of the sequence came in graduate school with a course called "introduction to algebra" (rings and fields, as in van der Waerden). In the early stages of the process I learned, in high school, about complex numbers, in two different courses. They didn't seem especially mysterious. To my astonishment and dismay high school students do not learn complex numbers nowadays, possibly because high school teachers don't know them. The students I met in a recent graduate course never heard of De Moivre's theorem; even absolute values and complex conjugates made them feel insecure.

Euclidean geometry was capital fun. We were challenged to discover the proofs of "obvious" statements and to write them down: step 1, reason, step 2, reason, etc. I loved it and I think I was better at the procedure than my teacher.

The best mathematics I learned was in physics class. Mr. Payne told us, for instance, about those famous rabbits that keep producing other rabbits with monotonous chronological regularity. (Surely everyone knows about them, but, just in case, here is how it goes. According to the story, each pair of rabbits, one male, one female, starts reproducing two months after they are born, and then they keep giving birth to another such pair regularly once a month. No rabbit in the story ever dies. Question: how many pairs of rabbits are there alive after a year?) The story leads to what I later learned to call the Fibonacci sequence. I didn't solve the problem, but it made a deep impression on me.

Was most of my high school mathematics a waste of time? Would I have been better off if someone had designed an intelligently organized curriculum with no waste motion? Should I have been exposed to nontrivial calculations, theorems, and proofs at the age of 14? Perhaps. I am not sure—but it's worth a thought.

Where was I to go to college? My father said he could afford to send me to one of two places. (I later learned that his income in those days was about $6000 a year.) I could go to the University of Chicago, right there in town, or to the University of Illinois, the state university in Urbana. I didn't need as much as a minute to decide that one: Urbana was 130 miles away from the parental home, Urbana meant freedom. I chose Urbana.

I have been kicking myself ever since. I enjoyed Urbana, and freedom, and I became a loyal Illini. The University of Illinois is not a bad school, but the University of Chicago had (and deservedly still has) the reputation of a great school. At Chicago my education would have been of a much higher quality, and the subsequent steps toward building a career (contacts, jobs) would have been much easier. Ah, well—next time I'll know better.

What did you want to be when you were growing up? Did you always want to be just what you now are? Not me. At 4 I wanted to be a carpenter, at 14 a pharmacist, at 16 a chemical engineer, and at 18 a philosopher. In between, and persistently ever since, I've flirted with the law. I bought thick books of cases (but read only a few pages in them), and I audited courses in evidence (at the University of Miami), both torts and criminal law (at Indiana), and Roman law (in Edinburgh).

When I was asked, in connection with the admission procedure at Illinois, what my proposed course of study was, I said chemical engineering. When I took high school chemistry, I was fascinated. Unlike physics and mathematics, it was all new to me, and I enjoyed it hugely.

The equations balanced in a very satisfactory manner; I really loved knowing that

$$4Zn + 5H_2SO_4 \rightarrow H_2S + 4ZnSO_4 + 4H_2O.$$

The engineer part of my decision came about because I like mathematics and didn't know that anyone except engineers had anything to do with it. (Correction: I liked what I then believed was mathematics.) My father was a medical doctor and my two brothers were medical students—medicine had to do with pharmacy, and pharmacy with chemistry. The combination P.R.H. (my initials) and R.Ph. (Registered Pharmacist) was good for a few feeble jokes. All these "reasons", about as profound as the ones that traditionally cause a 6-year-old to want to be a fireman, contributed to my declaration that my major subject was to be chemical engineering, and in September, 1931, I so registered at the University of Illinois.

CHAPTER 2

A college education

Move to Chambana

The train ride from Chicago to Champaign took a little less than three hours; it cost $4.56. Champaign and Urbana are twin cities, sometimes called Chambana. The University of Illinois is mostly in Urbana, but when you're there you can't tell where you are—you cross the street and you're in a different town. The two city administrations manage to live together well enough, but sometimes they generate confusion. In the days before the Uniform Time Act, for instance, it could happen, and it did, that Champaign would adopt Daylight Saving Time one summer but Urbana would not. If you were invited to dinner at 7, you usually had to make one extra telephone call to make sure what time you were really expected.

To get used to living in a strange town among strangers, speaking a language that has not yet stopped being somewhat strange, and doing the strange things required of freshmen during orientation week—these are non-trivial sources of strain even for a precocious 15-year-old. My second evening there I was herded into a gigantic room in the gym, a basketball court, with hundreds of others, to sing "There are smiles, that make you happy . . ." under the phony enthusiastic goading of Ray Dvorak, assistant bandmaster. I hated it, I looked down on it, I felt embarrassed to be doing it—but I sang, I enjoyed it, and I was warmed by it. Ray Dvorak achieved his purpose—despite me. When it was over I went home to Newman Hall—"home"!—and went to bed, across the cell from Stanley Fry, the gangling fellow freshman from Farmer City, Illinois, who was assigned to be my roommate. Next morning I stood in long lines to get each of dozens of cards initialed by a bored faculty

P. R. Halmos, 1931

representative so as to be permitted to register in Rhetoric 1, Chemistry 2, and Mathematics 4.

It was all strange. Newman Hall was strange—long, bare, cold, impersonal, jail-like corridors, the bathroom far down the hall—Stanley Fry was strange, and the meal tickets were strange. I moved out of Newman Hall at the end of the semester. Stanley and I had become friendly, but I didn't want any roommate, not even him. As for the meal tickets, in those depression years they were quite a bargain. You paid $4.50 for a ticket with $5.00 worth of punches in it; on that you could live for a week.

It took me a while to meet people, and I never really had a large circle of friends. I moved from one dorm to another and from one apartment to another, more than is usual, and that helped to enlarge the number of my acquaintances. I find their names in my diary of those days, but often I have no idea of who they were. Felix Giovanelli? —oh, yes, he was the chap who looked up a new word in his dictionary every day. I tried to emulate him, but even though I was already in love with words and was bent on mastering them, I couldn't stick to it systematically. Incidentally, I didn't really have a diary, to record the past; what I had was a small pocket-size appointment diary, to alert me to the future. I started the custom in 1932, and I never threw away any of those little books. With the exception of one or two that were unaccountably lost, I still have them all; they are a big help now in looking back.

How not to be a freshman

A semester hour is the amount of "credit" you get by attending one class, usually 50 minutes long, for each of the approximately 15 weeks that make up a semester. To graduate with a Bachelor's degree you need 120 credits, which means 15 semester hours for each of the eight semesters that it usually takes to get there.

Looking at the transcript of my record at the University of Illinois, I shudder at what a dull and bad general education I got there. That's not to say that I was bored—I didn't have time to be bored. I made myself be in a hurry. One semester I took 22 semester hours, and once, in addition to 16 hours worth of courses, I prepared for and passed special examinations to receive a total of 28 hours of credit. That sort of thing, plus one summer session, enabled me to finish four years in three.

In the first year a course in something called hygiene was compulsory. The term "auto-intoxication" was in vogue then, and we were told not to. I was in Dr. Judah's section (a *real* doctor, not a Ph.D.), and I think he was ashamed of the stuff and nonsense he had to tell us. At the end of the academic year I left a postcard with him, stamped and addressed to Budapest where I was to spend the summer— that was the usual way to find out what grade you got in a course. He was impressed; that was the farthest he ever sent an A, he told me later.

Physical education was compulsory for two years. Because I have a bad foot (injured in a streetcar accident when I was 2 years old), I couldn't do track and baseball or any of the usual things—couldn't, or said I couldn't, and for sure didn't want to. The second semester I was transferred to "corrective" gymnastics. There was no way to correct missing toes, so all I had to do was attend the class a decent number of times and do a few push-ups. My attendance became more and more spotty; my grades in the four semesters were B, B, C, and D. I got only one other C during my days as an undergraduate, in a history course. I liked the course anyway. It was called "The ancient world", and it compared the social standing of the Christians in Rome with that of the communists in the U.S.

I took a course in German the first semester, but I didn't like it. It was too easy, therefore I didn't study, and therefore it was too hard.

I bluffed my way through Professor Geissendörfer's course, persuaded him to allow me to take special examinations in three additional semesters of German, and thus managed to fulfill the two-year language requirement. I think I forgot more German in the process than I learned.

Then there was rhetoric. Freshman rhetoric (U. of I. language for English composition—the comma fault, split infinitives, weekly themes) was useful and interesting. I had never really studied English till then, and I liked it. My instructor, Mr. Peterson, was an old maid of perhaps 32, a thin, dry man, seemingly a permanent graduate student, a conscientious drill-master, and a clear explainer of the mysteries of punctuation. He was a bit of a cultural snob. I remember one morning after a symphony concert he let us know of his disapproval—contempt?— of those who didn't attend. He seemed to like me. He gave me A's both semesters, and (oh thrill, oh pride) he got one of my themes into the college magazine, the Green Caldron, that printed the best of them.

Inorganic chemistry had a large overlap with my high school course. The chief thing I learned from it was how to make beer out of cranberries. So help me! The recipe was given to me by the friendly graduate student (Shilz) who was in charge of my lab section. On the off chance that somebody really cares, here is the recipe. "Boil one quart of cranberries with one quart of water, for about five minutes, till cranberry begins to pop open. Strain to get rid of skin and seeds. Add one cupful of sugar and half a cake of yeast. Let work for 36 hours. Bottle, and let stand for a week. Drink." I never had the nerve to try it.

Speaking of drinking: those were prohibition days. Beer was hard to get; the home-brew that I was sometimes offered was foul. Hard liquor was easy to buy but bad. When I wanted "gin" or "bourbon" (powerful but otherwise unconvincing imitations), I called Barney the bootlegger, told him what I wanted, and identified myself by my code number. (It was 3-3-11: my birthday, translated back by five years.) A half hour later Barney met me at a prearranged street corner and in exchange for greenbacks gave me bottles. Both gin and bourbon cost $1.00 a pint.

There was a better way to get whiskey, but for that you needed a friend in the chemistry department. From that friend you got a couple of liters of good pure lab alcohol and a couple of liters of distilled water. At the drugstore you could buy a vial of "essence of bourbon", about as big as your little finger. The essence was non-alcoholic; what it contributed is the color. You put the alcohol, the water, and the essence in a carboy, corked it tight, and rolled it across the floor. This was known as "aging it". The result was a generous gallon of healthy whiskey.

Anyway, back to chemistry: laboratory work seemed to me an uninspiring and messy waste of time. I was never told and I never caught on that a laboratory could teach you new facts and insights; I regarded it as just one of those chores (like irregular verbs in German and finger exercises for the piano) that the world assigns to apprentices before allowing them to become journeymen. I knew in advance what each experiment was intended to prove, and I proved it; I cooked the books mercilessly. Before the year was over I knew that chemistry was not for me, and I arranged to be transferred to the general liberal arts curriculum. I shilly-shallied about my major; it would probably be either mathematics or philosophy, I said.

Hygiene, phys. ed., German, rhet., and chemistry—they, plus math., were my freshman courses. I was too near it then to see how shallow it all was—it didn't challenge me, it didn't stretch me intellectually, it contributed very little to culture, wisdom, or understanding.

Trig and analyt

Most of what I learned in my first year of college was not in courses. I learned, for instance, to beat the registration system; the second time at it I stood in no lines and went through the process, from beginning to end, in seven minutes. The secret of my success was that I read and understood the college catalogue and I observed and understood the registration bureaucracy. I knew which courses were required and which were allowed, and I knew that the initials on the registration cards were intended mainly to avoid ridiculous conflicts and egregious stupidities. (They were intended to avoid overcrowded classes too, but if you arrived early, that was never a problem.) I made out a sensible program, forged reasonable looking initials on my cards, and went directly to the window where we paid the fees—seven minutes.

What I did not learn was how to study, or, for that matter, what studying meant. I was just barely smart enough to know that it did not mean memorizing—but I did think that to have learned something meant mainly to remember it. Everything I had ever been taught was done by "here is how it goes—now you do it". That's how I was taught to count, to write, and to solve algebra problems. I didn't understand what it meant to understand something, and what one should do to get there.

The mathematics I was taught in my freshman year was algebra (old stuff), trigonometry (too much of it), and analytic geometry (a revelation).

My trigonometry teacher was a graduate assistant whose name got away from me. He taught me about adjacent over hypotenuse (all new), and "solving" triangles by logarithms (a crashing bore). He also taught me about identities (capital fun, like easy jigsaw puzzles). Much later, when I started teaching trigonometric identities, an ingenious student told me of his foolproof way of getting a perfect grade almost every time. If you're told to prove that some expression A is equal to a different-looking B, you put A at the top left corner of the page, B at the bottom right, and, using correct but trivial substitutions, keep changing them, working from both ends to the middle. When they meet, stop. If the identity you were given is a true one (it always is), everything on the page is true. To be sure, somewhere near the middle of the page there is a gigantic step, probably as big as the original problem, but very few paper graders will ever find it, or, if they find it, dare to mark you down for it—it is, after all, true!

The second semester was pretty much like the first except for day-dreams and analytic geometry. I wore the engineer's uniform (brown corduroy trousers, with a slide rule, called a slipstick, dangling from the belt), I crammed for and worried about exams, I read Corneille, and Fielding, and Racine, and I tried to find and date girls. I lived in Illini Hall, the ancient firetrap precursor of the subsequent gigantic Union building, and I even managed to get a part-time job as janitor there (sweeping the stairs and cleaning the water fountains). The idea was to save enough money to be able to make a plausible cost-sharing proposal to my father about a trip to Europe in the summer.

The daydreams were about going away, to a university somewhere else, to a glamorous place. I dreamt about France, Germany, Russia, and Argentina, about Madrid, Edinburgh, London, and Los Angeles. I studied college catalogues in the library, and wrote off and asked for many that I couldn't find. (All the universities I wrote to answered me, and all but one sent me a catalogue. The exception was Edinburgh: in reply to mine of the 7th ult., if I would transmit a postal order for two shillings, the catalogue would be sent. Are the Scotch jokes true after all?) Nothing came of the daydreams.

Analytic geometry was great. It began with a description of Descartes' great victory, the insight that made algebra out of geometry and vice versa. It was all about graphs, and mainly about conics. The conic sections were defined three ways: as plane sections of cones, in terms of foci and directrices, and by quadratic equations. The conics had eccentricities and latera recta (and we were expected to remember that that's the plural of latus rectum). There were also lemniscates and limaçons, and most phenomena had three-dimensional versions (but that got short shrift near the end of the course). The biggest mysteries were called simplifications and rotations. I never caught on that they were meant to be synonyms: we were supposed to simplify something by rotating it. Linear algebra (vectors and matrices) was never mentioned. I thought it was all great stuff and in my letters home I wrote enthusiastically about my mathematics course; it was a beauty, I said.

Calculus was considered much too sophisticated for freshmen in those days—it was postponed to the second year. (No one seemed to notice that virtually the only part of analytic geometry that calculus really uses is the idea of a graph, and most of us knew about *that* before we entered college.) Another procedural rule, different from to-day's, was that graduate assistants were not allowed to teach calculus, and they were not allowed to teach anything at all in their first year of graduate work.

Most of the content of my first-year mathematics courses is no longer taught in respectable universities; a part of it is relegated to high schools and the rest is forgotten. Is that good? I don't know, I really don't.

Calculus, and is there a doctor on the faculty?

In the summer of 1932 I went to Hungary. It took seven days to cross the Atlantic. I went third class on the S.S. Aquitania, and resolved to go first class from then on. I haven't always been able to keep that resolve, but it wasn't for want of trying. In September I was back in the U.S.; the next time I went to Hungary was thirty-two years later, in 1964.

I have some records of my college days—transcripts of grades, appointment diaries, and a few, very few, letters, but even with their help, I remember curiously little of daily life. The documentation helps me to recall that I read the doggerel of Robert Service, I read Pearl Buck, and I bought, enjoyed, and lost *God's Man*, that glorious "novel" in woodcuts by Lynd Ward. *Heavenly Discourse*, by the witty iconoclastic socialist named Charles Erskine Scott Wood, made a big impression on me—all of his names still stick in my memory fifty years later. But who, for example, is the Chuck Edward with whom I seem to have been in touch at least once a week for quite a while? I unremember him with pleasure—he was friendly, I think, he was someone I liked—but who was he?

My sophomore year was the year of calculus. It wasn't a big part of my life then; it was just a chore. I couldn't for the life of me understand it; B's were the best I could do. I could differentiate and integrate everything, but I had no idea of the meaning of something called the "four-step rule" in the text. (I know now: it means "apply the definition to find the derivative.") The text was the infamous Granville, Smith, and Longley that, according to rumor, brought each of its authors a royalty income of many thousands of dollars for at least 20 years. It was very bad. The explanations were not explanations—they were neither clear nor correct—they were cookbook instructions, no more. The selling virtue of the book was that it had many exercises, almost all of the routine mechanical kind. I was horrified to learn, years later, that it had been translated into French—the language of Goursat and other superb Cours d'Analyse.

My calculus teacher was Henry Roy Brahana. He was a tall man with a craggy face with his mouth usually dangling slightly open. He looked like a bewildered farmer in the big city. He tended to be incoherent, and he never really finished a sentence in all the years I knew him. He worked hard at trying to be a mathematician. He took a swing at the four-color problem, but mainly he examined in great detail many special metabelian groups of order p^n and type $(1, 1, \ldots, 1)$. I admired his ability to compute the square of a 3×3 matrix by looking at only one copy of it; to this day I cannot do the row-by-column multiplication without putting two matrices side by side (or the same matrix twice, if squaring is involved). Brahana was a friendly man, he took an interest in guiding and advising me, and (God forgive me!) I used to call him Doc (before I learned better and before we became friends and I could call him Roy).

Isn't the sociology of academic manners of address curious? At the non-first-rate places (Syracuse and Hawaii are examples from among the places where I have been a member of the faculty) "Doctor Jones" is the way to address Doctor Jones: he and his institution don't want you to forget, not for a minute, that he has earned that great distinction. (Not, please, "Doctor" or "Doc"—that rule applies to first-rate and tenth-rate places equally. "Doctor" means physician, and both it and "Doc" reveal an outsider's ignorance of the mores.) Some instructors may not have a doctorate; they and the janitors are called Mr. Jones. (Mutatis mutandis; a female instructor with no doctorate is Miss or Mrs. or Ms.) An assistant professor, low though he may be on the academic totem pole, is to be called Professor Jones if he is not a doctor, and the same is true of associate professors and the "full" (that is, adjectivally unrestricted) professors. In sum: at a low level place, "Doctor" is as high as you get, "Professor" is a poor substitute, and "Mister" is regrettable.

At a university on a higher level (Illinois, or, for that matter, most state universities) it is taken for granted that everyone on the academic staff has a Ph.D.; to call someone Doctor Jones would distinguish him from the janitor only. There you are called Doctor Jones till you are promoted to a professorship (with an adjective, presumably), and Professor Jones from then on. Remember that the army doesn't distinguish (not in the vocative case anyway) between an oak leaf colonel and a chicken colonel; just so, a university doesn't distinguish between an assistant professor and an associate professor.

The highest level of snobbishness (and the one I myself prefer to all others) is reached at a few ultra-elite universities such as Chicago. There it is taken for granted that the doctorate is too unimportant to give you any status, and, for that matter, all ordinary status symbols shrivel in comparison with the mere glory of being *there*—so none is used. You are Mr. Jones at Chicago, whether you are a young instructor, an elderly professor, the chairman of your department, dean, president, trustee, or janitor—you are Mr. Jones to your secretary, to students, and to Abe the newsboy on the street corner. (Poor Chuck MacCluer. I taught him all this, and I asked him to call me mister—he started out by calling me doctor—and then he called one of my Michigan colleagues mister too. He got a bad dressing down from that unreconstructed uninverted academic snob.) You are Mr. Jones to everyone except your colleagues—among them it is a hallowed tradition that whether you are a 25-year-old beginner or a 65-year-old quitter, at tea and at department meetings you must be Bill. That's easy downward in the chronological scale and hard in the other direction, the first few times, but since it would be insulting to make a point of avoiding the first name, we all soon learn not to.

Elementary mathematics and culture

Mathematics is the center of my life and has been that for a half century, but during most of my second year at Illinois my attitude was still sophomoric. I took courses to satisfy requirements, and spent a lot of time on social life and culture.

The only mathematics course I was taking, in addition to calculus, was called solid analytic geometry—it was probably intended to be a slow, gentle first glimpse of linear algebra. It was taught by Victor Hoersch, a tall, portly man, with wavy gray hair, and voice and bearing like a Texas elder statesman; and it was dull, very dull. Hoersch wrote 4×4 matrices on the board, all sixteen a_{ij}'s of them, often and carefully. A girl sat next to me—I had a good kidding relation with her —and she pretended to be upset because I didn't do my homework but copied hers during the first part of the hour (while Hoersch was writing subscripts on the board), so as to be ready to hand it in at the end.

There were many like Hoersch at Illinois then, people whose life work was not mathematics but the teaching of mathematics. Henry Miles was one (he was my freshman algebra teacher); he seemed middle aged to me when I was 16 (he was in his 30's), and he seemed middle aged when I last saw him some 40 years later. He never changed. He was a fine man, a friendly man. He had a small moustache (in Hungary we used to call them English moustaches) such as people's fathers had in old-fashioned photograph albums. He had a soft but clear voice, a ready wit, and more than enough wisdom and character to be able to put into his place any rambunctious undergraduate who thought he could outsmart him. He was sharp with one classmate who tried to stall off a quiz by asking questions the answers to which he neither wanted to know nor, obviously, could understand; the quiz came on schedule. Henry was not a mathematician, but he was great at his job: he liked people, he liked to explain things, and he knew how. He grew old gracefully. Sometime in the 1950's he was pronounced incurably ill (leukemia) and was given a year to live; he lived more than 20 years longer in good health and good spirits.

As for social life, all I mean is that I remember many parties, bull sessions, dates—many ways of wasting time in pleasant company. I got to know several graduate students, and for a while a bunch of us rented an apartment together and lived in luxury. We rotated housework duties (including cooking) weekly, and there was no nonsense about it: the carpets were vacuumed; the dinners were not hamburgers and ice cream but roast beef and home-made banana cream pie; and the dishes were washed and dried after every meal. Al Burton and Ralph MacCormack were in chemistry and Kiki Conder in zoology. I was the only mathematics student in the crowd and the only undergraduate. Al's thesis research frequently forced him to get up at 5 a.m.—he had to rush to the lab to turn on a switch. That's all he had to do, turn it on; with that his morning's work was done. Poor old Ralph, a second generation chemist, had a congenitally bad heart that killed him in early middle age. Kiki studied turtles. He had a checkered past, including a stretch as a preacher. He drank a lot. Before baking a pie, he would remove his false teeth and use them to trim the crust, chop-chop, neatly around the pie tin.

Culture meant literature and music. James Stephens (the Irish influence of one of my former roommates), and St. Augustine (my lifelong ineffectual flirtation with the Catholic Church), Tiffany Thayer (the legal limit of pornography in the 1930's) and Pirandello, James Branch Cabell and G. K. Chesterton, Molnár (after all, I *was* born in Hungary) and Plato, Aristotle, and Shakespeare (I loved *Henry IV* and *Henry V*), Lewis Carroll, and Evelyn Waugh—my tastes were not focused. Some of the readings, a small part, were suggested by Bill Templeman, my teacher in two courses in English literature, 18th and 19th century. (We became party acquaintances later and had several meals and drinks together—that's why "Bill".) He was a "good teacher"—clear, uninspiredly inspiring (had he, I now wonder, taken courses in Education?), sharper than his students.

In music I had been a romantic (Brahms) till I met Warren Ambrose. I do not remember the meeting, but it was the beginning of my longest and closest friendship. We were in several classes together, and we talked about philosophy and mathematics and sex and politics and music. In mathematics he was a semester or two behind me, and I felt old and wise when I told him about the mysteries of differentiation that were still in store for him. He introduced me to Bach's Little G-minor Fugue (via Stokowski's orchestral transcription). Since then (unlike Ambrose) I have become a thoroughgoing musical reactionary. Roughly speaking my attitude is that if it's after Mozart, it can't be any good. I don't really mean that, but I mean most of it. Some Beethoven is good and even some Prokofiev, but my ears are closed to Debussy and Stravinsky, and, of course, to Hindemith, Schoenberg, Bartók, Kodály, Copland, and, horrors!, Cage. The chief exception to my reactionism is that I like certain kinds of schmaltz: Hungarian gypsy music, Johann Strauss, John Philip Sousa, and Scott Joplin.

The word "culture" is sometimes used to refer to the works of art that humanity has accumulated in the course of the centuries: poetry, drama, literature of all sorts, painting, sculpture, architecture, music —have I left anything out? To be "cultured" in this sense is to be in touch with human beings long gone, to share a common heritage with those still here, and to have, in addition to the pleasure that the art gives, a kind of sensitivity and insight that cannot be got any other way. To stay culturally alive, we must keep looking and listening throughout life—but, I firmly believe, no one can become culturally alive unless he is made so in his childhood. Eighteen years is too old, and so is fourteen; I believe in Verdi for the 6-year-old, every 6-year-old, and Voltaire for the 10-year-old. The bus driver and I would have a lot in common if we had both heard about Falstaff when we were 12—and there is no reason we couldn't have. Music and poetry are more important than carburetors and calculus, because both the bus driver and I would be better human beings if we had more in common and because we could then collaborate better to live in a saner world.

Mathematical daydreams and BARBARA

Slowly, slowly I began to wake up. In the middle of my sophomore year I wrote the formula that summarizes the fundamental theorem of the integral calculus in my pocket diary. Near the end of that academic year I made one of the few truly diary-like entries: "I'm getting more and more convinced that I've found my right field by majoring in math."

Among my other notes from the same period I find these daydreaming questions that I thought profound at the time. "How many diagonals does a polygon of n sides have?" (I haven't the faintest idea, but it is a respectable, though, if memory serves, elementary combinatorial question; a discussion of it can surely be found in any one of the proliferating books on the subject.) "Is there a binomial theorem for fractional exponents?" (Of course: Newton and Professor Moriarty didn't live in vain.) "Are there coordinates that do not define position, just define—that is, is there a definition of a physical object without space?" (Am I trying to invent manifolds?) There is also an unexplained reference to non-Euclidean billiards.

I was 17. Girls were important, dates, dances, and parties. I attended several football games and a Homecoming Prom, and I enjoyed solitary walks in the cornfields and cemeteries near Urbana. My grades were mediocre that year; logic was the only course in which I got an A.

Logic was not mathematics—not in those days, not at Illinois—it was the pedantic, taxonomic, scholastically superficial study of syllogistics boiled down to the intellectual level of college sophomores by textbook writers. It was very easy for me. Although I couldn't seem to memorize the names of the valid syllogisms, I knew by inspection which they were, and the pointless classification schemes (like the square of opposition) amused me. In and through logic I discovered philosophy, and from then on, for the next two and a half years, more than half my academic energy was spent on that.

In logic I immediately became a pro: I registered with the Hobart tutoring agency and earned a few dollars by explaining BARBARA and the other orthodox syllogisms to students to whom they couldn't be explained. Mr. Hobart started his agency because he couldn't in those depression years get a job teaching mathematics. He was a heavy gent with beautiful white hair and an impressive manner. His advertisements in the Daily Illini appeared almost every day; referring to the A, B, C, D, E grading system in use, they announced that E + Hobart = A. He was known to us sharecroppers as A − E.

The only member of the philosophy department who had any interest in logic was Oscar Kubitz (an assistant professor at the time). He taught me formal logic, and, much more important, he told me about Russell and Whitehead's *Principia Mathematica*. Of course I never read PM, but I held all three volumes in my hands at various times, I riffled through the pages, I tried to pay close attention to the prose at the beginning, and I worked through the p's and q's and boldface propositional connectives with which it all gently begins.

All Gaul

Mathematics at Illinois in the 1930's was, like Julius Caesar's Gaul, divided into three parts: algebra, analysis, and geometry. Applied mathematics either didn't exist, or, in any case, news of it didn't trickle down to us undergraduates. Statistics did exist, but not much, and it wasn't very important. I began my third and last undergraduate year by diving into all of Gaul in the summer session of 1933. (That was a busy summer. I managed to get a full semester's worth of hours in the short term of eight weeks. In addition to algebra, analysis, and geometry I took a course in philosophy and a credit exam in English literature. My only A was in philosophy.)

Algebra, taught by James Byrnie Shaw, was a soft course. Groups and fields were mentioned, but mainly it was a relaxed meandering through parts of what used to be called the theory of equations. Polynomial equations of degree n have n roots; the elementary symmetric functions of the roots are good things; cubics and quartics can be solved.

Shaw was just retiring. He was an algebraist of some note, a courtly gentleman of the old school, and, as amused, non-malicious, whispers had it, he was much more fond of teaching freshman algebra to a classroom full of girls than to a classroom full of boys. One whisper was more pointed: it said that every girl in his classes got an A and every boy a B. He gave me a B. I was present at his last lecture in his last class. He ended with a small bow, a gentle smile, and a soft statement, half proud and half wistful: "And that's the end of 50 years of teaching".

Analysis was taught by Steimley, who taught like a marine drill sergeant. He prepared detailed notes for his advanced calculus course and used them over and over again. He graded homework and exams promptly and fussily. Your grade was not likely to be just B or 80 or 85, but something like 83. The digit to the right of the decimal point in your average could play an important role in determining your course grade. He was quick and sharp and compulsively organized—his grade books, his stamp collection, and his phonograph records were filed, entered on cards, and cross-referenced. Since I am inclined to have similar compulsions, we found each other sympathetic; he enjoyed telling me and I enjoyed learning about his classification schemes. I'll never forget one of his homework problems: use a photograph to draw your own profile on graph paper (facing upward) and approximate the graph by a Fourier series to within a prescribed degree of accuracy. I tend to blame Steimley for my never having learned advanced calculus. His course made me jump to the conclusion that classical analysis was nothing but picayune bookkeeping—the whole circle of ideas around Green's theorem and Stokes' theorem has always been a foggy mystery to me.

Geometry was based on the then famous book by Graustein; it was mainly projective, and it was kept on an elementary level. The ingredients were homogeneous coordinates, points at infinity, cross ratios, and the theorems of Desargues and Pappus. The approach was partly synthetic (i.e., axiomatic) and partly analytic (i.e., coordinatized), but nothing as modern and fancy as the fundamental theorem of projective geometry was ever mentioned.

Harry Levy was my geometry teacher, and he more than anyone else showed me the way out of undergraduate mathematics. He was a good geometer who just happened not to be outstandingly good, and, I later heard, one who missed the boat very sadly. He had a chance to discover something important in differential geometry (symmetric spaces?), he almost had it, but he made an oversight. When he came to the path that could have led him to glory, he dismissed it out of hand, saying that such examples obviously couldn't exist. He was an engagingly ugly man with a large nose, a pockmarked face, and an inviting broad grin. When I learned Cantor's proof that α is always less than 2^α, I rushed to him eagerly to share my new knowledge. He listened genially and then he waved it away—sure, sure, it's another of the paradox arguments.

He taught me a lot. I once heard him lecture on the point at infinity in the complex sphere, and I was impressed by his refusal to use the symbol ∞—he used an asterisk instead. Above and beyond the call of duty he taught me library technique. We walked down to the mathematics library, and he showed me the Revue Semestrielle, the Jahrbuch, and the Zentralblatt, and various other sources of mathematical information and wisdom. From time to time since then I did the same for students of mine, and I always felt secret pride as I was doing it— look here!, now I'm grown up, now I'm doing what Harry Levy did.

Harry Levy directed my bachelor's thesis too. I elected to write one (it was no longer required in my day) because it made me eligible for an honors degree. The topic was projective geometry, and I remember the plot. The idea was to take two books (Veblen and Young was one and Coolidge was the other), each of which gave an axiom system for projective geometry, and to compare the two systems. Since the subject was the same in the two books, the meaning of "compare" became to prove that each axiom of either one is a theorem in the other.

A Bachelor of Science

The summer of 1933 might have been busy, but the two semesters that followed it were maximal. The first of them was the one during which I acquired 28 hours of credit, dabbled in Esperanto, began to read *Imitatio Christi*, concentrated heavily on philosophy, took a beautiful course in Norse mythology, and still kept my hand in mathematics.

The heavy dose of philosophy consisted of three courses, of which the best was philosophy of science. The following semester I took three more philosophy courses, of which the best was metaphysics. Both bests were taught by D. W. Gottschalk, a small, wiry man with a Roman

nose and fierce, intense eyes. I loved Gottschalk's lectures. He spoke
fast and loudly, with fiery and contagious enthusiasm. I tried to listen
and understand, and, at the same time, to take careful, complete notes;
at the end of each lecture I was both exhilarated and exhausted. Gottschalk had developed an intricate and "precise" cosmological theory,
Spinoza-like, with definitions and proofs. (I put "precise" in quotation
marks because later, from the point of view of a professional mathematician, I came to think that it wasn't precise at all.) The following
summer, after I already had my Bachelor's degree, I spent several weeks
organizing and typing my notes on Gottschalk's metaphysics; it was a
labor of love with no conceivable ulterior motive. It made quite a hefty
book. The book was lost and its contents forgotten long since.

Professor Flom taught me Norse mythology, and, the next semester,
Ibsen. I only got a B in the first, but I loved him. He was a dour
Norseman with dry humor and dry scholarship. He not only explained
what Jove, Zeus, and Thor had to do with one another on the one
hand, and Venus, Aphrodite, and Freya on the other, but hinted at the
etymological connection between love and liberty. (Freya is the goddess
of love and the similarity between her name and the word "freedom"
is not just coincidence.) I noted that Thor's day (Thursday) belongs to
Jove in Spain (Jueves) and Freya's (Friday) to Venus (Viernes), and I
found it all heady stuff. Flom was precise; he never bluffed; and I loved
his courses. I wrote him a fan letter on my final examination paper ("the
two best courses of my undergraduate career"). I have never heard
about him since, and I have forgotten much of what he said, but to the
extent that I can be said to have any culture, he deserves a large
measurable amount of credit.

Aside from the work on my undergraduate thesis, the way I kept my
hand in mathematics was through three feeble courses: advanced aspects of Euclidean geometry, fundamental concepts of mathematics,
and probability. The latter could have been good but wasn't; it was just
a bunch of puzzles about permutations and combinations, no theorems.
The law of large numbers was never mentioned, but it seemed to hover
in the background as an experimental fact. The central limit theorem
was never mentioned, but we looked at some of the properties of the
Gaussian distribution, and regarded it as a mysterious act of God.

Professor Lytle taught the Euclidean course. I remember him as lame,
the kind of lameness that polio could leave one with, but much better
than that I remember his white hair, and his kind, peaceful, grand-
fatherly face. He usually taught off-beat courses designed for the cheery
but slow-witted students who were preparing to teach high school in
Peoria. I learned more about the nine-point circle from him than
I now want to know, but I enjoyed it then. I drew a gigantic triangle
with everything on it—the inscribed circle, the Morley triangle, and, of
course, the nine-point circle, and lots more. I liked Lytle and he liked

me. When I was applying for a fellowship, I asked him to recommend me. He looked sad and kind and told me whom I should ask instead—a recommendation from him, he said, wouldn't be of much help.

And that's it—that's a college education with a major in mathematics (or philosophy if I had chosen to call it that—I satisfied the requirements for either one). In mathematics I was not only uninspired, but shockingly ignorant. Weierstrass, Hausdorff, Poincaré, Galois, and Cayley were mere rumors; I knew nothing about honest epsilontic analysis, either set-theoretic or algebraic topology, algebra, or even linear algebra (other than how to multiply 4×4 matrices). But I was a Bachelor of Science—University of Illinois, class of 1934—and a Phi Beta Kappa at that.

CHAPTER 3

Graduate school

Statistics

The Bachelor's degree didn't even break my stride. I never stopped to think—just took it for granted that I'd go on to graduate school. Finances were a problem but not an obstacle; my father continued to help till I became 22 and started earning a little. What to study?—that was a problem. I was split between philosophy and mathematics. I had an A in every undergraduate philosophy course, and I thought I understood philosophy better than mathematics—so I decided on philosophy, but, once again, I wanted to keep my hand in mathematics. The course I chose was called "Theory of statistics".

Statistics was not a major industry in the 1930's, and I cannot off-hand remember whether any major departments of statistics existed. The people who were to influence the future of statistics were not trained as statisticians: Wilks and Wald, for instance, were mathematics students in their day, and Tukey (pretty much my contemporary) acquired early fame by writing about general topology and ascribing Zorn's lemma to Zorn. Much later I had a hand in pushing for a separate statistics department at Chicago (successfully) and still later at Indiana (unsuccessfully); the important statistics departments at Berkeley, Columbia, Princeton, and Yale are among the ones whose births came later during my professional life.

Statistics was not a popular subject among graduate students at Illinois; the registration in the courses on algebraic geometry was much larger. To be sure, algebraic geometry was part of the rigidly fixed requirements for "prelims" (the Ph.D. exams) and statistics was not. The embryonic form of statistics was taught by Crathorne, whose own training was in the calculus of variations. His main claim to fame was a well-selling freshman algebra book. He was near the end of his career by the time I took his courses, and he was not full of verve and fire. He spoke softly, reminisced a lot, and was known as an easy grader.

Crathorne told us about Student's *t*-test, and chi square, and he made us do quite a bit of computation. The computation was done on desk machines; to multiply by 37, you punched in the other factor, turned a handle 7 times, flopped a shift lever over one place, and then turned the handle 3 more times. Division was a mildly complicated process of repeated subtraction (the handle had to be turned backwards), and square roots were even more ingenious—I knew how to find them when I had to.

At the end of my year of statistics I made a diary note: "Crathorne shaved off his moustache and became full professor." He must have been about 60 at the time.

The end of the affair

That first year of graduate school was a busy year, and it did not have a happy ending. I made a weekly schedule for the first semester, from Monday through Saturday, from 8:00 a.m. to 8:00 p.m. The main entries, to each of which I allotted an appropriate number of study hours, were French, Ethics, Plato, Statistics, and Typing.

I took typing seriously. I bought a teach-yourself-the-touch-system pamphlet at Woolworth's (for 10 cents) and practiced half an hour after lunch every day for several months. I became a good touch typist—not super, but good. I made mistakes, but I could copy fast, and for the next 25 years or so I did all my own typing—correspondence, class notes, exams, papers, and books. When a prospective Ph.D. student approached me, in later years, I asked him, first thing, whether he could type, and told him to learn if he couldn't. That was regarded as eccentric for a while, and then as passé, but now, with computers on the desks of many executives, it might become a sensible notion again. I read once that the true mark of a pro—at anything—is that he understands, loves, and is good at even the drudgery of his profession. The apothegm impressed me, and I patted myself on the back; I was sure that my attitude to typing helped to prove that I was a real pro.

French was in preparation for the Ph.D. reading exam. German didn't worry me (although I had to bone up on it quite a bit), but I was a Francophobe (and stayed one all my life). I had never studied French, never lived in a French-speaking country, and regarded most Frenchmen with suspicion and mistrust. I find notes in my pocket diary like

"look up French irregular verbs that change the stem". I did it all myself (no course); I just looked up enough grammar and memorized
enough of the short words to get by (the long ones were trivial)—and
I got by. The exam was simple: translate into English something new,
something you hadn't seen before. There were two parts: 45 minutes
with a dictionary, and 45 minutes without. The examiner was
Nicholson, in the philosophy department, and when he told me that I
passed, he added that mine was the freest translation he had ever seen.
But I passed, and I passed German too, and I've been bragging
ever since that I passed them when I was a philosophy student, translating hard prose, and not when I was a mathematics student. Mathematics is much easier; the formulas on the page, or, for that matter, in the
sentence, give you a strong clue to what's going on.

I didn't study all the time, of course—I read (*Autocrat of the Breakfast Table, Das Kapital*) and I played a lot of chess. Diary note early in
1935: "Morphy must have been a genius." I helped organize a small
chess club, and I was an enthusiastic member of the CCLA, the Correspondence Chess League of America. All that's gone—I have forgotten
what Oliver Wendell Holmes had to say, I am no longer impressed by
Marx's ponderous prose, and I am not quite sure anymore how to
capture a pawn en passant.

Most important, however, was not French, or statistics, or typing,
or chess, but philosophy—Ethics and Plato the first semester, and
Locke–Berkeley–Hume and German Idealism the second. I wanted to
take it seriously, and I worked hard at it, but, somehow, the love affair
between us, between philosophy and me, started to go sour quite early on.

I kept a real diary in those days—well, almost a real diary. It was a
little book, about the size of a pocket paperback, and it allotted
four short lines to each day for five years: from January 1, 1935
to December 31, 1939. My entries are perforce telegraphic and sometimes, memory being what it is, incomprehensible. Many of the entries
in 1935 are about philosophy courses, and I was astonished, when I reread them recently, to see how sour they were. "Plato class mediocre...
Plato exam farcical and boring... Locke–Berkeley–Hume boring... Un-
exciting Kant class... Kant class getting beyond control." There are
also some less narrow expressions of unhappiness, such as the grumble,
"Is philosophy going to the dogs?", and the jejune gibe, "To be a philosopher you must have a mind that is agile, fertile, and futile." Which
came first: my hostile attitude or the philosophy department's low opinion of me? It's all far enough away that I am no longer emotionally involved, and I am quite prepared to believe that as a philosophy student
I was just no good at all.

The villain of the piece, and at the same time the hero, the
axman, the philosopher I had most to do with, was Glenn Morrow. He
was a dignified man, a neat man. I think of him as brown—brown

moustache, brown necktie, and a polished, comfortable, brown leather, Ivy League deportment. A diary note after one of my conferences with him says, "Morrow is a gentleman".

Morrow assigned me to give a class report on Kant's "Critique of practical reason". I studied for that, I prepared it, and I organized it very carefully. I wrote the major cue words on 3 × 5 cards, along with the pertinent references and quotations. The set time arrived, the five of us students and Morrow were sitting around a seminar table, Morrow caught my eye and nodded, and I pulled out my stack of cards. Morrow frowned: "Is that all you have?" "Yes, sir." "Haven't you got it written down?" "No—I thought I could present it better this way." "Very well—go ahead." Not a good start, but my report went over pretty well, anyway. I think.

I was inexperienced, of course. I have learned since that in some subjects when a scholar "reads a paper" at a meeting of his scholarly society, he literally reads it—that is he reads out loud the words typed on the paper on the lectern before him. Philosophers often proceed that way, and I think that's dreadful—it's hard for the audience to stay awake during such a presentation, and even harder to learn something from it. I have heard mathematics papers delivered that way, but only very rarely, and each one was a catastrophe. One of them was by a philosopher talking about a borderline topic—it could have been called propositional logic (philosophy) or Boolean algebra (mathematics). He read every word, every letter, every p and q, every propositional connective. He had the good grace to omit the more computational proofs, but when someone in the audience raised a question, he said, slightly offended, "Well, I could read you the proof, you know."

In the spring of that academic year I wrote a paper about axiology (the theory of value). It was not for a course—it was from the heart. It was an attempt to get off my chest my worries about philosophy and face them squarely. In effect I was asking whether philosophy had any value at all. I would probably be ashamed of that paper now, but I would still like to see it. At the time I showed it to some of my philosophical friends, including Gottschalk. They didn't like it. I wrote in my diary: "I have every reason to guess that this philosophy department will never give *me* a Ph.D."

Early in May there was a meeting of the American Philosophical Association in St. Louis, and I hitched a ride with Glenn Morrow and his charming wife Vic. It was great. I met important people, I liked many of the lectures, and I relished the personal contact with Glenn and Vic. "They're swell people", my diary says.

That's how it went, up and down, pro and con, till the crash came, near the end of May. The crash was the oral comprehensive exam for the master's degree. I knew it wasn't going well, of course. The questions I was asked were not only about philosophy itself (which I thought I knew), but also about the history of the subject (which I have always been very bad at). I failed, I flunked, I was out.

Glenn Morrow gave me the bad news. We were in his office, standing; occasionally I paced back and forth a few steps, walked to the window to look out, and then came back to examine very carefully the backs of some bound journals in the book case. I tried to keep the tears of disappointment, frustration, anger, and bitterness from coming to my eyes, but I was not successful. Morrow was trying to sound reasonable, and I hated him. "Well, Halmos, you know, after all, there are many fields of human endeavor other than the intellectual ones". The diary entry for that day says: "Flunk comprehensive. Home, get tight, go to bed dejected."

Twelve years later, in 1947, I got a Guggenheim fellowship; Glenn Morrow got his two or three years after that, and I couldn't help feeling victorious, self-satisfied, and smug about having "beaten" him. I have tried from time to time to locate him and establish contact, but I never succeeded.

Matrices

Even the philosophy failure didn't break my stride. I never burned my mathematical bridges, and the formal transfer from being a graduate student in philosophy to becoming a graduate student in mathematics was easy. Less than a month after the catastrophe, summer session classes began, and I was proceeding full steam ahead with "Introduction to higher algebra" (Brahana), "Introduction to higher analysis" (Steimley), and "Theory of numbers" (Carmichael). Algebra and number theory occupied most of my attention. I was sure I wanted to be an algebraist, like Brahana, and Brahana tentatively accepted me as a Ph.D. student. The algebra course was hard and I worked at it furiously; number theory was a revelation and a joy and I worked at it enthusiastically.

When I say furiously, I mean furiously. Brahana didn't know how to be clear, the text was Bôcher's book (which I thought was a mess), and my dominant emotion during much of the time that I spent on the subject was exasperation reaching to anger. Since the greatest part of my subsequent professional life was spent on linear algebra and its generalizations (the part of mathematics that I have come to like more and know better than any other), that anger might seem strange now. I don't think it is. Part of the explanation might be the cliché that hate and love are not far apart. Another possibility is that I was instinctively attracted to the subject and I was upset by the needless obfuscation that kept me from getting to the heart of the matter.

Vector spaces were not in the common mathematical vocabulary of Bôcher's time, and linear transformations, barely mentioned, do not play an important part in his book. Orthogonal matrices are brusquely defined ($P' = P^{-1}$) in the last chapter, without any hint of their geometric significance. The classical necessary and sufficient condition for similarity (invariant factors and all that) is made to depend on the theory of "λ-matrices"—they are matrices whose entries are polynomials in one variable. I organized a rump class of those who, like me, were having trouble with λ-matrices; after one of its meetings I smugly and proudly wrote in my diary, "I can teach". Some of Brahana's course was based on Dickson's *Modern Algebraic Theories*, a badly written but tightly organized book, in which various canonical forms get their due share of emphasis. Bôcher, Dickson, rump sessions, and some glimpses of inspiration from David Netzorg (I'll tell about him later), and somehow I survived my introduction to linear algebra. I didn't really begin to understand what the subject was about till four or five years later, after I got my Ph.D. and heard von Neumann talk about operator theory.

The Dean

Bob Thrall was a graduate student at Illinois in those days, and he was still pure. Later he became an activist proponent of applied mathematics, but then he was a Ph.D. student of Brahana's. We had that in common, and our interest in Carmichael's course, and chess; other than that we agreed to differ on almost everything. Bob looked dignified— his hair was white when he was 19, he always wore a tie, and he was a teetotal, politically conservative, church-going Methodist. I wore my hair long (for those days), didn't believe in neckties, drank like a fish, was beginning to lean more and more to the political left, and never went to church. We were together a lot, working on algebra and number theory, and playing chess. He beat me regularly. We also worked on Carmichael's group theory book; we read every word of it at least three times (galley proof, page proof, final proof). No pay; labor of love.

R. D. Carmichael was one of the outstanding members of the Illinois department. He told me once that for a period of several years there were only three mathematicians on earth who published more than 100 pages a year: G. D. Birkhoff, N. E. Nörlund, and himself. His lectures were supremely organized, clearly delivered, inspiring. When I learned from him that $2^{2^5} + 1 \equiv 0 \bmod 641$, I rushed home and entered that in my diary. He wrote several books and kept publishing a lot (mainly differential equations). I admired him and wanted to be like him in as many ways as I could. His handwritten p, for instance, was idio-syncratic (the vertical stroke extended almost as far above the level of the loop as below), and I adopted it; to this day my p's look odd.

I fell in love with number theory in Carmichael's course. He made us prepare a table whose row headings were the first 400 positive integers, and whose column headings, about 25 of them, were items like factorization, sums of squares, sums of primes. Our instructions were to fill in the table, and then proceed to guess (and if possible prove) as many theorems as we could.

My first research was inspired by Carmichael. He told us about a peculiar question (inspired perhaps by the four square theorem): for which positive integers a, b, c, d is it true that every positive integer is representable by the form $ax^2 + by^2 + cz^2 + dt^2$? ("Representable" means via integral values of the variables x, y, z, t.) The answer is that there are exactly 54 such forms, and Ramanujan determined them all. My question was: which forms of the same kind represent every positive integer with exactly one exception? I found 88 candidates, proved that there could be no others, and proved that 86 of them actually worked. (Example: $x^2 + y^2 + 2z^2 + 29t^2$ fails to represent 14 only.) The two I couldn't decide were later settled by Gordon Pall: both $x^2 + 2y^2 + 7z^2 + 11t^2$ and $x^2 + 2y^2 + 7z^2 + 13t^2$ fail to represent 5 only. No in-spiration was needed for this, my first published research, only patience and diligent application of the techniques Carmichael taught me, but the work gave me a feeling of accomplishment and (badly needed) the confidence that I could do research. I was very proud. I bought 200 re-prints; it took me many years to find enough people to give them to.

Carmichael was dean of the graduate school in my day; the mathe-matics department usually referred to him as the Dean. He liked me and he loved to talk. I used to go to his graduate school office, by appointment, with a prepared list of questions—but all the Dean needed was one question. On one question and topics related and unrelated he could and did give me a lecture for about an hour and a half (no exag-geration), at the end of which the rest of his afternoon appointments, waiting outside to see him, were up in arms, and I had to leave in a hurry. He told me about his days as a graduate student, his life at Princeton, his family of six, the poetry he had written, and the papers he was and would soon be writing. He complained about the time that his work as dean took—ever since he took the job, he said, he could never find more than 20 hours a week for his research.

One day, a couple of months after my philosophy exam, I received an official message from the graduate school office: the dean wanted to see me. He seemed almost embarrassed when he told me what it was about: he urged me to get a Master's degree. I was not eligible for one in philosophy, but, what with my statistics courses and my summer load, I had more than enough for one in mathematics. I was surprised but agreeable; the only difference it made in my life was the $10.00 I had to pay for the diploma. I didn't guess the reason for the dean's urgency till later. My plan was to go for the Ph.D.—the in-between degree was folderol. From the official point of view, however, I was a student with a failure on my record in one subject and too many B's in the other. The chances of my making it to the Ph.D must have seemed slimmer to the all important Miss Voigt, the graduate school secretary, than they did to me—and, I am guessing, she suggested to the dean that I be advised to aim for the customary consolation prize. In my naiveté I viewed the plan as a small pat on the back, and my friends and I celebrated the extra degree.

First class

If you were chairman of the mathematics department, would you give a person with my record a fellowship, or a teaching assistantship, or at least a tuition scholarship? Coble didn't. Should he have? At the time I was angry and felt discriminated against; I *knew* I was better than most of the fellows and assistants. Now I can see Coble's point: the record just didn't look good enough. I was a paying customer all the way through graduate school, except for one semester only, when there was an emergency and I was given one class to teach.

Arthur Byron Coble was a big name in those days: an algebraic geometer, the author of a book in the AMS Colloquium series, a member of the National Academy, an important and influential man on the American mathematical scene, and head of the Illinois mathematics department for many years. His yearly salary was $10,000—a figure representing wealth beyond the dreams of avarice to most people during those uninflated depression days. In 1933 the postage for a domestic letter was 2 cents, in 1935 you could buy a new Ford for $600, and in 1938 the standard salary of an instructor with a fresh Ph.D. was $1800. Roughly speaking the inflation factor over 50 years is 10; Coble's salary is the equivalent of $100,000 in 1983.

I knew Coble as an administrator only; I never heard him in a course or even in an isolated lecture. There is a story about him as a Ph.D. supervisor that may be apocryphal but is true in spirit. Allegedly he had an infinite sequence of Ph.D thesis topics. Having proved a theorem in dimension 2, he had his next Ph.D. student extend the result to 3, and the one after that to 4. The story has it that Gerald Huff was the spoilsport; he settled not only 5 but every greater dimension too, and thus put an infinite number of prospective Ph.D.'s out of business.

It was the day before classes were to begin at Illinois in September 1935 that I learned that this time there was a job for me—I was given a freshman algebra class to teach, 8 o'clock in the morning (beginning next morning), five days a week, for one semester only. The pay was $45.00 a month—which I thought was glorious, a big help. I was living in an old-fashioned, comfortable, large, 5-room apartment, within 5 minutes walk of the campus; the rent was $45.00 a month.

I had no fear of teaching. Stage fright, yes; fear, no. Stage fright, in the sense of being keyed up, slightly nervous, and temporarily unable to concentrate on anything, is something that has always been with me six minutes for each new class: five minutes before it first meets and one minute after it starts. The same is true of colloquium talks and all other kinds of public appearances. "Stage fright" may be the wrong name for the phenomenon—it may be just a subconscious automatic revving up that gives one the impetus needed to overcome the large inertia of every beginning.

Even though there was still a lot I had to learn about teaching, I knew I could cope with it, and I did. I have always been surprised by the beginners who say they don't know how to teach at all yet. Haven't they already spent almost two decades under the influence of teachers, haven't they noticed that some techniques seem to work well and others are just annoying, and haven't they ever muttered to themselves "I could explain that a lot better"? I had had enough bad teachers; I knew, I thought I knew, what not to do, and I marched to my first class confidently, with my head held high, and with the stage fright barely noticeable under my eagerness to get on with it.

I got on with it well enough. I taught at 8, and three days a week that made a full morning: at 9, and at 10, and at 11 I had classes, the graduate courses I was taking. I took the job seriously; many of my diary entries refer to my class. "Trip on factoring..., quiz on logs..., bawl out class with millions of conferences as the result..., get applause, [I don't remember why, but I'm pleased]..., teach determinants..., make out final."

The ages of my students were around 17 and 18; I was a wise and experienced graduate student aged 19. They believed what I told them. Some of them were good, and some of them were hopeless. The only one whose name I'll never forget (let's call him Drossin) is one of the hopeless ones. Attendance spotty, homework missing or weak, midterm exams around D minus, the final exam might have helped him pass the course, but I didn't have much faith and neither did he. The Saturday evening before the final a small party at my house was interrupted by the doorbell—it was Drossin. Could he speak to me a minute, privately. Somewhat surprised, I took him to an unused room, and asked him what's up. I was a graduate student, wasn't I?, and perhaps I wasn't too well off, was I?, and this course was important for him, so he'd appreciate it if I would help him pass it, and he'd make it worth my while—he'd give me five dollars. (Half a week's rent, a week's food.) I was too surprised to be angry, but I told him to go away. I returned to the party and told them that I had learned how much I was worth. Next Monday, Drossin's paper was the first I graded. He flunked, with no margin for error.

Hazlett and Netzorg

The courses I was taking during the academic year 1935–1936, especially in the first semester, were completely typical of the graduate program at Illinois; I was going by the book one hundred percent. I was taking algebra, analysis, and geometry—what else was there?

Algebra was taught by Olive Hazlett who was, by our lights, a famous and important mathematician: she published papers and she taught advanced courses. I remember three other women in the department, which may be surprising now but seemed perfectly natural at the time. There were no Negroes (the dignified and polite term for Blacks in those days) on the mathematics faculty. (A brilliant black freshman who appeared in an elementary class was quickly discovered, encouraged, and allowed to proceed at his own pace. He was David Blackwell, who has since become perhaps the best black mathematician alive.)

Hazlett's course was based mainly on the first volume of van der Waerden, with, of course, some deletions and additions. She enjoyed telling us about one of her papers whose title she gave as "Embedding a ring in a field", and she enjoyed telling us how her colleague Shaw teased her about it—it evoked a picture of nefarious agricultural activities, he said.

Despite her non-trivial standing in the mathematical world, Professor Hazlett had at least one limitation that I have never forgotten, a narrowly parochial view of mathematics and its parts, especially her part, algebra. When she came in the course to the part of algebra in which formal power series play an important role, she showed us how they worked. David Netzorg, a brilliant student, was in the class with me. He was far ahead of me in training and talent, but he had never seen this particular theory before, and he had a difficulty, an objection. You can't, he said, talk about infinite series without talking about convergence. He was wrong, you can and you should, but that's not what Hazlett's answer was. Her answer had a sharp tone (just as David's question had a querulous challenging tone): no, no, she said, you must not, you cannot talk about such things as convergence in algebra. She repeated one of her favorite phrases, "purity of method", and she seemed to think ill of David for wanting to make her algebra unclean. David didn't understand her answer, and didn't have sympathy for her point of view. The discussion ended when the next class started to arrive and elbowed us out, and the issue was not resolved.

Now, nearly 50 years later, I understand the controversy. David was wrong, formal power series make good sense without any analysis, but he knew or sensed something that, apparently, Professor Hazlett didn't know. He knew that a mixture of structures is what frequently makes modern mathematics richer than "purity of method" can make it. Topological algebra, and algebraic geometry, and geometric measure theory are important mixtures—without subjects like them modern mathematics would not exist. Netzorg was wrong, but Hazlett was wronger. To correct him a missing definition had to be supplied; to correct her an emotional prejudice had to be abandoned.

I have known many mathematicians who were both bright and disorganized, as David Netzorg was, but I was unusually fond of him. I didn't like him at first because he seemed too unsociable; in fact he was shy and his occasional gruffness was just awkwardness. I thought he was snobbish too and wouldn't associate with me because I wasn't as good at mathematics as he was, but I misjudged him. He wasn't a clear explainer but he knew a lot and loved to explain what he knew. Bit by bit I learned more and more from him, and he liked me for that, and I liked him. Mathematically he mothered me, and I looked after him in some of the bureaucratic ways associated with student life, such as getting fellowship applications in on time and supplying the right number of reprints of his thesis to the graduate school.

I remember complaining to him about my difficulty understanding the rational canonical form of a matrix, and I remember the spirit of his explanation. He might get his symbols mixed up and his words, but he knew the idea, and that's what he tried to communicate. It's what always happens, he said, it's the typical linear phenomenon. Given a linear transformation, you just choose a vector to start a basis with, and keep working on it with the transformation. When you stop getting something new, you stop, and, lo and behold, you have a typical companion matrix. I didn't know what he meant then, but I do now, and it's an insight, it's worth knowing.

We became close friends. He frequently spent his evenings in my apartment, reading my magazines and occasionally raiding my refrigerator. I would get sleepy first and go to bed; David stayed on. In the morning he'd either still be reading or asleep on the living room couch; in either case he'd accept some breakfast and then go home.

I accidentally eavesdropped once on a conversation between David Netzorg and Ted Martin. David had just introduced me to Ted, and then we got separated—Ted and David walked ahead and I was with others a few steps behind. Ted asked, motioning toward me with his head: is he good? So-so, said David, but he's a plugger—and then he happened to glance half backward, seemed to realize that I might have overheard, and looked embarrassed.

I didn't mind, much. At least I now think that though I minded being characterized that way, I minded it only a little. David had high standards and I was insecure; I never expected that he would call me just plain good. Of course I hoped I was "good", and perhaps subconsciously, in secret even from myself, I thought I was, but I didn't really believe it. Anyway, there it was—I never told David that I overheard.

David returned to the Philippines (where his parents were living) just before World War II, and he was there when the Japanese came; he was interned and he died.

Good morning, analysis

I became a mathematician in April 1936. Or was it March? I have that familiar "it's almost on the tip of my tongue" feeling—I can almost say which day in April (or March) it was. Before that day I was a student, sometimes pretty good and sometimes less good. Symbols didn't bother me, I could juggle them quite well, and I understood the clean, finite reasoning of algebra. (I still feel as if I were an algebraist by birth, and only the accidents of environment stopped me from becoming a full-fledged one.) I did not understand (never even dreamt of) the idea of a "structure", in the sense in which, later, Bourbaki used that word, and I was stumped by the infinitesimal subtlety of epsilontic analysis. I could read analytic proofs, remember them if I made an effort, and reproduce them, sort of, but I didn't really know what was going on.

The year before, when I was being introduced to "higher analysis" by Steimley, the sad diary entries outnumbered the cheerful ones quite heavily. Usually it was "Steimley's analysis is worrying me", and "analysis has got me down again", and "analysis is out of control"; only rarely was it something like "worked some Fourier series problems —correctly!" and "A in analysis, 39 out of 40".

Matters did not immediately improve when I started graduate school. Pierce Ketchum tried to teach me some analysis, but I just didn't catch on. Ketchum would say, let's consider a function defined in the unit circle, and Ted Wang, a bright showoff in class would ask, "open or closed?". I didn't see any point to the question, I didn't see what difference it made or who cared, and I was annoyed by the interruption. Another time Ketchum assigned me a topic to read and report on to the class—it had to do with uniform convergence. He showed me just what to read—certain specific pages in Hardy's *Pure Mathematics* and certain specific pages in Titchmarsh's *Theory of Functions*. I memorized (!) my report, and I so little understood what was going on that when I had to change from the Hardy part to the Titchmarsh part, I blithely changed the notation from Hardy's to Titchmarsh's. So far as I was

W. Ambrose, 1948

concerned, I wasn't talking about one concept, but about two sets of pages of symbols. I didn't like analysis.

The day when the light dawned—I remember the circumstances and the scene—Ambrose and I were talking in a seminar room on the second floor of the mathematics building, and something he said was the last candle that this blind camel needed. I suddenly understood epsilons and limits, it was all clear, it was all beautiful, it was all exciting. I spent an hour in exhilaration and joy going through Granville, Smith, and Longley, nodding my head impatiently and happily. Yes, yes, sure, I can prove this!—yes, that's obvious— how could they possibly have botched this up so badly? It all clicked and fell into place. I still had everything in the world to learn, but nothing was going to stop me from learning it. I just knew I could. I had become a mathematician.

Why geometry?

The designers of the graduate program in mathematics at Illinois in the 1930's seemed to have had no doubts: they knew exactly what it was that every mathematician had to know. Matrices and permutation groups and Galois theory were in; Hausdorff spaces and homotopy theory were out. (The "out" subjects were sometimes available, and in my last year, as I was finishing my thesis, I availed myself of the opportunity to take an elementary course in general topology.) Uniform convergence and Cauchy's theorem were in; Banach spaces and Hilbert spaces and probability were out. (Let's be fair, however; let's remember that Banach's and Stone's books appeared in 1932 and Kolmogorov's in 1933.) The most curious aspect of the program was that 19th century Italian algebraic geometry was in; to get a Ph.D. in mathematics at Illinois you had to know all about the 27 lines on a cubic surface.

I took my geometry course from Arnold Emch, a round little Swiss with a white goatee and a heavy German accent. Spiritually he definitely belonged to the Italian school. The algebraic geometry developed by that school was, according to some critics, the only part of mathematics in which a counterexample to a theorem was considered a beautiful addition to it. Emch mentioned all the right names (such as Clebsch and Severi) and tried to teach us how to count everything; the subject was sometimes known as counting geometry. I thought it was fun for a while ("Severi is very nice reading", my diary says). We counted double points and cusps and intersections and multiplicities, and there were combinatorial tricks of the trade: first you multiplied a couple of integers, but that wasn't the answer because some constituents occurred twice in the product, so then you had to subtract those, but the obvious way of doing that subtracted some of them more than once, so you had to add them back in—it was hard (for us) to know when we were finished, when we could stop. We were scared to ask, and when Felix Welch did ask (how do you know when to stop?), Emch was upset. Emch told us about generic points, but he used the word informally and we were expected to understand what it meant. When he learned that we didn't, he started the next class by telling us: generic, he said, means nothing in particular.

Our curves and surfaces were real—they must have been because we drew pictures (some of them consisted of more than one component), but they were complex—they must have been because polynomial equations always had solutions. In any event no words like "field" and "ideal" were ever mentioned; the only way we could see that the subject could possibly be made rigorous was by fussing with the epsilons of analysis, but we didn't really know how to do that.

I was just as confused as the others in the class, but, like several others, I treated the subject as a game—you had to catch on and then play according to the rules you guessed. Some of us got pretty good at it. Ambrose didn't—he insisted on knowing what the sentences meant. A couple of weeks before a crucial exam he gave up and asked me for help. We reviewed the course together, and I told him everything I knew. When the exam came, I understood all the questions (I thought), and I wrote down their answers. Ambrose did not understand the questions (he confessed afterward); he just wrote down everything he learned from me. I got a B, he got an A.

CHAPTER 4

Learning to study

Doob arrives

That was a great year, the academic year that began in the fall of 1935, that was a watershed year, and the most memorable event that made it so was the arrival of Joe Doob.

Ambrose and I were blasé graduate students; we knew everything about the department, we knew everyone, and we could be trusted to deal with anything that was likely to come up one early September afternoon —the department secretary left us in charge to watch over the main office. This boy came in, looking like a new graduate student, crew cut, shirt sleeves, and all. He was 25 years old at the time, I later learned, but he looked 19 or 20. What do you want?, we asked. Is Coble here?, he wanted to know; my name is Doob, D, O, O, B. Ambrose and I had heard of the bright young hot shot who was coming to Illinois from a fellowship at Columbia; we yanked our feet off the desk and told him who we were.

It didn't take long for our lives to become intertwined. A few days after Doob's arrival my diary starts having many entries such as "shoot bull with Doob", "thirteen games (!) of squash with Doob, 7:6 his favor", and "Doob's class good". [I wasn't in his class—I wasn't ready for it—but I dropped in to audit every now and then.] In one of our early squash games I missed a shot, and swore in exasperation. "Just call me Joe", he said. We talked about mathematics, not to the exclusion of politics and music and professional gossip and many other human concerns, but more than about anything else. Is there a set in the plane such that... ?, ... but if the derivative itself is not differentiable, does it still follow that... ?

The mathematics curriculum at Illinois was skimpy in those days and not perfectly up to date. It was probably no worse than at most other universities; the number of exceptions (such as Chicago, Harvard, and Princeton) must have been very small. Doob was the first well-informed modern mathematician in the department; Martin and Baer, and subsequently many others, started coming later. Although I spent my professional life thinking about ergodic theory, topological groups, mathematical logic, and functional analysis, subjects that were hot news in the 1930's, when I got my Ph.D. in 1938 I had been exposed to hardly any part of any of them, and I certainly never had a course in any of them. A few of us graduate students organized a seminar based on Banach's book for a short while, but except for that what went on was routine course material that did not come near the frontiers of research. Doob was not a missionary, and he did not set out to make a revolution by designing a new curriculum, but just by being himself he began to instill a new spirit in the place.

The graduate students he had to face had the algebra, analysis, and geometry that I've been telling about. Second and third year graduate courses on borderline subjects such as homological algebra, differential topology, and functional analysis, which became quite common later, were unknown then; the only thing remotely resembling them was a rarity like a course on elliptic functions that Carmichael offered once. During my last year Doob tried something new: he gave the course in topology that I mentioned before. It was mainly set-theoretic, but it included some homotopy and the classification of 2-dimensional compact manifolds. For many years after coming to Illinois, while he supervised several Ph.D. theses on stochastic processes, Doob was never given a chance to teach a research course on his specialty; his students learned the background by reading the original literature themselves. I am not trying to imply that that's bad—maybe it's better than the predigestion system—but nowadays it would certainly horrify many graduate students and many would-be teachers of prestigious research courses.

Incidentally, though I admire Doob and think he is great, I do not think he is a good lecturer. He is too flippant. He seems to be so worried about appearing self-important that he plays down everything—nothing is important, nothing is hard, no one thing is worth a lot of time (enough time?). In personal conversation his insight wins—but even there you must believe how he looks and how he sounds rather than listen only to the words he says. He looks alert and interested, and he sounds pleased and excited—but what he says is something like "this is pretty trivial really, all you do is...". He is best when he gets stuck and you can watch him battle his way through a difficulty—then he can be determined and serious.

He has the one absolutely necessary quality of a teacher: he can see what it is that the learner is not seeing. His very flippancy and brevity can then be the source of great insight. I remember one time when he said

J. L. Doob, 1967

just the right two words to me. I had been trying to understand rectifiable curves, and the books I looked into never mentioned the crucial two words. (Thinking back on it all, I am surprised and skeptical—probably they did, but I missed them.) In any event, I was confused by the sums and absolute values and inequalities and upper bounds all over the page, and I was confused by the differences and similarities between absolute continuity and bounded variation. In a conversation among a bunch of us rectifiable curves came up again, and I must have looked vague and troubled. Joe must have sensed it. He turned to me and said: "has length". That did it. It made the reading I had been doing fall into place.

All work and politics

The advice I give to graduate students isn't always what I did when I was one. I advise them, for instance, never to stay on in graduate school at the same university where they did their undergraduate work. It's a bad idea: you don't learn enough, you don't meet enough people, and your point of view becomes too parochial. To change from Illinois to Michigan, and even from Princeton to Illinois, and certainly from Los Angeles to Chicago exposes you to new views. It may teach you mathematics you haven't seen and is certain to show you approaches and arrangements that are new to you. The faculty at the new place, and their professional friends, enlarge and enrich your sources of scientific contact. Even the mundane level is important; we should all know that there is more than one way to go about organizing courses, seminars, libraries, and departmental governance in general. That is senior wisdom that almost everyone shares; when I was a would-be junior mathematician I paid no attention to it, but now I wish I had.

Another way in which I wish I had then followed the advice I give now has to do with the old adage about all work and no play. I don't believe in it; I think all work and no play is the only way to get anything done. Having made my point in as shocking a way as I could think of, I'm ready to take it back and modify it. I am not talking about a lifetime of torture and slavery, and I am not excluding the relaxing tennis game, detective story, dinner in Chinatown with a bunch of friends, or Saturday night movie. What I am saying is that the work of a scholar is not torture that would be insupportable without distraction, and that, for most of us, two consuming passions are one too many.

We mathematicians are lucky, and we share our good fortune with many artists and craftsmen: we get paid to do what we enjoy doing. We are not in the state described by the old labor chant "we go to work, to get the cash, to buy the food, to get the strength, to go to work, to get the cash,...". The restaurant cashier, worn out by a day of work he hates, might study law in extension school, for pleasure or to improve his life, or he might go bowling every night, because he needs a hobby. We don't need hobbies. I might go so far as to say that we shouldn't have any. Most of us ordinary mortals don't have enough psychic energy to split between two passions; we cannot be good at both mathematics and cello, we cannot be satisfyingly creative in both mathematics and cabinet making, we cannot be effective in both mathematics and politics. To have but one passion, to channel all one's energies into one major activity, is especially important for a professional at the apprentice stage of his development. A graduate student in mathematics doesn't—shouldn't— have time to do anything except be a graduate student in mathematics.

I violated that rule. I violated it not by auditing a course in music appreciation, not by reading the odd book or two about philology (a lifetime flirtation), and not by getting in with a poker-playing crowd of chemists—those are just minor relaxations, like the tennis game, the detective story, and the dinner in Chinatown. I violated my rule by a temporary but frightfully time-consuming attention to political ideals.

From a patriotic Hungarian royalist I changed into a typical activist leftist of the middle 1930's. Hitler was rising to more and more power in Germany, and, in America, demagogues like Huey Long and Father Coughlin were acquiring more and more followers. The socialists, Norman Thomas and Leon Blum, and then the communists, Earl Browder and even Joseph Stalin, looked good in comparison. Influenced in great part by my admiration of Joe Doob, who leaned to the left long before I did, I became a convert and a joiner. I joined, I attended meetings, and I collected petition signatures and money for organizations that were later described as communist front. I don't even remember what they all were. My diary mentions the ALWF often, and that one is easy: it's the American League against War and Fascism. I belonged to something called the Youth Commission, and I seemed to be a member of the Entertainment Committee of some organization and the Peace Committee of another—the diary doesn't say which. I distributed leaflets advertising left-wing speakers, and I collected money for the Spanish loyalists. I helped to organize a Co-op, a part of whose purpose was to run a restaurant where Negroes [sic] and whites would be equally welcome.

Even my social life was affected. Most of it came from mathematics, but, for a while, a large part came from elsewhere on the campus, motivated by the zeal to spread the word. Joe Doob organized the Friday evening discussion group. One person would be persuaded to "volunteer" to read up on something of social significance and give a talk about it to the dozen or so people who came. The subjects included the Supreme Court, life insurance, dialectical materialism, Mexico, freedom of the press, and such obvious villains as the temporarily famous British self-styled fascist Oswald Mosley. After the talk we had coffee and cookies, and then, presumably, a deep intellectual discussion of the topic of the day. That part was unpredictable; the discussion topic could veer off to the latest local divorce scandal or to mathematics. Ambrose came sometimes, and, of course, he and I would include politics among the subjects that we enjoyed arguing about. I remember him as being originally just as conservative as I, or even more so, and my diary records, with surprise and glee, the day I got him to join the ALWF. Ever since then he has been conservative (I mean, consistent, persistent, faithful, unchanging) about being a radical. I gave up long ago, in disillusion, disgust, and impatience, but he stayed loyal and often active on the right (meaning the left) side of most of the social issues that the world has faced in the last forty years.

While my honeymoon with politics was still on, it affected everything I did (except, of course, mathematics). My reading, for instance, jumped from John Reed's *Ten Days that Shook the World* and Anna Louise Strong's *I Change Worlds* to Alexander Woollcott and P. G. Wodehouse, and then back again to the New Masses and Upton Sinclair's *No Pasarán*. I started studying Russian and had daydreams about emigrating to that land of milk and honey.

Most of this was in 1935–1936. Late in 1936 I began to realize that there just weren't enough hours in the day for both mathematics and politics; I resigned from committees and curtailed my attendance at meetings. My opinions were still heavily tinted by party-line ideology, and I'd make a note such as: "This guy Plato is a rank reactionary—a fascist". A year or so later fatigue, boredom, impatience, and the all-consuming eagerness to do mathematics began to change my views. My notes began to say things like "Laborious ALWF meeting with everything wrong as usual—I don't want any", and "intolerable evening—I want to quit", and "no more for me after this year". For me, as for many other intellectual radicals of the period, the end came suddenly, with the Nazi–Soviet pact of 1939. I was surprised, of course, and horrified—and in a way relieved. I told my friends that I was committing "political suicide", and I did. I resolved to support nothing, sign no petitions, and contribute to no cause ever again. I have always been too self-conscious to be a proselyting missionary—it seemed undignified, like hard-selling patent medicine or, from another point of view, like shameful beggary—the resolution to refrain from converting others to my views was easy to keep.

What are my views? In a discussion focused on mathematics, the question should be irrelevant, but having come this far, I might as well try to answer it and then get on with the main plot. The phrase for what I probably am used to be "tired liberal". Economically I am conservative, but I favor responsible collective bargaining in industry. I am against unionization of university faculty—I am convinced that it makes for a deplorable inattention to and consequent lowering of academic standards. I am for complete fairness to blacks and other minorities, and I regard women as members of my club—the human race. All of us, including Hispanics, honkies, and Hungarian Jews, should have the same opportunities to live in safety, comfort, and dignity. I am, however, opposed to what has been called affirmative action. To make an incompetent Hispanic, honky, or Hungarian Jew a senator or a professor, just because at the moment of the search there are not enough competent ones to make their representation in Capitol Hill or University Village proportional to their part of the census—that is, I think, a step backward in humanity's striving toward spiritual and intellectual perfection.

I was a Roosevelt democrat in 1932, in 1936, in 1940, and in 1944—and I am one still. Roosevelt the man had flaws (is there someone who doesn't?), but Roosevelt the leader pointed the way to the right goals. I don't really think it's sinful to vote for a Republican president—but I never have. The lawyers and car salesmen I play poker with think I am a flaming radical, and my younger colleagues know that I am an incontrovertible reactionary.

Born again

Even after I finally mastered epsilons, I didn't immediately begin to like analysis. I didn't like it at all; Doob's influence was insidiously slow-acting. In retrospect I can see the same pattern as I encountered in linear algebra: I didn't understand it, and therefore (a) I didn't like it, and (b) I had to work hard to learn it; having worked hard I came to understand it better than the subjects that seemed easy, and therefore ultimately I came to like it more.

Early in the academic year 1935–1936 life looked dark ("Discouraged; what if I don't get through?"). I was still a dyed-in-the-wool metabelian groupie and had many long conferences with Brahana. More and more, however, as the year rolled on, I kept struggling with analysis, and occasionally achieving a small victory. I began reading Whittaker and Watson in December, de la Valleé Poussin in January, and Townsend's *Real Variables* in February. God forgive me!: I wrote "Townsend is good". Later I came to think it was terrible, neither rigorous nor clear, but at the time I must have been attracted by how un-Whittaker-and-Watson-like it was. Much less fuss, many fewer epsilons—mathematics becomes easy if you dismiss the difficulties with an airy wave of the hand. I kept going back to Hardy's *Pure Mathematics*, which I found difficult but honest—if you kept your wits about you, you could soon see that progress was being made.

In the second semester Carmichael offered a course in elliptic functions. He said I couldn't take it because I hadn't yet learned the requisite parts of complex function theory. I begged and pled and asked to be allowed to learn about complex variables concurrently. He reluctantly agreed, and I worked my head off. I taught myself complex functions by writing out a complete English translation of both volumes of the little Knopp book, and I got an A in Elliptic Functions. Dreadful stuff, I now think: long identities between power series, from which you can read off, by examining sufficiently many coefficients with great care, in how many ways a prescribed integer can be represented as the sum of three squares.

In June I told my diary that "Knopp's treatment of periodic functions is beautiful", and, a couple of weeks later, that "Riemann surfaces are terrible things". Early in July "Function theory is great stuff"; in the middle of July "My mind is almost made up to major in analysis", and near the end of the month "The die is cast: I am an analysis major". That did not mean, of course, that I was at the end of anything—just the opposite. But work was rewarding. "More enthusiastically busy day than any recently; Lambert series in the morning, Carathéodory afternoon". In August: "Carry Hahn around with me most of the day; futile. Still I feel that I *am* mastering the stuff slowly."

While I was being born again in analysis, algebra and number theory were still very much on my mind. There was a book on groups by Mathewson that I tried to read, and the famous Miller, Blichfeldt, and Dickson (was the title *Linear Groups*?) was giving me trouble. (Miller was the prolific American finite group theorist, old, stooped, frail, dogmatic, mathematically narrow, and financially increasingly rich. I saw him in the mathematics building often; he came to meet his stock brokers. When he died he was a millionaire and he bequeathed his fortune to the University of Illinois.)

In number theory I was trying to read Landau. It was like walking up a mountain road; the incline may not seem too steep, but after half an hour you're puffing breathlessly. Around this time, too, I was finishing my almost-universal-form work that I mentioned before, and some of the diary entries refer to the excitement of the progress. July 1: "The mountain labored and brought forth a proof for the even case. The odd looks harder." Two days later: "Got the odd numbers this morning! Very screwy proof! Carmichael himself vaguely amused."

Carmichael was teaching number theory again that summer, but by then I was too far advanced for his course; I just dropped in every now and then for fun. His audience was mainly summer students—school teachers who took a course or two each summer toward an ultimate Master's degree. They were not good. I recall one detail. Carmichael was reminding the class of the one-to-one correspondence between complex numbers and points of the Cartesian plane. One hopeful Master's candidate raised his hand. "I can see what you did to your a and to your b. But would you mind telling me where the i went?" The Dean looked sad.

The most exhilarating experience that summer was my introduction to set theory. I would sit on the swing on the front porch, drink iced tea, and devour Hausdorff's *Grundzüge der Mengenlehre*. I was like a child with a new toy: I thought it was beautiful and astonishing. The most elementary ideas took me by surprise, and yet I was sufficiently prepared for them to be able to appreciate their significance when they came. I loved cardinal arithmetic, and I eagerly computed the number of elements of every set I could think of. Ordinal numbers were another revelation, and the strange order types, the non-well-ordered ones, were like music: intricate, but just right.

All work and no play? I was 20 years old, I had just learned to ride a bike, and in the middle of August I proudly recorded that I could turn corners.

Other forces, other tongues

While my "conversion" was gradually taking hold, other influences, both personal and professional, were also very much present.

A minor influence, but non-zero, was Trjitzinsky. I liked him immediately: "he's quite human", my diary says. He had a dark movie-actor kind of handsomeness and a thick Russian accent; he and his blonde and generously built wife Varvara (that's how the Russians say Barbara, we were told) were generally regarded as likeable but somewhat

eccentric middle-aged children. The most child-like thing about him was
his love of driving. He and Varvara would go to the west coast often (to
visit an offspring, as I seem to remember). He would put four new tires
on his powerful Hudson, and then drive for 20 hours at 85 m.p.h.
whenever he could get away with it, stopping only to fill the tank. He
would pull over for an occasional nap (Varvara didn't drive); I think
he said that he made it to Los Angeles in two days.

He was an important man in Illinois in those days. He was a prolific
publisher of papers and an indefatigable producer of Ph.D.'s. He attracted
many weak students; rumor had it that his way of helping you was in
effect to write your thesis for you. Some of his better students complained
to me years later—they wished they had been made to work harder and
learn more. Triji (the obvious nickname that everyone used) did "hard
analysis" (differential equations, integral equations), but at the same time
he was deeply enamored of the modern set-theoretic concept of transfinite
classification (Borel sets, Baire functions). Several of his papers, and of his
students' papers, mixed classical analysis and Baire classes—whatever it
is, arrange it in a transfinite sequence and then stir multiple integrals into
it. Triji thought that he was very good, and he resented that he wasn't
suitably recognized. One time when Chevalley was nominated to a post
on the Council of the AMS, Triji mumbled "what has he ever done?"
and wondered grumbling why *he* wasn't nominated.

Other influences? Yes, mathematics in all shapes and forms. It was all
around me and whatever part I wasn't struggling with, and learned to like
later, I was already liking. In geometry I learned about invariants, in
the 19th century, pre-Hilbert sense of the word, and I noted "study invariants all day—the stuff is getting to be exciting". For number theory I
read Euler's proof of the four square theorem—in Latin! There were
many words I didn't know, but I could guess enough from the context so
that there were no linguistic hurdles between me and the proof.

Languages continued to be a source of joy and reward. English is my
best language by far, for all purposes, and, in particular, for reading
mathematics. Although I can speak Hungarian pretty well, Spanish so-so,
and German very badly, my second best language for understanding
mathematics is German. (Why is that? A part of the reason could be that
my childhood exposure to it made German seem a "real" language, not
like French, and another part could be that most of the best books in my
student days—such as van der Waerden, Seifert-Threlfall, and Carathéodory—were in German.) The Latin of the Gymnasium and the
French of the Ph.D. requirement made it possible to read mathematics
in Italian (easy), Portuguese and Rumanian (hard), and Esperanto (trivial
—I read an entire article once, with the greatest of ease, without even
bothering to ask myself what language it was in). English and German
form a conspiracy and make it possible to read Dutch, but only with a

lot of inspired guessing; they are not enough, I was sad to learn, to make Danish, Norwegian, and Swedish accessible.

Dutch is funny. I don't mean strange or peculiar; I mean humorous. I asked a Dutchman once whether he understood why his language causes speakers of English to smile or to chuckle, and he said he did. It looks like baby talk to them, he said, just as Afrikaans looks like baby talk to him. I love the challenge that a page of Dutch presents—it is like a crossword puzzle—and many of the cognates that I finally recognize give me the kind of fun we all get from recognizing a friend, long not seen, behind a bushy new beard.

Russian is not funny—it's just plain hard. I never really learned it, but I took a part of a course once or twice, looked into two or three grammars, and tried to stick to good resolutions about a page a day. The technique that did the most for me was the Raikov translation project. I took his 90-page paper, "Harmonic analysis on commutative groups", and settled down with a dictionary. I had to look up almost every other word, and the first page took me an hour and a half. Now I can read Russian mathematics—with a dictionary and with pain, to be sure, but I can read it.

More influences? Perhaps the Lieber books should be mentioned; I enjoyed them so much that they made a deep impression. I don't know how many there were; I still have three. They are thin (34, 64, and 73 pages). The title pages say "Text by Lillian R. Lieber, drawings by Hugh Gray Lieber". (Married? Siblings?) The ones I have are about non-Euclidean geometry, Galois theory, and relativity. The pages look like poetry—lines of uneven length, each beginning with a capital letter—but the preface explains that.

> "This is not intended to be
> Free verse.
> Writing each phrase on a separate line
> Facilitates rapid reading.
> And everyone
> Is in a hurry
> Nowadays."

They are beautiful exposition, addressed to T. C. Mits (The Common Man In The Street). To read them, Mr. Mits would need more mathematical patience than he usually has, but his mathematically curious children in high school are sure to enjoy them and profit from them.

Plugging away, reading, studying, and worrying about prelims—that sword of Damocles was always swinging over the heads of my friends and myself—I was moving along, but if it looks like plain sailing in retrospect, it didn't always look that way at the time. Diary entry: "There seems to be no limit to the amount of mathematics I can forget in six months."

Prelims

The preliminary examination for the doctorate consisted of four parts. Three of them were written (one each on algebra, analysis, and geometry, given one at a time, spaced out), and the fourth was a comprehensive oral. The oral covered the "minor" also—usually one of the hard sciences but in my case philosophy.

In retrospect I like the system. It provides sufficient information (to both sides, the victims and the faculty), and it makes for a coherent curriculum. The old adage is that a written exam tells the faculty what the candidate knows, and the oral reveals what he doesn't know. As for coherence, that's the desirable opposite of the current tendency to provide each student with a custom-made, privately tailored education, instead of a large body of shared background that everyone can count on everyone else knowing. There always was and there still is a basic minimum of mathematical knowledge that every mathematician should have, and I think it's better to guess at what that basic minimum is incorrectly than not at all. In current departments of "mathematical sciences" it's difficult to find any single subject that is required of all students; more is excepted than accepted. Why should a computer scientist know calculus?, some people ask; why should a numerical analyst learn number theory?; and why should a statistician be informed about homology groups? The result of this practical attitude is intellectual chaos, which at some universities even makes it possible for a Ph.D. in pure mathematics to be totally innocent of the theory of analytic functions.

I studied hard for the prelims, and I worried, of course, but I didn't really think I would fail. The writtens were a breeze; my diary says "took analysis written; pass, of course". The oral was a positive pleasure. The committee that I remember consisted of Brahana, Carmichael, Doob, Emch, and McClure (the dean of the College of Liberal Arts, representing philosophy), and I even remember some of the questions. Doob asked about the proof of the classification theorem for 2-dimensional manifolds, and I thought I answered him, but he didn't look pleased. Later he told me that he was trying to get the intuitive background out of me, and what he wanted to hear me say was "cut and paste". Carmichael asked me who Sturm was and what he did; I floundered. Brahana made me write $\begin{pmatrix} a & b \\ -b & a \end{pmatrix}$ on the board, told me that a and b were real, and asked me what the set of all matrices of that kind was. I felt relaxed, friendly, and cocky, and I said, "if you dictated it right, it's isomorphic to the set of all complex numbers".

[I am reminded of a story I heard once and am fond of telling; it establishes an acquaintanceship thrice removed between me and one of the all-time greats of mathematics. I knew Arthur Rosenthal slightly, when he was at Purdue. When he was a student, many years before, he roomed with Landau. One day Landau came home and complained about the long and boring afternoon he had just spent with old man Dedekind. Beginning to lose his grip, all Dedekind would talk about was the old days—he complained bitterly about those hard, unkind, and unfair questions he was asked on his Ph.D. exam—by Gauss.]

When my exam was over, I was sent out of the room while the committee deliberated. I strolled up and down the corridor, taking deep drags of my cigarette, and a few minutes later they all trooped out. Emch happened to be first. He smiled, shook my hand, and said "Congratulations! It was fine—just a small condition in philosophy". That is: I passed, in fact Doob said I did well, but the pass was conditional on my submitting to and passing one more written exam in philosophy. I grumbled about that, but there was nothing else I could do; about three months later I took that exam and removed the condition.

The year was 1936–1937, my last year as a bona fide course-taking student. My courses were not arduous: one semester of differential equations (Carmichael) and two of topology (Doob), plus the accounting formality called "Seminar and thesis", Math. 100, just to make the number of semester hours come out right. The following year was really my last one, and I registered for its first semester only. During the spring semester of 1938 I wrote up my thesis and saved my fees. There it is: as far as courses are concerned I have described my entire mathematical education. How does it stack up with what goes on nowadays?

For example

Having "finished" my education, I settled down to study. The prelim pressure was over, and the thesis pressure was just beginning; in the interim days of relative peace I had time to read *Tobacco Road* and lose money playing pea pool (same as bottle pool) with Ambrose and Dick Leibler. When I asked Joe Doob if he would be my thesis supervisor, he said "I think we'll be able to find something for you to do". He had a pocket-size dime-store notebook with some possible research problems scribbled in it, a dozen or so. What he originally had in mind was that I apply modern, high-powered, measure-theoretic methods to the statistical ideas that R. A. Fisher was promulgating. He had already started to "clean up Fisher" with his work on the method of maximum likelihood, but there was a lot still left to do. The first step was for me to learn as much as I could of modern probability and statistics.

By now, having been a student for over 60 years, registered in schools of various sorts (from grade school through Ph.D.) for 16 of them, I think I know how to study. After prelims I didn't, not yet, but I began to grope toward finding a method that would be right for me. If I had to describe

my conclusion in one word, I'd say *examples*. They are, to me, of para-
mount importance. Every time I learn a new concept (free group, or
pseudodifferential operator, or paracompact space), I look for examples
—and, of course, non-examples. The examples should include, whenever
possible, the typical ones and the extreme degenerate ones. Yes, to be
sure, $\mathbb{R}^2$ is a real vector space, but so is the set of all polynomials with
real coefficients, and the set of all analytic functions defined on the open
unit disc—and what about the 0-dimensional case? Are there examples
that satisfy all the conditions of the definition except $1x = x$? Is the set of
all real-valued monotone increasing functions defined on, say, the unit
interval a real vector space? How about all the monotone ones, both increasing and decreasing?

[I cannot resist the temptation to mention a pertinent puzzle. A
routine examination question, frequently encountered near the beginning
of linear algebra courses, asks whether the set $\mathbb{C}$ of all complex numbers
is a vector space over the field $\mathbb{R}$ of real numbers. A precisionist would
object to the formulation: what should have been asked is whether there
is a way of defining multiplication of complex numbers by real ones so
as to make the additive group $\mathbb{C}$ a vector space over $\mathbb{R}$. Legend has it
that a flustered student once misread the question: he thought it asked
whether $\mathbb{R}$ can be a vector space over $\mathbb{C}$. That is: is there a way of
defining multiplication of real numbers by complex ones so as to make
the additive group $\mathbb{R}$ a vector space over $\mathbb{C}$? That's one of my favorite
puzzles. If you give up, turn to p. 64.]

When Reinhold Baer first arrived at Illinois he gave a course that I
audited for a while. Early on he defined what it meant for a subring of a
ring to be integrally closed. The definition sounded natural enough—too
natural—the only example I had ever heard of satisfied the condition.
I asked Baer to give us an example where the condition did *not* hold—
and he wouldn't. He certainly could have, but he didn't; he made an impatient sarcastic comment putting the question down and went on with
his lecture. That was typical Baer, by the way: he was not only brilliant
but also arrogant; a prolific mathematician, a forceful "Herr Professor"
in the autocratic German tradition, and a cultured and charming conversationalist.

A sample of his majestic manner of operation (which I have envied
but never had the nerve to try to emulate) is the way he used his
assistants. One of their duties was to keep scanning the new journals as
they arrived in the library, and pick out the articles of interest to Baer.
They had instructions, of course, and, even so, they had to use mathematical insight in making decisions; in any event, the result was that Baer
could look at what he wanted to look at without wading through masses
of irrelevant material.

At Illinois he was a welcome ambassador from the great world of cosmopolitan scholarship. He stayed a long time—long enough to retire,

in his middle 50's, and start collecting a very reasonable pension. Except for the pension, however, all that retirement meant for him was a change of job: the official German governmental guilty conscience toward people maltreated by the Nazis resulted in his being appointed professor at Frankfurt. He threw himself into that renewed career with vigor; he churned out papers and Ph.D.'s, and was an active participant in many conferences. When he reached Frankfurt's mandatory retirement age, he started collecting his Frankfurt pension along with his Illinois one, and started still another career in Zürich.

He and I became friends at Illinois, and we kept meeting in many countries for many decades. But I didn't go back to his course; I thought his attitude toward examples was bad.

It has become a cliché, based on Eugene Wigner's elegant phrase, to express wonder at the unreasonable effectiveness of the application of mathematics to the "real world". The other side of the coin is the maladroit ineffectiveness of mathematics at matching precise definitions to our intuitive preconceptions. (The truth might be that our intuition is vague and inconsistent, and nothing precise could possibly match it.) We all think we know what continuity means, but some of us, surely, are inclined to wriggle uncomfortably when we learn of a function that is discontinuous at every rational number but continuous at every irrational one. (A much deeper and much more famous monster is a curve that turns continuously everywhere but has a tangent nowhere.) We are all sure we know what connectedness means, but aren't we inclined to be skeptical when we learn of a curve that consists of two pieces (such as a circle and a spiral winding infinitely often around it) that is nevertheless called connected? In any event, I maintain that a student does *not* know what continuity and connectedness mean till he knows examples such as these.

A good stock of examples, as large as possible, is indispensable for a thorough understanding of any concept, and when I want to learn something new, I make it my first job to build one. Sometimes, to be sure, that cannot be done immediately: to prove the existence of even one example might need more of the theory than the mere definition reveals. When we first learn about transcendental numbers, for instance, it is not at all obvious that such things exist. Just the opposite happens when we are first told about measurable sets. There are plenty of them, there is no doubt of that; what is less easy is to find a single example of a set that is *not* like that.

Counterexamples are examples too, of course, but they have a bad reputation: they accentuate the negative, they deny, not affirm. The difference is often just a question of language and attitude, and while it has some intellectual validity, it is more a matter of emotion. Is the Cantor set a beautiful example (a perfect set with empty interior), or is it a nasty counterexample (a set of measure zero that is not countable)?

Affirmative or negative, it's examples, examples, examples that, for me, all mathematics is based on, and I always look for them. I look for them first, when I begin to study, I keep looking for them, and I cherish them all.

Do you remember the question on p. 62 (can $\mathbb{R}$ be a vector space over $\mathbb{C}$)? Much to most people's astonishment, the answer is yes.

Here is a sophisticated way of rephrasing the question: is there a ring monomorphism from $\mathbb{C}$ into the ring of group endomorphisms of $\mathbb{R}$? The answer to this formulation is obviously yes. Reason: the additive group $\mathbb{R}$ is isomorphic to the additive group $\mathbb{C}$. No? Still don't see it? Then think of Hamel bases and let them remind you that the additive group of $\mathbb{R}$ is the direct sum of uncountably many copies of the additive group of the field $\mathbb{Q}$ of rational numbers. Consequence: the additive group of $\mathbb{R}$ is isomorphic to the additive group of $\mathbb{R}^2$.

Statistics, no

Where can we find examples, non-examples, and counterexamples? Answer: the same place where we find the definitions, theorems, proofs, and all other aspects of mathematics—in the works of those who came before us, and in our own thoughts. We find them, that is, in books, published articles, preprints, courses, colloquia, seminars, personal conversations, correspondence—and, first, foremost, and above all, in ourselves, by creative thinking. At the beginning of my post-prelim and pre-thesis study I relied heavily on books—old and new—and, as I began gradually to catch on, on recent journal articles. (As for preprints: the word didn't exist yet.)

I looked up "measure" and "integral" and "probability" and "statistics" in the card catalogue in the mathematics library, made a list of attractive titles, and then went to the parts of the stacks where those books lived. I didn't find all the books on my list in the stacks (does anyone, ever?), but I found several. Scattered among them I found several others that looked interesting and that I hadn't noticed in the catalogue. I took down and looked at perhaps 30 or 40 books. Does this one look as if I could read it? Does the preface of that one exclude me (for not knowing enough, for not sharing its specialized purpose, or for any other reason)? Is the third one in a language that I have difficulty with? In less than an hour I found a dozen promising books, I staggered to the checkout counter with them, and took them home.

I didn't read all those books. I didn't read any of them. Except for the books I have written, I can think of only two mathematics books that I have ever read, in the sense of having read every word, the way I read

Evelyn Waugh and Lewis Carroll. (They are Knopp's *Theory of Functions*, which I translated, and Carmichael's *Theory of Groups*, which I proofread.) I did read the first 10 or 20 pages of all those books, and I dipped into other parts, skipping back and forth among them. (I wish I had read the first 10 pages of many more books—a splendid mathematical education can be acquired that way.) Then I returned most of them to the library and came home with another huge armload.

Hobson was on my desk, frustratingly, for a long time. Saks's *Theory of the Integral* was the bible—or perhaps the old testament— Kolmogorov's *Grundbegriffe der Wahrscheinlichkeitsrechnung* was the new. I was very enthusiastic about Saks's treatment of abstract integrals; Kolmogorov's conditional probabilities I found troublesome. Some of the volumes of the series published under Borel's editorship looked promising, but I found most of them disappointing—obscure or sloppy. I liked Levy and Roth ("an excellent book, but needs amplifying for the mathematician"), I almost got apoplexy from von Mises's ex cathedra dicta about Kollektivs, I found Khintchine difficult but enlightening, and I never learned the trick of understanding Paul Lévy. [The Levy of Levy and Roth is not Harry, but Hyman Levy, an Englishman, and Paul Lévy, a Frenchman, is one of the greatest names in modern probability theory. He combined deep insight with almost total incomprehensibility; he seemed to have no notion of what it meant either to explain or to prove anything.]

My study was meant to be directed (learn measure theory to learn probability to learn statistics), but my reading did not proceed in a straight line. I loved Saks, so I went off on tangents, dipping a lot into Fundamenta Mathematicae. That great journal was at the height of its glory in the 1920's and 30's; in it I met Tarski, and Steinhaus, and Banach, and many other great Poles. That kind of serendipity (read the "wrong" article in a volume you took out for another purpose) always works: I urge all students to give it a chance. [It is, incidentally, an argument against specialized journals. If the volumes you check out have articles on nothing but C^* algebras, you are not giving yourself the opportunity to bump into the best paper on Lie groups that appeared in the last 25 years.]

At the same time as I allowed myself to be distracted from building a solid logical foundation for my understanding of probability, I tried to go the historical way: just what did Chebyshev and Markov do? As for statistics: an extensive literature simply didn't exist. I looked into Czuber, I borrowed some notes by Hotelling, I had the chance to listen to a couple of lectures by Neyman, and I did my best to forgive Fisher for not being a mathematician. That was not easy. Diary entry: "Try to study statistics out of Fisher: impossible." My patience gave out toward the end of the academic year (May 1937): "Try Fisher again and Doob's papers. I'll take probability, and to hell with Fisher."

Readings and ratings

In addition to the Colloquium, there was the Graduate Club, a sort of colloquium junior grade. The speakers at the graduate club were graduate students and faculty, and the talks were expository. There was also a private clique, the Real Variable Club, with only six or seven members: I remember Don Kibbey, Andy Lindstrum, Phil Maker, Felix Welch, and, of course, Ambrose and myself. We were like the Graduate Club, except that all our talks were about or near real analysis (Ambrose on the Cantor function, Welch on non-separable spaces, I on "chance variables") and except that (we snootily thought) our talks could be on a higher level. I first met Hilbert space in the Real Variable Club, and I was pleased ("very cute").

Along with all that, I tried to keep adding to my general mathematical culture by, for instance, reading Felix Klein's *Elementary Mathematics* ("from the higher point of view out"), and by dropping in on an occasional lecture in Bob Martin's course. R. S. Martin came from Princeton, a different world, with an up-to-date, active, exciting mathematical atmosphere, inhabited by glamorous people with famous names. Though he was kind, he had a sharp wit; he's the one from whom I first heard the judgement "there's less in it than meets the eye" (speaking of Tarski's celebrated paper on the meaning of truth). He understood mathematics, but he couldn't explain it. He spoke well about literature and gossip and politics and history, but the mathematics he knew was too easy for him, and he would stutter and stumble trying to clarify the obvious. Students in his classes couldn't follow him, he was called a bad teacher, and he was on the point of being fired when he died. (He had an infected heel. Several of us were tested for blood transfusion, but only Ambrose turned out to be acceptable, and the blood didn't help. Penicillin could have saved his life, if it had been discovered four years before it was.) Bob is the one who discovered the famous Martin boundary in potential theory, but because he published very little and died young, other Martins frequently get the credit.

Graduate students spend their time studying (they had better!) and griping (which they don't have to be taught). A frequent subject for our gripes was the Ph.D. rating system and its consequences. Course grades don't mean much (many teachers seem to find it difficult to look a student square in the eye and give him a low grade), and, in particular, they are of almost no help to a faculty trying to make an intelligent estimate of a student's chances of getting the Ph.D. The rating system was designed to be more helpful. The idea was that each member of the graduate faculty rated each graduate student whom he knew, and reported his rating to the chairman's office in confidence. Students were told their averages only—not the individual ratings. A rating of 1 meant "top Ph.D. caliber"; a rating of 3 meant "hopeless".

We, the students, didn't think much of the faculty's estimate of us: we were sure we knew better. By now I am not so sure. My experience since then suggests that students tend to overrate students; each is so sure that he himself doesn't know enough that he is too much impressed by even a small gleam of what looks like intelligence in someone else.

The ratings were the main base for the award of assistantships and fellowships. Ever since that lucky first semester in 1935, I was eagerly hopeful every September and every February (another emergency?, another chance to teach?), and anxiously prayerful before April 1, the day that next fall's fellowships were announced. The diary reports the bad news each September, each February, and each April—"no job". The last such entry is in 1937: "And for what seems to be the last time: I did not get a fellowship."

Reprints: Doob's and others'

Xerox didn't exist yet and the use of preprints, while technologically possible, was not widespread—how then did you assimilate the currently available scientific information? Answer: you could subscribe to the journals, or use the library copies, or ask authors to send you reprints. Journal prices were not outrageously high. For a while, for instance, in the early 1940's, there was a special deal: as I remember it now, if you joined the MAA (the Mathematical Association of America, called just "the Association" by the in-group), you could subscribe to the Annals of Mathematics for $4 a year. As for using the library copies, that was much more feasible than people nowadays seem to think. If the paper you wanted to look at was in a journal that did not circulate, no great harm was done—usually in half an hour (or even in five minutes) with the volume before you on a library table, and paper and pencil handy, you could learn what you needed. (Doesn't the modern tendency to Xerox first and look later tend to be rather wasteful?) If the paper was long and you were slow, the time would grow—you would have to keep going back to the library and spend many hours over that volume at that table. Not so comfortable as lying on your own living room couch with a few sheets of Xerox, but not too bad. Besides, if you wanted, nothing stopped you from copying the article, or the crucial parts anyway, by hand. That's not necessarily a waste of time—you learn as you go.

Writing away for reprints was a common practice. Almost everyone was pleased to be asked and sent what was asked for. Sometimes the printed matter was accompanied by a more personal communication. I was thrilled, for instance, when I received a postcard addressed to Dr. Paul R. Halmos (six months before I got my degree) by Khintchine, the great Khintchine. He saluted me as "Sehr geehrter Herr Kollege!", and he apologized for no longer having any copies of one of the papers I asked for. He sent some others on related subjects, and he expressed the hope that they might interest me. Pride and glory!

I was an avid collector: I wrote to France, Germany, Denmark, Russia, Poland, and Japan, and, of course, all over the U.S., and I accumulated a splendid collection. I had (and still have) reprints of papers by Paul Lévy, Eberhard Hopf, Børge Jessen, Marcinkiewicz and Zygmund, and Yosida and Kakutani, as well as Bochner, Copeland, Feller, Tukey, van Kampen, von Neumann, Wiener, and Wintner. I treasured them, and in many cases went so far as to get bunches of them bound together, in good quality, long-lasting library buckram; four such volumes, red with gold lettering on the spines, are still in my bookcase. They may be, or they may someday become, valuable collector's items: they contain Hopf's earliest works on ergodic theory, and what may be Tukey's first paper on statistics.

Once I got them, I read them and, in many cases, studied them in attentive detail. I enjoyed Steinhaus on the probability that a series such as $\sum \pm \frac{1}{n}$ converges, and on the probability that a power series such as $\sum \pm \frac{z^n}{n}$ has the unit circle for a natural boundary. I struggled with the mystic approaches that several Germans (Reichenbach, Tornier, von Mises) took to probability. Part of what they wrote was pure nonsense, and I couldn't help wondering whether that was intellectually related to the motives that made some of them active and proud Nazis. One of the earliest papers on probability via measure on infinite product spaces is by Łomnicki and Ulam. It's a good paper, but there is a mistake in it, a big one, involving non-uniform convergence. I struggled with it, but couldn't see what was wrong. The volume of Fundamenta that contained it *was* allowed to circulate, and I had it at home. For a period of two or three weeks Joe Doob stopped at my house almost every other day to look at the Łomnicki–Ulam paper: he too was having difficulties. Ultimately he pinpointed the error, and, in a subsequent paper of his own, showed how to save the pieces.

Joe knew about the ergodic theorems that von Neumann and G. D. Birkhoff published in the early 1930's, and he was one of the first to exploit the connection between them and the so-called strong law of large numbers. I spent a lot of time with Joe, without consciously realizing the extent to which he was educating and inspiring me, and I spent most of my study time reading and swearing at his papers. Diary entry, April 25, 1937: "Struggle with Doob's 'Limiting distributions' all afternoon and evening. See Doob about it. I am a dope." Diary entry two days later: "Work on Doob paper afternoon. Doob is a dope; win 5¢."

Doob has an outgoing, relaxed personality; he seems to get along easily with everyone. He seems to take nothing seriously, but he takes everyone equally seriously. He is never intellectually dishonest—well, hardly ever. There was the time when Ambrose and I insisted on pinning him down. He said something about time and motion (could it have arisen from a discussion of homotopy?) and when we demanded precise definitions, in terms of sets and functions and parameters, he said "you can't do that—you can't make the dynamic static". Pure baloney.

He has no feeling for elegance in organization, exposition, language, or notation. Here is an example. In some of his early stochastic process papers he used expressions such as "the space of all functions $x(t)$" [naughty—the name of the variable shouldn't be there], and a few pages later he discussed the assumption that "$x(\tau)$ is a measurable function on $T \times \Omega$". The latter expression was meant to refer to the mapping $\langle x, \tau \rangle \mapsto x(\tau)$, the one that sends the pair consisting of a function and a point to the value. You were expected to have guessed that t was a "variable", and that the use of the Greek letter τ meant something else. For reasons such as this, and worse, his papers are not easy to read. At the same time, however, he has great feeling for elegance of idea and great insight into depth. Are he and I, I sometimes wonder (at times when I am not maximally satisfied with myself), exactly complementary in our talents?

Study

Here you sit, an undergraduate with a calculus book open before you, or a pre-thesis graduate student with one of those books whose first ten pages, at least, you would like to master, or a research mathematician (established or would-be) with an article fresh off the press—what do you do now? How do you study, how do you penetrate the darkness, how do you learn something? All I can tell you for sure is what I do, but I do suspect that the same sort of thing works for everyone.

It's been said before and often, but it cannot be overemphasized: study actively. Don't just read it; fight it! Ask your own questions, look for your own examples, discover your own proofs. Is the hypothesis necessary? Is the converse true? What happens in the classical special case? What about the degenerate cases? Where does the proof use the hypothesis?

Another way I keep active as I read is by changing the notation; if there is nothing else I can do, I can at least change (improve?) the choice of letters. Some of my friends think that's silly, but it works for me. When I reported on Chapter VII of Stone's book (the chapter on multiplicity theory, a complicated subject) to a small seminar containing Ambrose and Doob, my listeners poked fun at me for having changed the letters, but I felt it helped me to keep my eye on the ball as I was trying to organize and systematize the material. I feel that subtleties are less likely to escape me if I must concentrate on the bricks and mortar as well as gape admiringly at the architecture. I choose letters (and other symbols) that I prefer to the ones the author chose, and, more importantly, I choose the same ones throughout the subject, unifying the notations of the part of the literature that I am studying.

Changing the notation is an attention-focusing device, like taking notes during lectures, but it's something else too. It tends to show up the differences in the approaches of different authors, and it can therefore serve to point to something of mathematical depth that the more complacent reader would just nod at—yes, yes, this must be the same theorem I read in another book yesterday. I believe that changing the notation of everything I read, to make it harmonious with my own, saves me time in the long run. If I can do it well, I don't have to waste time fitting each new paper on the subject into the notational scheme of things; I have already thought *that* through and I can now go on to more important matters. Finally, a small point, but one with some psychological validity: as I keep changing the notation to my own, I get a feeling of being creative, tiny but non-zero—even before I understand what's going on, and long before I can generalize it, improve it, or apply it, I am already active, I am doing something.

Learning a language is different from learning a mathematical subject; in one the problem is to acquire a habit, in the other to understand a structure. The difference has some important implications. In learning a language from a textbook, you might as well go through the book as it stands and work all the exercises in it; what matters is to keep practising the use of the language. If, however, you want to learn group theory, it is not a good idea to open a book on page 1 and read it, working all the problems in order, till you come to the last page. It's a bad idea. The material is arranged in the book so that its linear reading is logically defensible, to be sure, but we readers are human, all different from one another and from the author, and each of us is likely to find something difficult that is easy for someone else. My advice is to read till you come to a definition new to you, and then stop and try to think of examples and non-examples, or till you come to a theorem new to you, and then stop and try to understand it and prove it for yourself—and, most important, when you come to an obstacle, a mysterious passage, an unsolvable problem, just skip it. Jump ahead, try the next problem, turn the page, go to the next chapter, or even abandon the book and start another one. Books may be linearly ordered, but our minds are not.

Does your spouse believe everything you say? Mine doesn't always accept my expert opinion. Once when my wife wanted to read some mathematics, I gave her the advice (about skipping) that I just expressed, and she looked skeptical. The next day, however, she showed me, pleased and excited, the preface of a pertinent book: "look, you were right, they say just what you said". Sure enough, the preface said that a reader must not "expect to understand all parts of the book on first reading. He should feel free to skip complicated parts and return to them later; often an argument will be clarified by a subsequent remark". What my wife didn't notice was that "they" were not exactly a source of independent confirmation. "They" were an editorial panel, with their names listed on the title page; I was a member of the panel, and the words "they" said were in fact written by me.

As long as I am musing on how I learn mathematics through the eye, I might muse a moment on learning through the ear. Lecture courses are a standard way of learning something—one of the worst ways. Too passive, that's the trouble. Standard recommendation: take notes. Counter argument: yes, to be sure, taking notes is an activity, and, if you do it, you have something solid to refer back to afterward, but you are likely to miss the delicate details of the presentation as well as the big picture, the Gestalt—you are too busy scribbling to pay attention. Counter counter-argument: if you don't take notes, you won't remember what happened, in what order it came, and, chances are, your attention will flag part of the time, you'll daydream, and, who knows, you might even nod off.

It's all true, the arguments both for and against taking notes. My own solution is a compromise: I take very skimpy notes, and then, whenever possible, I transcribe them, in much greater detail, as soon afterward as possible. By very skimpy notes I mean something like one or two words a minute, plus, possibly a crucial formula or two and a crucial picture or two—just enough to fix the order of events, and, incidentally, to keep me awake and on my toes. By transcribe I mean in enough detail to show a friend who wasn't there, with some hope that he'll understand what he missed.

One-shot lectures, such as colloquium talks, sometimes have a bad name, probably because they are often bad. Are they useful to a student, or, for that matter, to a grown-up who wants to learn something? The something might be specific (I need to know more about the relation between the topological and the holomorphic structures of Riemann surfaces) or only a vague but chronic pain in the conscience (my mathematical education is too narrow, I should know what other people are doing and why). My own answer is that colloquium talks, even some of the very bad ones, are of use to the would-be learner (specific or vague), and I urge my students and colleagues to support them, to go to them.

Why? Partly because mathematics is a unit, with all parts interlocking and influencing each other. Everything we learn changes everything we know and will help us later to learn more. A good colloquium talk, well motivated, with a sharp central topic, well organized, and clearly explained is obviously helpful—but even a bad one can be helpful. My favorite case in point is a bad talk I heard once on topology. I didn't understand the definitions, the theorems, or the proofs—but I heard "stable homotopy groups" mentioned and, a few minutes later, "Bernoulli numbers". I had only the dimmest idea of what stable homotopy groups were, and an equally dim one of Bernoulli numbers—but my knowledge grew and my ability to understand mathematics (and, in particular, future colloquium talks) became greater just by listening to that odd, surprising, shocking juxtaposition. (It has become commonplace to the experts since then.) It's worth 55 wasted minutes to learn, in 5 minutes, that the theory of the zeroes of meromorphic functions has a lot to do with the distribution of prime numbers.

Seminar talks are thought to be different, but they are not really. The usual notion is that the audience in a seminar consists of experts who know everything that has been proved in the subject up to the day before yesterday and are just there to see the last filigree. That notion is false; a good seminar talk is good and a bad one is bad, the same as a colloquium talk. The idea that the speaker's technical incomprehensibilities become conceptual insights if the occasion is called a seminar instead of a colloquium is just not realistic, just not true. The "experts" present, be they colleagues working on problems similar to the speaker's (but remember: "similar" is almost never "identical"), or graduate students trying to work their way into the arcane technicalities, get lost and impatient and bored only 30 seconds later than the so-called "general" audience at a colloquium. In my opinion good seminar talks and good colloquium talks are interchangeable—and even the bad ones are worth going to.

The best kind of seminar has two members. It can last five minutes—a question, followed by a partial answer and a reference—or it can be a strong collaborative bond that lasts for decades, and it can be many

A. L. Shields, 1969

things in between. I am very strongly in favor of personal exchanges in mathematics, and that's one reason I am in favor of colloquia and seminars. The best seminar I ever belonged to consisted of Allen Shields and me. We met one afternoon a week, for about two hours. We did not prepare for our meetings, and we certainly did not lecture at each other. We were interested in similar things, we got along well, and each of us liked to explain his thoughts and found the other a sympathetic and intelligent listener. We would exchange the elementary puzzles we heard during the week, the crazy questions we were asked in class, the half-baked problems that popped into our heads, the vague ideas for solving last week's problems that occurred to us, the illuminating comments we heard at other seminars—we would shout excitedly, or stare together at the blackboard in bewildered silence—and, whatever we did, we both learned a lot from each other during the year the seminar lasted, and we both enjoyed it. We didn't end up collaborating in the sense of publishing a joint paper as a result of our talks—but we didn't care about that. Through our sessions we grew... wise?... well, wiser, perhaps.

CHAPTER 5

Learning to think

Optional skipping

Ray Wilder has been president of both the major mathematical organizations in the U.S.; he was a member of the National Academy, and the author of several books and many articles. When he was 80, fifteen years after he retired from the University of Michigan, he told me that his days of study were definitely not over (and neither was the feeling of pressure that makes one study): he was still reading mathematics, going to colloquia, and trying to keep up with what was going on.

In one sense or another a scholar learns all his life, but absorbing the knowledge that others discovered (study) is a very different kind of learning from discovering a part of the truth yourself (research). Some time in the spring or summer of 1937 the studying that I had done got me to the point where I could see into a certain narrow tunnel as well as anyone, and the time had come to stop studying and start digging. Doob certainly suggested a direction in which I could look for a thesis (the tunnel), and I don't remember the extent (if any) to which he also suggested the first theorem or two.

The basic idea was to extend Doob's mathematization of the von Mises *Regellosigkeit* (irregularity) principle. What does "random" mean? A part of the von Mises answer is that a sequence of observations of an experiment, in a scientific laboratory or in a gambling casino, doesn't deserve to be called random unless it is totally unpredictable. In the language of the casino, unpredictability means the non-existence of a successful gambling system. For a simple but typical example, consider an experiment with only two possible outcomes (yes and no, on and off, 0 and 1, or heads and tails) repeated regularly, once a minute. Doob's proposed definition of a gambling system in this context is a sequence of functions $f_1, f_2, f_3, \ldots$, the nth one a function of n variables, attaining the values yes and no only. The idea is that the gambler bets on experiment number $n + 1$ if the result $f_n(x_1, \ldots, x_n)$ of feeding the first n outcomes $x_1, \ldots, x_n$ into the function f_n is yes; if it's no, he sits out that deal. (Doob describes these goings on by the phrase "optional skipping".) Possible concrete special case: $f_n(x_1, \ldots, x_n)$ is yes if and only if the last ten x's are tails. Doob's formulation of the ingenious but imprecise *Regellosigkeit* principle is a theorem: the transformation that converts each infinite sequence $x_1, x_2, x_3, \ldots$ of outcomes into the sequence $y_1, y_2, y_3, \ldots$ of the outcomes of the games that the gambler actually bets on preserves all probability relations. It follows, in particular, that with probability equal to 1, the limiting average of the x's is the same as the limiting average of the y's. In other words, there can be no "successful" gambling system; no matter what system you use to decide when to bet and when to wait, your chances are the same as if you had decided to bet every time—neither better nor worse.

Doob's theorem was about independent repetitions of an experiment; my problem was to see what could be said about not necessarily independently performed experiments. Typical of the sort of conclusion that I was eventually able to prove is that optional skipping cannot change the "fairness" of a game. Suppose, to be a little more explicit, that the "random variables" $x_1, x_2, x_3, \ldots$ are not necessarily independent, but they describe a fair game in this sense: the conditional expectation of the gambler on game number $n + 1$, given the outcomes $x_1, \ldots, x_n$ of the first n games, is always 0. (Note: therefore the conditional expectation of the casino is 0 also; the game treats both sides of the table equitably.) Example: the gambler bets on the total rainfall each day; he wins if it is over 5 millimeters and loses otherwise. The observations are not independent—it is more likely to rain Tuesday if it rained Monday—but the payoff is arranged to make it a fair gamble. That is: on a rainy Tuesday following a rainy Monday the payoff is small and on a dry Tuesday following a rainy Monday the gambler's loss is large, with the numbers arranged so that, on the average, the conditional expected win is 0, no matter how much or how little precipitation there was the preceding week. Under these circumstances the gambler using optional skipping may increase the frequency of his wins, but, because of the rigged payoff, will still come out financially even in the long run.

Many years after I wrote my thesis, Jimmie Savage and I collaborated in teaching a course on stochastic processes at the University of Chicago. (We alternated weeks, more or less.) One week I was supposed to lecture on gambling systems, and I got mixed up in class, badly confused. By then I had worked on several other parts of mathematics, and probability no longer came naturally to me. Since Doob's book was our official text, I tried to get myself unconfused by looking at it. To my astonishment, I found that Doob generously described the work of the pertinent couple of pages (309–311) as "following Halmos"—the stuff that was the hardest for me in the 1950's was the same as the stuff I struggled with in the 1930's.

Roller coaster

Theses aren't produced by instantaneous inspiration—at least mine wasn't. The process was a less than merry merry-go-round, a roller coaster, with rare good times mixed in with many discouraged days. June 1937: "It's fun to have ideas, even if they don't pan out right"; two days later: "it's much more fun to have ideas when they *do* pan out right". Next day, exasperated: "nobody knows anything about measure preserving transformations". Many days: "work on research, ... stuck on research, ... wish I knew the answer, ... , the more I get into my thesis reading, the less I know". Later, in October: "unusually cheerful Doob conference, ... making headway with Doob's paper, ... start thesis!" Three days later: "stuck on thesis, ... slow hammering", and, a week after that: "get hot—finish theorem—exult to Doob". The roller coaster continues on and on this way, slumping to one particularly bad low, followed by a premature high, and then a maddening, goading, spurring low, as the months of my last student year keep rolling by.

Doob was good to work with. He did not hound me, sometimes he encouraged me, and he was always ready to listen. We weren't working in Paradise though, and we were human—there were times when he lost his patience with me, and there were times when I thought he could help more. At the time of the particularly bad low, I must have been pestering him too much, asking him too many petty, detailed questions, and that was when he lost his patience. I forget the exact words he used but I sure remember the idea: "This is supposed to be your paper, you know, but if we go on this way, it's going to have to be a joint one, and that won't look so good for you."

That rocked me. I went to work, with a vengeance, and the tone of the diary entries becomes almost cheerful. "Get continuous case afternoon, ... work on thesis all day—capital fun—can't get the damn series to converge, ... find the desired Gegenbeispiel, ... maybe there is a theorem on the horizon—asymptotic theorems look hopeful, ... get invariance of law of large numbers, ... start blue-pencilling thesis—hard work, ... , type and fill in thesis (as much as there is of it)—see new possibility in continuous stochastic processes."

[The meaning of "fill in" is probably no longer known; in those days it was an inevitable part of mathematical typing. If you were lucky, your typewriter had a plus sign and an equal sign, but Greek letters and inequalities, not to mention integrals and the other weird symbols that mathematicians are fond of, were unheard of. I doubt if any typewriters are produced now without a half-space ratchet; then it was an expensive luxury. The result was that mathematical typing consisted of typing the prose and skipping enough spaces to leave room for letters from other alphabets and for weird symbols—once that was done you went back over the page and by hand, with a pen, *filled in* the subscripts and the rest that you had left out. When, about ten years after my thesis, I typed my book *Measure Theory*, several hundred pages of an original and five carbons, I filled in each of the six times several hundred pages—short of using the expensive and unsatisfactory photostat, there was no other way to get six copies.]

My exultation was premature: what I anxiously referred to as my thesis didn't make the grade. One afternoon in January my diary says "Thesis showdown with Doob. I'll show them, if it takes me ten years! Superficial, eh?!" The professional difference of opinion didn't hurt our friendship; later the same week the diary says "Beat Joe in very good squash games."

From then on things got better. Further, successive, diary entries: "Getting somewhere?;... Work a little all day—not worried, because I enjoy working;... stuck at 8:30 a.m.—unstuck at 4:30 p.m.;... got somewhere today—consolidate position and make great advances;... good work today—two major additions to thesis;... work on Gegenbeispiel— finish it;... exultant and optimistic—things are going splendidly." In the middle of March (1938) I spent four days copying what I again hoped was my thesis, and then turned it over to Joe. Next day: "Joe irritating about footnotes", but, a week later, "I'm as good as a Ph.D.".

Have I conveyed my perception of the frightening-glorious up-down-up phenomenon of research? First you are excited because you understand the problem and are eager to get on with it. There is a lot of underbrush to be cleared that keeps you busy and makes you think you are getting somewhere—and then you get stuck. You keep working, you try a different avenue every week (every day?, every hour?), you go round and round, you think about it as you eat, as you walk, as you dream, but you are on a plateau and nothing seems to be happening.

As I gradually realized what was going on, I began to see that I just couldn't attack the theorems head on, in their ultimate general form— I had to try to break them and their proofs down to their simplest special cases. I kept simplifying the problem. I made the measure spaces finite sets, I restricted the optional skipping functions so as to make each one depend on only one variable, I replaced infinity by 2, and I replaced convergent sequences by constants. The need for that sort of activity is obvious to anyone who has ever tried to do research. It's all said by Pólya's dictum: if you can't solve a problem, then there is an easier problem you can't solve—find it!

I found it, eventually, I reached the triviality that I could see through—and that was the breakthrough. The down was over and I was on the way up. The rest (for me then, and I think for everyone always) is just work—exhilarating, rewarding, backbreaking work. Squeeze the proofs, simplify the techniques, extend the results, harvest the corollaries, and finish in a blaze of glory. A month after my thesis glory the official version was typed and handed in to the graduate school; another copy of it, an identical copy, was mailed, submitted for publication in the Duke Journal. Its descriptive and only slightly rambunctious title was "Invariants of certain stochastic transformations: the mathematical theory of gambling systems". The glory, by the way, lasts a few days only, perhaps a week—and then it's "why haven't I been able to do anything for such a long time?".

Jobs, no

I didn't spend every minute of my last student year on the thesis—but the measure of the set of exceptional minutes was very near to zero.

At the beginning of the academic year, in September 1937, I attended my first American Mathematical Society meeting—AMS or just "the Society"—in State College, Pennsylvania. I didn't know a soul, everything was strange, I didn't learn much, and I felt lonely.

As the year progressed, I relaxed with frequent games of snooker, and many sessions with Banach's book. The latter was an addiction. The diary has many entries such as "study Banach all day,... I'm a fiend for it,... Banach is not too easy,... Banach morning conquered with aid of Vaughan."

Oh, yes—Herbie Vaughan. He was a young and still ambitious instructor, recently arrived from Michigan, a topologist. We talked a lot about mathematics, and other things, and, twelve years later, we collaborated on a tiny 2-page paper, "The marriage problem". The paper made us famous in combinatorial circles; it could be the most frequently cited paper that has my name on it. The marriage problem assumes that each of a possibly infinite set of boys is acquainted with a finite set of girls; it asks for additional conditions that make it possible for each boy to marry one of his acquaintances. Herbie contributed a formulation of the induction proof in the finite case that was clearer than any earlier one, and I contributed a topological proof, via Tihonov's theorem, of the infinite case. Herbie then went on to deduce an ingeniously generalized corollary, which he dubbed "the celebrated problem of the monks"; for its historical background he cited a nonexistent part of the *Droll Stories* of Balzac, "Des moines et novices".

The reference impressed many readers; I still get a letter every now and then asking for help in locating it.

On my active reading list in 1937–1938 were the several papers Nikodým kept publishing on what came to be called the Radon–Nikodým theorem, Julia's *Théorie Quantique*, and every issue of the Zentralblatt as it came out—several of us almost snatched it out of each other's hands in our eagerness to get the latest news.

Another important and highly time-consuming non-thesis activity was typing job applications. Xerox did not exist, and secretarial service was not available to starving graduate students. I typed 120 letters of application, mailed them out, and checked my mailbox several times every day for the answers that did not come. A few did—more than the two I remembered the last time I told this story, but not very many more—my diary reveals five (or, if one unclear entry refers to an answer, then six). The diary refers to Harvard several times: on March 8, "wait for mail from Harvard"; on April 1, "no news is good news from Harvard"; and, finally, on April 7, "Professor Walsh is sorry". Later in April "nibble from Michigan"; three days later, "Michigan?"; on May 6 "wait for mail"; on May 11 "no job—wait for mail"; on May 20, "Professor Hildebrandt (of Michigan) is sorry". Columbia, Kansas, and Ohio State were sorry too. Total number of offers received: zero.

Looking for a job, I went, of course, to the spring meeting of the AMS in Charlottesville, and, as a recently joined member of the Society, I exercised my privilege of giving a 10-minute talk about my thesis. The Associated Press routinely sent a couple of reporters to see what all those mathematicians were up to, but the reporters were badly discouraged by the titles of the important talks—they didn't know what the words meant. That is, they didn't till they came to the second half of my title— the part about gambling systems—and then they became interested. J. R. Kline was secretary of the Society, and he asked me if I'd mind being interviewed. The answer was, yes, of course, I'd mind, but, why not, it would be interesting, and, sure, yes, it might contribute toward the solution of the job problem. The reporters and I had a pleasant conversation; they put a brief item on the AP wires; most of the newspapers in the country found it easy to ignore the item; but for the *Daily Illini*, back home, it was hot news. It was hot not only because of the "local boy makes good" aspect, but for another reason. It turned out that Champaign was waging one of its almost periodic wars on vice; the city fathers were against wine, women, and song—and, incidentally, gambling. My interview was timely; my name was large in the headline: "YOU CAN'T WIN, SAYS P. R. HALMOS".

The final Ph.D. exam was a formality, and, like everyone, I sailed through it with ease. I was glad and proud; and I am still glad and proud to have been Doob's student. [I was, by the way, his first student. He had four Ph.D. students within two or three years of each other, and then a gap—a few years went by before he started producing them again. The second, third, and fourth were Ambrose, Kibbey, and Blackwell.] As soon as the exam was over I heaved a sigh of relief and I thought "Never again! I never have to take another exam in my whole life!"— and I still wake up some mornings feeling good about that. I didn't attend the commencement exercises; I disapproved of such goings on. I had my Ph.D. diploma mailed to me and I filed it away. I still have it, but it's not framed, hanging on a wall—it's just filed, under old records.

The year I got my Ph.D., 68 people got Ph.D.'s in mathematics in the U.S.; their names are listed in the 1939 volume of the Bulletin of the AMS. Twelve of the 68 were from the University of Illinois. One of the twelve was Netzorg; except for him, I was the best of the twelve, I thought. I thought of all the fellowships and assistantships and jobs that I kept applying for and never got, but all the others did. It still rankles a bit.

On my own

The next year, roughly from July 1938 to July 1939, was a kind of interregnum. Life went on, things happened, but it seemed then, and it still seems that way in retrospect now, that I had to run as fast as I could just to stay in the same place.

Early in August I had a minor operation on my bad foot. ("Minor" may have been the surgeon's word for it; the next two months of first pain, then discomfort, then annoyance, and worry all along, seemed to me to call for a more impressive term.) In the convalescent stage I was advised to use a walking stick—and I have been using one ever since. I still have my first one; worn smooth, the paint long gone, still as good as new.

Later in August I got my last two pieces of job news: no from Duke, yes from Illinois. (Doob told me, many years later, that J. J. Gergen, the chairman at Duke, told him why I wasn't hired. "I don't want any refugees", he said.) How the yes from Illinois came about is fuzzy in my mind by now—partially through pity, perhaps, as a consolation prize? I wasn't looking that gift horse in the mouth, but my diary is curiously unjoyful; it just records the facts. The pay was $1800 per year; the duties were 15 hours of teaching per week. Those terms were standard. And, glory!—I had it made, I was a grownup; I was allowed to teach calculus.

Classes have personalities. Some are cooperative, some are surly, some are friendly, and some are dull. Two of my classes in the autumn semester of 1938 were trigonometry. They were 2-hour courses, back to back: one was at 9 in the morning, Tuesday and Thursday, and the other, after a ten-minute cigarette break, at 10. I loved the 9 o'clock class and I hated the 10 o'clock. The differences between them could have been my fault, at least in part. For the 9 o'clock I was fresh and I was curious to see how the explanations I had figured out would go over. I told jokes, and they laughed. We loved each other. By 10 o'clock I was jaded, and I gave the explanations much more patly, smoothly, quickly. I told the same jokes, and they stared at me. We didn't like each other.

Fifteen hours a week looks like a heavy teaching load now, but we all learned to live with it then. On the days when I had three classes in a row, I grumbled—that's work!—but I still had time for research. I didn't have a lot of time, but I had some, and, of course, not having any committee duties, or any administrative duties whatever, my 15 hours of teaching left me more research time than it would have later.

As far as research goes, I was on my own—a scary way to be. Doob was still around, of course, but he wasn't "teacher" any more—he was just a slightly older and much wiser friend and colleague. Bob Martin was around, and he was helpful during those weeks when I was trying to prove that a series converged, and I kept proving, time after time, that it was dominated term by term by the harmonic series. Ambrose wasn't around; he accepted a temporary job at Alabama. We corresponded furiously—two or three long letters each week—but that's not the same as eye-to-eye conversation.

The diary entries for the year reveal not so much the roller coaster effect of not knowing the answer as the same effect, more frustrating, of not knowing the question. Here is what one sequence of them (stretching from July to December) looks like. "Get research idea—hope it works. ... Worry about details of problem all day. Feel extraordinarily good. I think I've really got something. Now if I can only find the answer.... Stuck tight on problem all day, but not feeling lost yet.... Stuck tight on research. Dark doubts about my being a mathematician.... Get hot and then run into blank wall on research. Worry about research future.... Giving up research problem.... Inequality in better form; still incomplete.... Chat with Doob: get good hint on inequality. Work, exult. Life is nice.... Putter futilely with my series. My research is discouraging, but I haven't given up yet.... Waste another afternoon on my inequality.... Snappish, irritable day. Write up results of last six months' work: six pages of shallow stuff."

Interlarded with that is a sequence of notes reporting the search for a way out. "I can't read mathematics for more than four hours a day. [I still can't. Can anyone?]...Worry about mathematical future; decide on algebra. The stuff is fun. Construct Gegenbeispiels all over the place. ...I feel that I am learning a lot of mathematics this year.... I have (again!) decided to change fields and see if I can't do research in some mathematics that *is* mathematics: Hilbert space and cognate subjects. To hell with the Paul Lévy's.... Capital fun reading Wintner's *Spektraltheorie* ... One damn day after another with only Wintner's

Spektraltheorie to sweeten life. . . . I *think* I've found my field, but it's going to be hard sledding. I hope that the knowledge axis is Archimedean."

I had a theory, first subconscious and later deliberate, that the way to stay young is to change fields often. I never did become an algebraist, but I did change from measure theory (under which I mean to include probability and ergodic theory) to Hilbert space, then to algebraic logic, and then back to Hilbert space, with some dabbling with topological groups and statistics sprinkled in. After about thirty years, around 1968, I realized that I wasn't likely to be able to change again, and, at about the same time, I started writing more expository papers (such as "Ten problems in Hilbert space") and sermons (such as "How to write mathematics").

The end of an era

It wasn't all teaching and research—I still had time, in those days, to practice the piano, at least a little, and I had time to read, to see a few shows, and I had to take time to keep applying for jobs.

I needed to read something by Pincherle; to be able to do so, I spent one strenuous day with an Italian dictionary and a grammar and "learned" Italian. I read about the "lines of Julia", normal families, and Picard's theorem; I tried to learn something about integral equations; and I spent quite a bit of time reading Uspensky's book on probability. In between I read several Charlie Chan books, Briffault's *Europa*, Dashiell Hammett, Nero Wolfe, philology (was it called linguistics yet?), Dorothy Sayers, Saki, Ellery Queen, Upton Sinclair's *Jungle*, and Hermann Weyl's *Space, Time, Matter*.

I listened to Toscanini on the radio Saturday afternoons, and went to see "Ruddigore" (local amateurs), "The lady vanishes" (a classic Hitchcock movie), Alec Templeton (a pianist-humorist-imitator, in a sense a precursor of Victor Borge), and "La Grande Illusion" (my favorite movie—I have seen it 14 times). I kept late hours—going to bed at midnight, or 1, or 3, or even 5 a.m., and, on rare occasions, skipping a night's sleep.

And I typed letters. I don't know how many application letters I mailed that year, but it was a lot. I applied for the super good things (National Research Council fellowship), the good things (Benjamin Peirce—pronounced "purse"—instructorship at Harvard), and the ordinary things (any teaching job anywhere). I got more answers that year, but they were no and no and no; they kept coming from the middle of February all through the spring. The NRC said no early in April, and Harvard a few days later; a nibble came from California, and fizzled out, early in May. The diary keeps referring to the mail. "Lots of mail morning; none of it good..... No mail.... It's not pleasant to live under this cloud of uncertainty.... I sure wish I had a job.... Bad news from Seattle.... Type letters all afternoon...."

Finally, in the middle of May, there was a flurry—a temporary job opened up at Oregon State College (as it was then called) in Corvallis, and, of course, I went after it as hard as I could. No interview—too far away, too expensive—but several letters and a few telegrams—and in the middle of June I had it. A month later, in the middle of July, I apologetically resigned from it!

What happened was that on July 7 Ambrose heard from the Institute (THE Institute—the Institute for Advanced Study in Princeton)—he had been an alternate for a fellowship there, and the first designate turned the offer down. Oh, no you don't, I said; you can't do that to me. If *you* are going to the Institute, then so am I—come hell or high water, I'll go. (Incidentally, Ambrose doesn't agree with this version of the story. He wasn't alternate, he says, and, anyway, how could we have known that he was?—that sort of thing isn't usually revealed to the person concerned. He may be right, but I can only tell the story the way I remember it. As for how we could have known: perhaps only from the lateness of the date—such fellowship-like appointments were usually announced on April 1—and possibly from an explanatory-apologetic cover letter— "we couldn't invite you sooner, because...". The main facts are uncontested: Ambrose had the invitation, and I was bound and determined to stay even with him.)

I made wild plans, and then I gave them up; I kept hoping, and then I became disheartened. I calculated probable costs, and wrote my father: could I borrow $1000? My hopes and plans were public knowledge in the mathematics building. Sam Campbell (a fellow graduate student— I'll always remember him fondly) offered to lend me $300 (an offer that, as things turned out, I didn't need to accept). My father came through. I wrote to Oregon and asked to be released; I was. The next month or so was spent in excitedly but efficiently rearranging the circumstances of daily life. I transferred the lease of my apartment (easy—apartments were hard to find); I sold my bike; I packed all my books and notes; and, on August 25, 1939, four days before the tenth anniversary of my first arrival in the U.S., I took the train to Princeton. (Young Ph.D.'s didn't have cars.) Nine days later World War II broke out.

CHAPTER 6

The Institute

The common room

For most people in this country (me included) the war was a worrisome, exciting, but remote series of news items—a couple of years were to pass before America became part of it and it started to affect our daily lives. In the meantime life went on as before. The depression was still with us, though it had begun to abate, bus drivers drove busses, doctors doctored, and the members of the Institute for Advanced Study did research.

When the Bambergers were persuaded to endow the Institute, their advisor, Abraham Flexner, wrote to the world's top six or eight people in each of three or four disciplines and asked them "who are the top six or eight people in your discipline?". Only in mathematics did he get anything like unanimous answers—and that's why the Institute began in 1933 with its School of Mathematics. The first six professors were Alexander, Einstein, Morse, Veblen, von Neumann, and Weyl.

The Institute was a small, cozy operation when I arrived in Princeton in 1939. The center of the life of all the mathematicians in Princeton, both University and Institute, was Fine Hall (the old Fine Hall), which is still my Platonic ideal of a mathematics building. Dark corridors, leaded windows, heavy furniture, worn carpets; the common room always open, always in use; the library up to date, complete, run with an iron hand by Bunny Shields, tiny, white haired, probably born looking as if she were in her late 50's, severe, but always helpful.

The mathematics department at Princeton extended its hospitality to the Institute for a few years; my first year there was the last of them. It was easy to get to know everyone. I felt awkward and overwhelmed for a while, but some of the graduate students (notably Arthur Brown and John Olmsted) took me under their wing, and before long I felt like a native and did my share of welcoming newcomers.

I virtually lived in the common room. (Not "commons room", dammit!; that's a careless mishearing that took place somewhere in the middle of the Atlantic Ocean, as the words were being imported from England. You eat at the Commons, and you lounge in the Common Room.) I loafed and read science fiction magazines on the comfortable soft leather couch, I tried to become good at Go and Kriegspiel on the game tables sprinkled around, and I wolfed chocolate cookies at tea every afternoon. Early morning or the middle of the night, there was always someone there, and some conversation about mathematics, the war, the quirks of the big shots, or the best nearby restaurant for oyster stew was always going on.

Everybody—well, almost everybody—came to tea. Eisenhart and Wedderburn sometimes (that's old Eisenhart, the differential geometer, not his son, the statistician), Lefschetz and Bohnenblust always, Bochner, Church, Fox, Tompkins, Tucker, Wilks—everybody in the Princeton mathematics department. The Institute professors were not quite so visible. Veblen yes, but Einstein no, and the others occasionally. The junior (but, of course, post Ph.D.) Institute visitors, such as myself, were there, and certainly, always, all the graduate students. The tea and cookies were free. They were endowed, we were told, as was the extra pay the janitor received for putting them out every day and washing up afterward.

Does everybody know about Kriegspiel? Perhaps not. It is chess, but it needs three boards. The two players sit back to back, each with his board, and the referee is between them, with *his* board. The game begins by White making a move on his board, and then Black on his; the referee duplicates both moves on the center board. Then it's White's turn again (to avoid confusion the referee calls out turns), and then Black's, and so on. Neither player ever sees what his opponent did; only the referee keeps track of all the moves of both sides. When one player, in his turn, tries to make a move that is impossible (because some of the squares his piece has to traverse are, unbeknownst to him, occupied) or illegal (because, say, he puts his own king in check), the referee says "no". From accumulated no's, bit by bit, the players begin to be able to guess where their opponent's pieces are. The purpose of the game is the same as in chess: mate the opponent's king.

"Blitz" is classical chess, but played with an almost vanishingly small time limit—something like two seconds between moves. Some adventurous souls have even tried Blitzkriegspiel.

"Go" is an old Japanese board game. The American Heritage Dictionary gives a curiously incorrect definition: "a Japanese game for two, played with pebble like counters on a board divided into 361 squares." What's wrong is the number. The board is indeed divided into squares, but the counters are placed not in the interiors of the squares (as the pieces in chess), but on their vertices. The number of vertices is $19 \times 19 = 361$, but that means, doesn't it?, that the number of squares is $18 \times 18 = 324$. The counters are all the same shape and size; half of them are white and the other half black. Roughly speaking, the idea is for White to place his counters so as to surround connected groups of counters of Black, and Black tries to do the same to White. The rules are not too complicated, but the grand strategy and the detailed tactics are horrendously so. I have heard chess masters who knew Go say that Go is by far the deeper and more difficult game.

Chess, bridge, Go, Kriegspiel—something like that was almost always going on in the common room.

The two major figures in Fine Hall were Lefschetz (university) and Veblen (formerly university, defected to the Institute). They were colorful figures, charismatic leaders. The irreverent young people in the common room added a few verses to the classical Princeton undergraduate song about the faculty.

> Here's to Veblen, Oswald V.,
> lover of England and her tea;
> he built a country club for math,
> where you can even take a bath.

Yes, Veblen was an ardent anglophile, and yes, there were showers in Fine Hall. There is another version of the song that credits Eisenhart with the design of Fine Hall. The alternative last two lines for Veblen in that version are:

> He's the one mathematician of note
> who needs four buttons to button his coat.

The jackets Veblen wore were distinctive indeed: long, straight, almost military looking, reminiscent of the ones sometimes called Nehru jackets.

Veblen was an effective operator. He knew the mathematical world, and he came to wield great power by having successfully placed many mathematicians junior to him in their jobs. He was a geometer, and he had a big hand in introducing "analysis situs" (= topology) in the U.S.

Lefschetz was the greater mathematician and the more theatrical character of the two. In his youth he was a chemist, till a laboratory accident blew both his hands off. That was the end of laboratory work; Lefschetz decided to become a mathematician. The natural hands were replaced by a neat-looking pair of wooden hands, covered by gloves. As prosthetic devices they were awkward, but they were good enough to hold a pen or a piece of chalk. The handwriting that resulted was rough and wiggly, but legible.

Dear Halmos:—

The proof of the J.M.G. was sent to you before I had a chance to check it up (I was ill). Cross out line −7, −8, ···, −13 of p.1.

P.2. Proof that K_1 is invariant is given a head.

P.2 line 11 $K_1 = K \oplus Fh_1 K_1$.

Hope that this is clear

Sincerely

Lefschetz

Letter from Lefschetz, 1964

Lefschetz was an irrepressible extrovert, a talker, a charmer. He spoke French and Russian and English perfectly; near the end of his life he learned Spanish and gave many lecture courses in Mexico. He ran the Annals of Mathematics with an iron hand—he rejected papers if he didn't like their subject, and solicited papers from authors known to be on a hot streak—he more than anyone else made the Annals a great journal. He saw mathematics not as logic but as pictures. His insights were great, but his "proofs" were almost always wrong. He interrupted colloquium speakers, often and pugnaciously, with what I then thought were stupid questions. I know better now: they were shrewdly stupid questions, asked to help the silent shy majority who were scared to ask them. His verse in the faculty song went like this:

> Here's to Lefschetz, Solomon L.,
> unpredictable as hell;
> when he's laid beneath the sod,
> he'll start right in to heckle God.

The center of the world

Memory of the size of things, such as houses, people, rooms, books, or crowds, is usually much less trustworthy than of their other sensable aspects, such as color, shape, smell, or sound, but, for what it's worth, I report that according to my memory the Princeton mathematics department was small in the early 1940's. There were about a dozen members of the faculty, and not more than a couple of dozen graduate students. As for the Institute, I can be sure—I have copies of the mimeographed address lists of the time in front of me—and it, too, was small: six professors, one "associate" (namely Walter Mayer, whose name is still remembered as that of half of a homology theory), roughly 20 visiting members (varying slightly from semester to semester), and one secretary, Miss Blake. (I am speaking of the mathematics people, of course; there were also the humanists and the economists, but, for all that most of us cared, they might as well not have been there.)

Five or six of the visitors were called assistants—theoretically one for each of the professors—but Alexander was so shy, so unwilling to be involved, so private, that he usually didn't have one. He belonged to an old family (Alexander Street was named after one of them), and he was rich. When he saw a ski-able hill once, the story goes, he bought it. He loved music, and he loved it loud—and there is another story about that: he designed a gigantic exponential speaker, reaching down to the basement, with its opening taking up most of the living room floor. (His wife said no.) He and von Neumann were friends, and enjoyed many parties at the house of one or the other of them; the parties were not dry.

One year Alexander announced a series of lectures on homology, and he drew a large crowd for the first one. I went, and I was both pleased and disappointed. The cause of both emotions was the same: the lecture kept its promise of beginning at the beginning, very abstractly, very axiomatically, and therefore I could understand it, but, for the same reason, it didn't get very far. At the second lecture, a few days later, Alexander, the perfectionist, announced that he was not satisfied with the axiomatic basis of the first one—he would now start all over again. The same thing happened at the third lecture. The fourth one was postponed—one week was skipped. There were more postponements and skips, but a total of six lectures was actually delivered—all starting all over again from page 1. After the sixth, Alexander abandoned the course—he just couldn't get the approach into the perfect shape that he wanted.

The administrative duties of the professors at the Institute were not heavy. They had meetings, but only a few—the main problem each year was to choose next year's fellows from among the applicants. Alexander found even that much administration onerous, and he resigned. There was some backing and filling, but ultimately the resignation "took", and Alexander succeeded in demoting himself from professor to "permanent member".

Einstein and von Neumann spent almost as little time on administration as Alexander. As far as contact with the organized university world of mathematics was concerned, Veblen was the executive officer. Back home in Princeton, however, both Morse and Weyl interpreted their jobs as having an educational function. Each junior visitor was invited to a private meeting with at least one of them, and sometimes with both. I was interviewed by Weyl, and I found the experience terrifying. He was formally polite, he asked what mathematics interested me at the moment, he asked about my long-range research plans, and he suggested that I prepare and present a talk at his current literature seminar. Yes, sir!

While the total mathematical community in Princeton consisted of only 50 or 60 people, it was, in one sense, infinitely large. It was taken for granted that almost every mathematician who arrived in the U.S. from abroad, to live here or to visit (and there were many arrivals then), stopped in Princeton for a day or for a year. The ones who were here already stopped in Princeton whenever they were on their way to Harvard or Yale or anywhere else. The ones who lived nearby, at Columbia, or NYU, or Rutgers, dropped in frequently to give a talk or to listen to one, or just to have a cup of tea, a game of Go, and a mathematical chat. The graduate students and the fresh Ph.D.'s who spoke at the colloquia and the seminars were competing with the most famous mathematical names in the world. Princeton was the center of the world, and not only the mathematicians in Princeton thought so.

Insignificant people

The greatest stimulation that we young fellows got at the Institute came neither from the professors, nor from the already famous visitors, but from each other. I was impressed when I met Fubini (a tiny little man, speaking bad staccato English, and apparently never again able to match the glory of the theorem he is immortal for), Kleene (a gigantic man, with a two-syllable name, a booming voice and a super careful approach to notation, not yet at the height of his career but well established), and Gödel (a slight, sallow, shy man, shrinking away from human contact, a legend, a magician)—yes, I was impressed, but I didn't learn much from them. The people I learned from, the people I exchanged ideas with, and was stimulated by, were the insignificant beginners such as Valja (spelled Valentine, pronounced Val-ya) Bargmann, Mahlon Day, Dorothy Maharam, and Ralph Phillips.

Others equally "insignificant" at the time, but, because of their different fields of interest, with less direct mathematical influence on me, were Garrett Birkhoff and Claude Chevalley (very impressive, sharp, heavy smoker, fast talker, always excited). Associated with the university, not the Institute (but who could tell?), were Ed Begle (later of SMSG fame), Leonard Eisenbud (a very mathematical physicist who is currently proud to be identified as the father of David Eisenbud, a lily white pure algebraist), Ralph Fox (a concert caliber piano player, among the top three or four best Go players), Gerhard Hochschild (with Cape Town English superimposed on a German accent, the pleasantest and kindest of all perpetual doom-sayers), Arthur Stone (who was part of the British team that had recently published a solution of the famous problem of squaring a square), John Tukey (who was busy writing Lefschetz's book), and Henry Wallman (of Hurewicz and Wallman fame, an addicted putterer with all electric equipment, such as radios; he subsequently emigrated to Sweden where he has lived ever since). Colloquium speakers during my first Princeton year included Mark Kac (recently arrived from Poland), Norman Levinson, Barkley Rosser, Paul Smith, and Oscar Zariski. How could we help being sure that we were at the center of the world?

One of the first people I met in Princeton was Paul Erdös. He was 26 years old at the time, had had his Ph.D. for several years, and had been bouncing from one postdoctoral fellowship to another. His English was then, as it still is, very good but heavily accented; my Hungarian was almost accent-free but getting feebler by the day. We agreed to speak English, and we usually do. Though I was slightly younger, I considered myself wiser in the ways of the world, and I lectured Erdös. "This fellowship business is all well and good, but it can't go on for much longer—jobs are hard to get—you had better get on the ball and start looking for a real honest job." I doubt that he listened to me. He has kept on travelling, lecturing, seminaring, and, from time to time, for short intervals, even teaching, in a geographically totally unpredictable manner ever since then—living out of a suitcase, talking mathematics every minute that he can find a willing collaborator, and writing many hundreds (literally, hundreds) of papers. Forty years after my sermon, Erdös hasn't found it necessary to look for an "honest" job yet.

I don't like the kind of combinatoric-geometric-arithmetic problems that Erdös likes, but he is so good at them that no one can help being impressed. It would happen that someone would ask him a question in a field that he knew nothing about; he then demanded that the basic words be defined for him, and if it turned out that his set-theoretic "counting" techniques were at all pertinent, he proceeded to find the answer. A typical example is the dimension of the set of rational points in Hilbert space, a problem that Hurewicz raised. Erdös was a little vague about what Hilbert space was, and he had no idea what "dimension" meant. Henry Wallman gave him the definitions—and before long Erdös produced the solution. (The answer is 1.) The paper came out in the Annals in 1940; it is a technically important contribution to a subject that a few months earlier Erdös knew nothing about.

Work

How did anybody—how did I—ever get any work done at the Princeton "country club for math"? My picture of the days there is full of happy memories such as a three-hour walk with Strodt and Ambrose; playing Go, Go, Go every day; taking time out to construct a travelling Go board (with holes drilled at each of the 361 points, and thumbtacks to be used for the counters); beating Edward Lasker at Go; inspecting the construction of Fuld Hall (the Institute building-to-be); and just plain loafing in the common room. The evenings were sociable; play hearts at the Foxes, and go have a cup of hot chocolate at the Balt at 3:00 a.m. (It was a gigantic fast food place, all white tile, looking more like a bathroom than a restaurant, but having the advantage of being open all night. As for Edward Lasker—he was not the same as Emmanuel Lasker, the great chess champion, but he too was a chess master, an expert gamesman, and a prolific and famous writer of books on games like chess and Go.)

In between other activities I had time to read Rolland's *Jean Christophe*, and Heinrich Mann's *Young Henri of Navarre*, and to fuss with the bureaucracy of an application for a certificate of citizenship. I had been a U.S. citizen all along, but I didn't have a sufficiently official-looking document to prove it—I thought that in times of war such a document might be useful. It wasn't—I never needed it. Much later, when I lived in Ann Arbor, and had frequent occasion to cross the Canadian border, I asked the Immigration and Naturalization Service to give me their standard U.S. Citizen Identification Card. It has my photo on it, 20 years out of date by now. It doesn't say that I *am* a citizen; it says instead that I have "claimed under oath to be a citizen of the U.S.". Tricky: if I slipped up anywhere on the paper work, the crime became perjury.

Sociability and bureaucracy notwithstanding, work did get done, by everyone, me included. I wasted several weeks trying to prove an extension theorem for measures, before Erdös located a counterexample in the literature. I doggedly kept looking for good new questions to work on, and I kept working. The counterexample made me feel disappointed, but, at the same time, relieved. Knowledge never hurts—what hurts is helplessness, the futility of banging your head against a brick wall without finding either proof or disproof. I have often spent weeks trying to prove a false statement—and when I learned that it's false, I felt victorious. Progress was made, knowledge was acquired, one more step toward the truth was taken.

The most beautiful step toward the truth was attending von Neumann's lectures. He was talking about rings of operators (they are called von Neumann algebras now), and, in typical von Neumann style, he began at the beginning with classical, finite-dimensional unitary geometry. At first I found his lectures illuminating but elementary, then I found them enlightening, getting better and better, and, before long, my diary records "von Neumann high-powered about spectra, ... von Neumann bewildering, but spectra and operators are awful good fun".

J. von Neumann, 1948

Hugh Dowker was von Neumann's assistant that year, and he wanted to think about homology groups, not operators. I on the other hand loved operators, attended every lecture, and took careful notes. Not having notes of his own, Hugh took mine to von Neumann—open, above board, with no intent to deceive—and it turned out that everyone was satisfied. Hugh didn't have to take time away from homology, Johnny approved the notes (everyone called him Johnny), and I—well I, possibly because of those notes, I was called in by Veblen and told that, as of the second semester, I would get a fellowship after all!

Work and between work

In the times between work and play I worked, and so did most other members of the Institute. The traditional "gentlemanly amateur" attitude existed to a small extent—you mustn't seem to be trying too hard, mustn't seem to be a greasy grind—but if we had to choose between not working and getting caught working, we chose getting caught.

My work habits, writing as I think, made me especially likely to get caught. I *can* think lying down, or taking a walk, or soaking in a tub, and some people prefer to do their thinking under such informal conditions. Steve Smale said (many years after my Princeton days) that he did some of his best work while he was deepening his suntan on Copacabana beach. For me, however, real work means to sit at a large table, with papers, and books, and notes spread out all over, and to write. It worried Slim Sherman. He saw me sitting at a library table in Fine Hall, hour after hour, writing. "How long is that paper, anyway?", he asked; "where did you get so much to write about?"

That's the real work, but what I was going to tell about is the other work, between play and work. I kept reading Stone's book (and I still haven't read all of it), I tried to study von Neumann's notes on continuous geometry, and, best of all, I read Pontrjagin's *Topological Groups*. The English translation by Mrs. Lehmer (usually referred to as Emma Lemma) had just come out, and it was an eye opener, a revelation, a thriller. Yes, a thriller—I read it almost as I would read a detective story, to find out whodunit. I thought it was a beautifully, excitingly written book, and each new aspect of the big duality theorem, as it was revealed to me and then exemplified and applied, was a surprise and a joy.

As for discovering new mathematics of my own—that went very much slower. To convince myself that I was still alive, I wrote up and submitted the "six pages of shallow stuff" that I produced during the preceding year at Illinois. The Annals printed it, but it turned out to be less than an overwhelming success. Feller's review in the first volume of

Math. Reviews was not unkind, but severe; my necessary condition for the strong law of large numbers is necessary even for the weak law, he pointed out, and, he went on to remark, it is a consequence of conditions that he himself had studied before. He was right, and though at the time I hadn't met him yet, I wrote and told him so. ("Dear Professor Feller: You are right....") We became friends a little later, and we consumed many lobsters and vodkas together.

W. Feller, 1964

Feller was an ebullient man, who would rather be wrong than undecided, and who preferred getting a hearing for his views to getting applauded for them. The circles we moved in were convinced that cigarettes helped cause lung cancer; that was enough to make Feller express his doubts about the conclusion. Not knowing all the facts, I cannot call him wrong, but I think it was irresponsible for him to make his statements: coming from a man in his position, a mild doubt could be interpreted as a strong defense. He spoke loudly, very fast, with a strong Yugoslav accent, with wit and charm and understanding. He enjoyed telling about the time when he and his wife tried to move a large circular table into the dining room. They pushed and turned and tugged it, but they could not get it through the door. Annoyed and tired, Feller took time out to diagram the situation, and, having found a workable mathematical model, he proved that what they were trying to do couldn't be done. In the meantime, as he was constructing his proof, Clara kept tilting and pulling the table; by the time the proof was finished, the table was in the dining room.

A weak paper and a pretty good book

My first stay at the Institute consisted of three consecutive academic years (1939–1942) and they tend to blend together and become one great nostalgic memory. (My diary stopped at the end of 1939.) My spectra paper belongs to the first year, "Elementary theory of matrices" to the second, and negotiations with Coble about returning to Illinois to the third—but usually the events do not seem to be arrangeable in a total order. One of the two summers in the interior of those years I spent in Champaign, collaborating with Dick Leibler part of the time, and the other I spent in Cambridge, teaching summer school at Harvard.

The "spectra" paper is probably the weakest one I ever wrote. It came out of the sudden realization that the set of real-valued bounded measurable functions (on, say, the unit interval) has an algebraic structure very similar to that of the set of bounded Hermitian operators on Hilbert space. (Of course people like Bochner and von Neumann had known that all along.) Since I was a probabilist, I saw merit in operatorizing stochastic concepts. The distribution function of a random variable corresponds to the spectral measure of an operator; what operator concept does the independence of two random variables correspond to? If, as editor, I now received a paper with as little hard content as that one, I would reject it out of hand, but then I thought it was all right. Not great—I had reservations—but all right. When I won the battle with my reservations and decided to enlarge my list of publications, I submitted the paper to a journal published by "our heroic ally", the USSR and, sure enough, it was accepted and printed (in English) by Mat. Sbornik.

The matrix business was something else again—it was, in a sense, the most exciting and most successful course I ever taught. I found von Neumann's 1939 lectures beautiful and revealing and inspiring, and in 1940 I was inspired to spread the word. The recipients of the word were to be graduate students at Princeton University. With no official pre-arrangement, I simply tacked up a card on the bulletin board in Fine Hall saying that I would offer a course called "Elementary theory of matrices", and I proceeded to offer it. I prepared for it carefully, a goodly number of students (something like a dozen) attended regularly, and a couple of them took notes. They were Ed Barankin (who subsequently became a statistician in California) and Larry Blakers (a topologist who returned to his home country, Australia). The notes circulated around Princeton for a while in mimeographed form. I found them invaluable; my first book, *Finite-Dimensional Vector Spaces*, was based on them.

As far as money goes, Princeton was agreeable to accepting my services free; I was slightly surprised when, at the end of the term, those services were officially recognized. I received an official request to assign grades in the course—so far as Princeton University and the students were concerned, the course carried graduate credit. From my point of

view, as teacher, the course was splendid. No one came unless he wanted
to; there was no nonsense about prerequisites and distribution requirements. There was no syllabus; we talked about what we wanted to talk
about. There was no homework, there were no exams. When I had
to give grades, I did so on the basis of subjective impressions acquired
during class and during between-class discussions. It was an ideal course:
information and ideas were transplanted into minds eager to receive them
and willing to work for them.

To write a book based on the notes of a course is a good way to
write a book. The most important single feature of good writing—of clear
communication of any kind—is organization. If you know the right order
in which a sequence of things should be said, and if you know the extent
to which you need to emphasize some parts and play others down, your
communication battle is more than half won. I organized my matrix
course before I gave it, of course, but presenting the prepared material
to a class of good students helped a lot when the time came to
re-organize it, to change the emphasis, to insert examples at just the right
places, and to polish the phrasing.

I enjoy writing. I enjoy organizing what I want to say, and, once that's
done, I enjoy looking for the *mot juste*, and feel good when I think
I have found it. Despite all that, I still find it hard work to write a book,
hard and slow work. It took a few months to prepare the matrix course
(or a few years, if I count my student struggles with Bôcher and Dickson,
and why shouldn't I?), and those months followed close on the heels of
listening to von Neumann's course and taking notes on it. During the
months the matrix course was going on, it cost me more time still in lec-
ture-by-lecture preparation. Once the course was finished and the
Barankin–Blakers notes were sitting next to my own notes on my desk,
it took me six months to write the book, at the rate of a relentless two or
three hours each morning.

When I submitted the final manuscript to von Neumann and asked
that it be considered for publication in the new series, Annals of Mathematics Studies (the orange series, still going strong), he was in favor of
accepting it, but a few days later he told me that he had encountered
some resistance. Shouldn't the Studies be more on the research level than
expository?—that was the question. The question was perfectly appropriate, and the history of the Studies before (number 3 was Gödel's
Consistency of the Axiom of Choice ...) and after (number 15 was
Morse's *Topological Methods in the Theory of Functions...*) shows that
the answer was yes. Nevertheless, it wasn't long before FDVS was accepted and published as number 7. The outcome made the Princeton
University Press happy. Reason: many of the volumes in the series were
money-losing prestige items; FDVS had no prestige but it sold well and
helped to keep the series afloat. During its 16 Princeton years, FDVS

brought me an income of zero dollars and zero cents; authors of Studies received no royalties. In 1958, when van Nostrand published the second edition, the Princeton Press was good-natured and sold the copyright to van Nostrand, for the traditional $1.00.

Collaboration

In the summer of 1940 I did a bad thing that I've been feeling ashamed of ever since.

For no special reason except auld lang syne I spent the summer in Champaign. Dick Leibler was there, and it was pleasant to be able to renew our friendship—we had been quite close for a while. We used to play pool a lot, we had once shared an apartment, and we even had a few mathematical interests in common. (Dick was Trjitzinsky's student.) We started, that summer, a sequence of mathematical conversations that looked like the beginning of collaboration. The problem was to find out when a measure-preserving transformation has a square root. When it comes to square roots the number -1 often plays a special role, and that turned out to be true here too; it makes a big difference whether -1 is or is not an eigenvalue of the induced unitary operator. The results we got were not deep—most of them held for the special case of pure point spectrum only—but they were new and we thought they were mildly interesting.

As our conversations proceeded I felt that Dick did not share my compulsive-obsessive eagerness to get the job done, my impatient ambition, my fierce insistence on "getting points". One time when we arranged to meet he didn't come, and at other times he didn't pull his weight, I thought. Near the end of the summer I typed up what we had accomplished, I gave him a copy, and I said something like: "Look at this, Dick, and tell me how you feel about it. If you think it's really joint work, we should make it a joint paper; otherwise I'll publish it myself." That was not a smart thing to say, and what happened made it worse. What happened was that a day or so later Dick said "Yes, I think it's joint work", and I said "I don't". (I don't remember the exact words on either side.) The paper was published with my name only.

We didn't have a fight, we didn't become enemies, and our relation did not suffer a discontinuous break, but things have never been the same between us. For a while I kept seeing Dick, and going to movies with him, and having him at parties and dinners at my house, but in the last 35 years I doubt if we met as much as a dozen times. When we do meet we are not cool and unfriendly, but we are polite.

I now think, I now know, that what I said was bad. It was bad
because it was an insult that gave pain (I am guessing, but how could it
be otherwise?); it was stupid because you get just as many "points" from
joint papers as from single ones; and it was almost certainly wrong, fac-
tually wrong—my evaluation of the extent of Dick's contribution to the
work was not the result of objective judgment.

As for the effect of joint papers on a scholar's reputation, at least in
mathematics, what I said is not 100 percent right—but it is near
that. Some mathematicians try to make a reputation on the coattails of
others; if most of your papers are joint, some cynical deans and chairmen
will wonder how much you yourself contributed. But twenty papers, of
which five are joint, are counted as twenty papers, and there exist outstanding, creative, original mathematicians most of whose published
work is collaborative.

The most important point, however, is not that, but the marriage principle of collaboration, a principle strongly supported by Hardy and
Littlewood, the famous pair of great collaborators. The principle is that
a collaboration once joined together shall not be put asunder. Once a
conversation that might lead to a collaborative result is begun, then anything it leads to, for better, for worse, is a collaborative result. The contributions of the partners to the final result might be numerically equal—
the same number of definitions, the same number of theorems, the same
number of proofs—or they might not; it must not matter, the count must
not be made. Perhaps one partner contributes the insight and the other
the technique; perhaps one partner asks the questions and the other
knows the literature well enough to avoid the waste of time that comes
from trying to re-discover already known answers; or, possibly, one is
active and the other is the foil needed to keep up his morale and inspiration. No matter—once a collaboration is begun (which is not necessarily
the same thing as formally declared), anything that comes out of it
is, must be, called a collaborative product.

Measures and Harvard

In 1940–1941 von Neumann lectured on invariant measures. He began with general measure theory and went on to Haar measure and
some of its generalizations. Kakutani was at the Institute that year, and
he and von Neumann had many conversations on the subject. The conversations revealed facts and produced proofs—quite a bit of the content of the course, especially toward the end, was discovered just a week
or two or three before it appeared on the blackboard. That year
(and also the next) I was von Neumann's assistant, so that taking notes

on his lectures was not only a labor of love but also my job. (There aren't many people on this planet lucky enough to get paid for doing what they enjoy doing.) To write notes for von Neumann was a rewarding activity. He read the handwritten version before it went to the typist, and sometimes scribbled comments on the margins. I am proud to remember the spirit (but not the words) of one of his flattering comments; by my proof of the monotone class theorem he wrote something like "good!—new?"

The notes were typed. I still have one copy, and two or three others were kept in the Institute—von Neumann had one of them and the Institute library had another. Since then, I have been told, a few xeroxes have been made, but the notes have never been published in any proper sense of the word. I doubt that they are of much value now, except as a historical curiosity; much of the material has become out of date and superseded.

My own research was tending toward ergodic theory (the morphism theory of the category whose objects are measure spaces). I proved a Fubini-like theorem about the decomposition of measure-preserving transformations into ergodic parts—the measure-theoretic analogue of the decomposition of a permutation into cycles. As a follow-up, Ambrose and Kakutani and I extended the result from single transformations to flows (one-parameter groups of transformations). Kakutani's English then was not as nearly perfect as it has become since, but it was very good, and his knowledge of grammatical rules was more extensive than Ambrose's or mine. When the time came to write up our results, Ambrose became the chief scribe and I the chief revisor. Kakutani did not just meekly rubber-stamp his approval of what we wrote; he wanted to change some of our idiomatic constructions, and, though he always remained soft-spoken and polite, he was not easy to shout down. Ambrose and I liked him a lot, but I was exasperated. "The nerve of the man!", I muttered.

The following summer a small emergency arose at Harvard, and along with several other young people I was hired to teach summer school. I had heard about the super undergraduates that Harvard has, and it's true: I have met a few of them since then. In summer school, however, or in the summer of 1941 in any event, there was nothing super about the calculus students; many of them were the weak ones who flunked in the spring and wanted to use the summer to catch up. The mechanical organization of the course was excellent. Sample: the first day when I showed up for class a stack of copies of the class roster was on the desk.

It was a pleasant summer. Several rooms at 17 Sumner Road were being passed from one visiting young-mathematician to the next; Kaplansky had lived there, and that's where I met Leon Alaoglu and Herb Robbins. The tragedy in Leon's life was that, much later, he married a woman who didn't love garlic.

Classical mechanics

Modern game theory started in 1928, when von Neumann's minimax theorem appeared in the Mathematische Annalen. In 1941 von Neumann and his economist friend Oskar Morgenstern started a collaboration that resulted in their gigantic book. The subject has become a well-established part of mathematics since then; in the fall of 1941, when von Neumann started his Institute lectures on the subject, it was still in its groping, formative stage. I was von Neumann's assistant, and it was my job to take notes on the lectures, but I didn't do it gladly and I didn't do it for long. The subject caught on, it was almost instantly popular, and someone (I wish I could remember who) played for me the role that I played for Hugh Dowker. Game theory left me cold, and I was hot on the ergodic trail; I was happy to leave my note privileges to the enthusiasts.

As for ergodic theory, my interests led to two collaborations, and I was pleased and proud to be a part of both. One was with Johnny von Neumann.

The main point of the paper with von Neumann was to prove that "all measure spaces are the same" (a statement to be taken with a grain of salt, of course). I had been brought up on abstract measure theory; so far as I was concerned the classical theory for Euclidean spaces was just a special case, with no special measure-theoretic virtues. Johnny was a mathematical generation or two older than I, and he was brought up more classically, but, of course, he was more than clever enough to understand the abstract version. Many other analysts, however, usually the older ones, were put off by the abstract stuff—they didn't trust it, somehow, or, I uncharitably thought, they didn't really understand it. When they needed measures in contexts more general than the Euclidean one (in probability, for example), they felt comfortable only when they could reduce the general case to the special one. The reduction took the form of a correspondence between the probability space of interest at the time (say, for instance, the one that arises in the theory of Brownian motion) and the unit interval; the correspondence carried Lebesgue measure from the interval to the space under study. The result of the desire for and the possibility of such reductions was that many papers on probability began with the (totally unnecessary) establishment of a mapping from some set onto the unit interval; one of the most prolific offenders along these lines was Norbert Wiener.

I saw a pattern in these reductions, and I hoped (and suspected) that there might exist a grandfather reduction that they all could be inferred from. Of course I talked to von Neumann about my problem and my beginning efforts, and he was helpful. At first I said "thanks", then I said "Gee, thanks a lot, that was a big help", and before long he was contributing half (at least half) of the results and of the techniques. I was pleased as Punch—look at me!—I am collaborating with von Neumann!

I am not ashamed of my role in the production of our paper; I think I pulled my weight. There is at least one result, however, that is 100 percent von Neumann—I am sure, because I remember the details of how that result was discovered. We needed to know whether or not a certain kind of "bad", pathological, set could exist. We were talking in Johnny's office, both of us at the blackboard; I was completely stuck, and Johnny was threading his way through a subtle transfinite double induction, making heavy use of the continuum hypothesis. At one point he interrupted himself to ask whether I didn't want to make a note of what was growing before our eyes. I was eager, and impatient, and cocky, and I said "no, no—I understand it and I'll remember". Well, you can guess what happened—the construction was successful, but at home that night I found to my horror that I didn't understand it—or that I just plain forgot it—and, in any event, that I couldn't for the life of me write it down. I went back to Johnny as soon as I could after that—he was out of town for a few days—with my tail between my legs. That's the only time I saw him angry. The anger lasted a few seconds only, but it was clearly visible; he stamped his foot and said "Oh, dammit...". (He didn't finish the sentence, but I could hear the end: "...I *told* you to take notes!") The trouble, the main trouble, was that he too forgot the construction, and had to start all over again, encountering the same obstacles and re-discovering the way around them. I took notes that time.

When the work was finished, I had the job of writing it up for publication. Johnny made many changes, and we kept passing the manuscript back and forth between us. I won a lot of the time, but I remember one battle that I lost. Tired of the repetitive reduction to the interval, I wanted to begin the paper with a snide crack at Wiener. The first sentence of my first version went like this: "The purpose of this paper is once and for all to map all measure spaces for which this is possible on the unit interval." Johnny struck out the "once and for all".

The title of the paper was my idea. I had read and admired Johnny's 1932 paper titled "Zur Operatorenmethode in der klassischen Mechanik". Our joint paper continued some part of the work of that one, in a slightly more abstract setting, and for me it was a dream fulfilled to be associated with that memorable predecessor. I hesitantly suggested and Johnny good-naturedly accepted the title "Operator methods in classical mechanics, II".

Birthdays

The year 1941–1942 had its share of glamorous Institute visitors, and then some—the astrophysicist Chandrasekhar was one of them, and so was the great analytic number theorist Carl Ludwig Siegel, and the logician-philosopher Tarski. The visitor that stands out in my mind, how-ever, was not glamorous at all, at least he wasn't so considered then. He is Hans Samelson, who quickly became one of my best friends, and who was my other collaborator in that vintage year.

The first I heard of Samelson was one morning when von Neumann stopped in my tiny office that was adjoined to his own luxurious one and asked me to be Samelson's first day host. Why not?—it's all part of the job—but I wasn't thrilled by the assignment. I met Hans and Renate, his pretty but flustered wife, and took them to lunch at the Institute cafeteria. (Her English was rather inexperienced, and she was sufficiently flustered that she couldn't remember that one of the German words for carrots was Karotten.) I showed them around the Institute some—here's the library, here's the common room—and drove them over to their apartment. Chore done.

Hans's thesis was about topological properties of topological groups, and my work with von Neumann involved measure-theoretic properties of topological groups—we obviously had something in common. We started chatting about monothetic groups (groups that can be topo-logically generated by a singleton), and we soon found that there were interesting unanswered questions in the subject. We exchanged ideas at lunch and at tea, and then, quite frequently, one of us would phone the other in the evening to report progress. In the course of these con-versations we discovered a powerful technique for answering questions

H. Samelson, 1977

about topological groups—frequently all you need to do is pick up André Weil's book on the subject and riffle the pages till you find what you're looking for. Sometimes the technique didn't work, and we could have some fun discovering our main result (the algebraic characterization of compact monothetic groups).

The human memory is a curious instrument. Forty years after Hans and I wrote our paper, Hans had occasion to tell the story. The way he told it, he didn't even know we were writing one till I showed him the manuscript I prepared with both our names on it! No, Hans, no!— you've got it all wrong. I remember it all very clearly. When we decided that the time had come to do the writing, we tossed a coin, and I lost— the coin said that Hans was to do the writing. I squirmed and I revised, but Hans's hand is visible throughout. The first sentence, for instance, begins with "A topological group G is called *monothetic* (following van Dantzig)..."—and, right or wrong, I regard "following van Dantzig" a Germanism, and I would never use it. So there!

Hans and I have stayed in touch, visited each other often, and exchanged birthday cards every year. There is a reason for the cards. Do you remember the birthday problem: What's the smallest number of people for which the probability is 1/2 or more that two of them have the same birthday? The answer is fun to find; it turns out to be 23. That means, of course, that any time you are with a group of 40 people, say, you should happily agree to bet even money that there is at least one birthday coincidence among them—the odds in your favor are, in fact, 10 to 1. If on the other hand you are with only 15 people, you would be ill advised to accept even money on coincidence. A couple of times in our lives Hans and I made a dishonest dollar by carefully guiding the conversation along the right channel and then accepting such a bet when we were together at small parties. We knew of course, we discovered early on in our acquaintance, that we were not only born the same year, but in fact within a few hours of each other on the same day, March 3. (Astrologists to the fore: does it have any significance that we share our birthday with Georg Cantor—1845—and Emil Artin—1898?)

[Computers and hand-held calculators can be good things and they can have bad effects. The birthday problem can be used to exemplify a bad effect. A good way to attack the problem is to pose it in reverse: what's the largest number of people for which the probability is less than 1/2 that they all have different birthdays? A few seconds' thought shows that all you have to do is multiply $\frac{364}{365}$ by $\frac{363}{365}$, then multiply the product by $\frac{362}{365}$, and continue with $\frac{361}{365}$, $\frac{360}{365}$, etc., till the first time that the answer gets below 1/2. In other words, the problem amounts to this: find the smallest n for which

$$\prod_{k=0}^{n-1} \left(1 - \frac{k}{365}\right) < \frac{1}{2}.$$

The indicated product is dominated by

$$\left(\frac{1}{n}\sum_{k=0}^{n-1}\left(1-\frac{k}{365}\right)\right)^n < \left(\frac{1}{n}\int_0^n\left(1-\frac{x}{365}\right)dx\right)^n = \left(1-\frac{n}{730}\right)^n < e^{-n^2/730}.$$

The asserted domination comes from the celebrated relation between the geometric and arithmetic means; the next inequality comes from the definition of the definite integral, and the last one from $1 - x < e^{-x}$. The inequalities are moderately sharp. The last term is less than $\frac{1}{2}$ if and only if $n > \sqrt{730\log 2} = 22.6$.

The reasoning is based on important tools that all students of mathematics should have ready access to. The birthday problem used to be a splendid illustration of the advantages of pure thought over mechanical manipulation; the inequalities can be obtained in a minute or two, whereas the multiplications would take much longer, and be much more subject to error, whether the instrument is a pencil or an old-fashioned desk computer. No longer: nowadays inexpensive and convenient calculators, especially the programmable ones, yield the answer for all probabilities (in place of 1/2) and for any number of days (in place of 365) in a matter of seconds. What they do not yield is understanding, or mathematical facility, or a solid basis for more advanced, generalized theories. A pity.]

Good and bad, it all ends some time. It was a great year, I met and enjoyed a lot of people who became great and famous (such as Jack Calkin, George Mackey, Deane Montgomery, and Jimmie Savage), and I did indeed negotiate with Coble and, quite early in the year, got an offer from him of a job at Illinois. Jobs were not easy to get; I didn't gamble on a better job at a better place but accepted Coble's offer immediately. I was glad to have it sewn up, and I gloated—"I've got Coble by the short hairs", I said. "He, you", said Ambrose.

CHAPTER 7

Winning the war

Back home in Illinois

Pearl Harbor day was December 7, 1941; for the next $3\frac{1}{2}$ years the U.S. was at war. The war affected everything and everybody—sometimes only a little and sometimes a lot. There was the draft, for one thing—all the men of my age group were affected by that. Only a few active research mathematicians were drafted, but enough that we were all aware of the possibilities. Most of us were deferred throughout the war; we were badly needed as teachers. I knew, I thought I knew, that if I were to be called, I would fail the physical exam (because of my bad foot)—for me the draft was only a vague threat on the horizon.

I was lucky. The war was a source of inconvenience (shortages, rationing, travel difficulties, housing problems), but not a personal tragedy; it was exciting, it ended the depression, and it promised to end fascism. I spent the first six months of the war finishing my Princeton years, and then moved to Illinois as an associate at a salary of $2200 per year. "Associate" was a curious vestigial rank at Illinois; it was between instructor and assistant professor. Its existence lengthened the climb up the academic ladder, and, I suspect, it was sometimes used as a shelf to put people on to forget about them. I know for sure that some people at Illinois retired and became Assistant Professors Emeriti.

For me, going to Illinois had all the advantages of going home—and all the disadvantages. I wasn't the conquering hero returning—but I was something a wee bit like that. Mrs. Walls, the tough old departmental secretary that I used to be scared of, welcomed me back ("sure we got *you* back—you've been doing a lot of research"); I knew the buildings, the campus, the streets, the towns, and, above all, the people; I could be with Joe Doob again. At the same time, returning to the same old places offered no glamor, no novelty, no excitement, and, of course, no new people. Princeton had spoiled me; Illinois turned out to be a letdown.

As far as Illinois was concerned I was a grownup now—I had been out in the great world and learned about people and places that the folks back home had never known. Old man Crathorne was still around and he was in charge of the colloquium; he asked me to give a talk on describing life at the Institute. I enthusiastically joined the life of the department. When, for instance, I noticed that the Proceedings of the National Academy did not come to the mathematics library, I started a battle that lasted through many months of memos and meetings. (ProcNAS was more important for mathematics in those days than now.) Soon after I left Illinois, a year after my return, we won.

There is not much more to tell about that year. I met an eager young MS candidate in Speech, who wanted me to be her thesis—she was going to eliminate my foreign accent. She taught me how not to roll my r's; in most other respects, however, she failed, but not for want of trying. She did get her master's degree. I kept working at research, but the result of that effort was only one tiny paper. (I am not ashamed of that one. It was about automorphisms of compact groups viewed as measure-preserving transformations. The paper answered some questions and raised some others, which are still not completely answered.) And, of course, I taught classes.

Most of my classes were routine, but one remains in my mind. It was linear algebra (once again I was a missionary preaching to the heathen), and it was remarkable because it had only two students. I remember Hundertmark and Langebartel fondly. I lectured at them as if there had been 20 of them. I would know better now—lecturing is always a clumsy way of teaching, but in a class that size it's extra awkward. I could—I should—have tried to draw them out, like a Socrates junior grade, and, almost certainly, I would have been more successful at implanting ideas more efficiently and lastingly. Live and learn.

Meetings

Bit by bit I was becoming a member of the mathematical family of America and the world. There were frequent meetings of the AMS at Columbia, and I acquired the habit of attending them. I met Leon Cohen (who later became Mr. NSF, the man who gave out the money), and Ralph Palmer Agnew (chairman at Cornell), and Norbert Wiener, and many others. I began to feel like one of the insiders. At first Ambrose and I were scared of Wiener—a pudgy man with a goatee, thick glasses, and impressively great fame—and we hesitated to approach him. He was,

however, despite his seeming bluster, extremely insecure, and before long
he didn't have to be approached—he chased us. He wanted to talk. He
wanted to tell us the mathematics he had been thinking about (he was a
notoriously bad explainer who thought that the way to make his
beautiful and deep ideas clear and rigorous was to surround them with
a plethora of integral signs); he wanted to tell us about the languages he
knew (German, Russian, Spanish, and, he fondly believed, one version
of Chinese—but at the Chinese restaurant, where he tried to show off,
the waiter had no idea that what Wiener was saying was Chinese); and
he wanted to ask us how good he was ("What do you think? Am I slipping? I just proved...—that's not so bad for an old man, is it?").

The AMS has two major meetings each year—they used to be called
the Christmas meeting and the summer meeting—and out of the 80 that
took place in the last 40 years, I must have attended at least 60. No—
probably 70. Why? Partly for the talks, but that's only a part. Yes, you
learn mathematics at meetings, but mainly you get a chance to meet
people informally over meals, to exchange vague but promising ideas
(instead of listening to precise ones polished to incomprehensibility) and
you keep up with the professional news (a conference of interest to you
is being arranged for next summer). You may learn some theorems and
proofs, but, more important, you hear about interesting conjectures, you
get reactions to your own tentative ideas, and you get helpful references.
Also, and that too is vitally important, you establish personal contacts.
You *can* write to a stranger and ask him for the history of an idea, or for
advice, or for a good place to look something up (and you should), but
it's much easier, pleasanter, and more effective to write to someone
you've met.

The summer meeting is usually on a university campus; you can
sleep in a dormitory, eat at the student cafeteria, and attend sessions in
classrooms. The winter meeting used to be during the week between
Christmas and New Year's; nowadays it is in January, so that instead
of grumbling about their ruined family holidays people can grumble
about the classes they had to dismiss or find substitute teachers for. The
winter meeting is more expensive (large hotels), easier to get to (large
cities), and less convenient (the sessions are in ad hoc meeting rooms
filled with undertakers' chairs, and the lecturer must use a rickety porta-
ble blackboard, or else that great modern invention, which will probably
be perfected someday, called the overhead projector).

What goes on at meetings, besides lectures and beer? Answer: the
wheels of the Society (and the Association, and several other ancillary
organizations) are kept rolling in the traditional smoke-filled back rooms
(that aren't quite so smoke-filled now). To arrange the meeting itself is a
long, complicated, and expensive piece of administration; transportation,
meeting rooms, hotel rooms, telephones, messages, books for sale—
everything has to be thought of in advance and provided for. Meetings

and publications—those are two major products of any learned society, and to keep the journals and books coming out is even harder work, more time-consuming, and more expensive than to arrange meetings. The chances are that many of the people that you came to the meeting to see and then cannot find are there all right, but they are invisible. They are in some small crowded room trying to figure out how many pages of the Transactions they can afford to put out next year, how large an editorial board is needed and who should be on it, and where can they find a printer that will do good quality work at finite prices.

Teaching at Syracuse

In 1943 Ted Martin became chairman of the mathematics department at Syracuse and he offered me an assistant professorship (to my mind a great elevation in rank—like going from a cadet to a full-fledged officer) and a munificent raise ($3000). Should I have accepted? Probably not, but I did. We negotiated about details for a short time, but I didn't even tell Coble about the offer so that he could try to persuade me to stay. My final telegram to Martin said LONG LIVE ARCHIMEDES.

I have always been restless—the new always looked preferable to the familiar—but from the sensible point of view the move was unwise. Illinois was a pretty good university; Syracuse was a pretty bad one. In Syracuse the professors of philosophy and of Latin were anything but devoted scholars, and the astronomy department was so weak it was sad—and weaknesses such as that affect the whole campus. Who cares, some might say, if the astronomers are no good, so long as the mathematics department is fine—but it doesn't work that way. That's almost like saying "it's your end of the boat that's leaking, not mine". Esprit de corps was low in the university, and (and therefore?) it was low in the mathematics department.

Martin's first four appointments were Lipman Bers, Abe Gelbart, Hans Samelson, and I. We were all young and enthusiastic, but we were pushing uphill a lot of the time. Part of the difficulty was the clash of the old and the new. Hans was, and is, quiet, soft-spoken, polite, but Lippa (Bers, that is) and I were brash, noisy, and more energetic than diplomatic. The old guard had been there for years and years, and they didn't appreciate the change of life that our presence implied. The rules were being changed midstream. They weren't treated unfairly, but they felt, I think, that they were regarded as second-class citizens, and such feelings don't make for high morale and good comradeship.

Martin worked hard to build a good department, and he succeeded in putting Syracuse on the mathematical map. People still look back on those years of first flowering and remember that there were many good

W. T. Martin, 1960

mathematicians in Syracuse—but they forget that only a few of the many were there at any one time. Those were years of high mobility, and good people (like Löwner and Paul Rosenbloom and Scheffé and Fox) kept coming and going all the time; it was part of Martin's administrative genius that he generally found good replacements for the ones who left.

I didn't like the climate (nasty, dark, cold, snowy, and long winters seemed to fill eight months out of every twelve), and I didn't like the town (too small for the conveniences, the shops, the restaurants, the theaters of a metropolis like Chicago, and too large for the charm and the academic domination of a college town like Champaign–Urbana). Teaching at the university was pretty discouraging. I don't have statistics to prove it, but subjectively I remember the students as being much worse than at Illinois. Stories circulated about parents irritably demanding that their children be given passing grades: "Your income depends on us, remember". There were many students who couldn't be talked out of $\frac{1}{2} + \frac{1}{3} = \frac{2}{5}$. From time to time a good student came my way, and I found it hard to remember where my income came from—my conscience made me advise him to go to Pennsylvania or Michigan.

Not everyone was so cynical. A wonderful example was one elderly woman schoolteacher whom Martin unearthed somewhere and, along with others like her, temporarily added to the faculty—he was desperate for warm bodies to put in front of classes. At the end of her first semester, when Martin examined her grade report (which he did routinely for every class), he was astonished and worried to note that 16 of her 34 students got A's, 13 got B's, and the remaining 5 got C's—there were no D's and

F's. Suspecting an error, he looked at the grades that she gave on the uniform departmental final exam—and, sure enough, those grades were consonant with the course grades. Was she too soft, too kind-hearted? Martin went to talk to her, and she showed him her grade book: the homework grades and the in-class hour exam grades were all high. He asked to look at the final examination papers, and he found that they were correctly graded—they deserved their high marks.

What happened? The class was not specially selected, and the possibility of a class full of semi-geniuses having been thrown together by chance could safely be discounted. Martin's curiosity was aroused, as well as his administrative responsibility, and he questioned the old lady in detail. Bit by bit, reluctantly, modestly, she told her story. She was a widow; she had nothing else to do; she loved to teach; and she loved her students. She had many extra help-sessions for them, she urged them to consult her during her all-day office hours, she invited them in small bunches to her apartment for an evening cup of hot chocolate. She corrected all the homework papers in great detail, explaining where they went wrong, what they should have done, and why the grade had to be what it was. When they cut class, she tried to reach them outside—"are you all right?, you're not sick, are you?, do you need anything?". They loved her, they worked hard for her, at the end of the term they, acting jointly as a class, sent her two dozen roses—and none of them got a D or an F.

There was a war going on. Meat and butter and sugar and clothes and gasoline were hard to get; you had to stand in line to buy things even if they weren't rationed; black market practices were whispered about. Despite all this, somehow, none of the people I knew suffered from any shortages—the goods were there, but the distribution system broke down. Some shops, for instance, sold cigarettes for only an hour a day; when the time started, the line would form faster than you could run to the end of it. I could no longer buy a carton of cigarettes at a time—I had to stand in line every day, sometimes twice a day. But, although I was a heavy smoker, at least as heavy as before the war, I never went without—the cigarettes were there, they were just hard to get.

To be sure, my cigarette problem was greatly relieved by my ASTP classes. Army Specialized Training Program is what that stands for, or the ASTP program as it was redundantly referred to. I taught several ASTP classes, and most of them were not much different from civilian ones. The boys wore uniforms, of course, and they marched to class, singing "Roll out the barrel". They had been ordered to stand up when I entered the room, but I put a stop to that. As long as we're in this class together, I told them, we can adopt one of two attitudes: either I am an honorary officer, or you guys are honorary civilians—which do you choose? That started them off with the hope that I was not the devil incarnate—and that, in turn, solved my cigarette problem. Soldiers could buy all the cigarettes they wanted at the PX, cheap too, and the non-smokers in my classes were quite willing to buy and resell to me a carton every now and then.

The most challenging course of my teaching carreer was an ASTP one that I was assigned. It was an educational challenge, not an intellectual one; it was a super course, the brainchild of an inspired military administrator.

The problem was that many enlisted men, bright ones, had had good educations, college degrees and more, that came near to satisfying the requirements for specialists of various kinds, but didn't quite reach. Possibly a mechanical engineer lacked a language course that the army curriculum deemed indispensable, or perhaps a chemist didn't have enough mathematics. Solution: send the boys to university for a term or two, in the ASTP, and have them make up what they lacked. It would have been too complicated, however, to teach Russian to the engineer and calculus to the chemist; the simple solution was to teach everything to everybody.

The bright and nearly perfectly educated soldiers were split into class-size platoons (I had about 30 of them), and then exposed to crash courses in several subjects. I can describe the mathematics course. Some of my students had been teaching assistants in mathematics and had taught the very topics that they had to endure hearing from me; others were language students whose only knowledge of trigonometry was that the word came from the Greek. The course I had to give to this heterogeneous group met 2 hours each day for 5 days a week. The syllabus called for covering algebra, trigonometry, analytic geometry, and differential and integral calculus in one semester. The most frequent complaint (of both students and teachers) concerned the speed of the course; the students swore that if someone dropped his pencil and leaned over to pick it up, he missed analytic geometry.

Research at Syracuse

Before Martin, the Syracuse mathematics department didn't offer many graduate courses—just enough for a gentle Master's degree. We had hopes and ambitions for a doctoral program, and so, of course, we started teaching some slightly more advanced stuff. I had a class on real function theory with two students in it: Tommy Aulbach (who died young) and Chuck Titus (who later became my colleague at Michigan). They were good students; some of our Master's candidates were not so good. I remember the oral examination of one whom we started out asking about Jacobians. That proved too hard—it had partial derivatives in it—so we dropped down to functions of one variable. No, that didn't go either, so we went back to two variables but made the functions linear—in other words, we were asking questions about 2×2 matrices. No, still too hard—a quadratic equation appeared, and the business of completing the square was an insuperable obstacle. The last question, before we gave up, had to do with the binomial theorem for the exponent 2. The candidate did get his degree, but not just then—a semester later.

We tried to run a seminar or two, and the usual sort of colloquium, but we didn't really have enough manpower. The four Martin appointees had at least three different kinds of mathematical backgrounds and interests; it was not easy to find seminar topics that would be attractive to all of us. Martin himself came, although administrative work was rapidly making an ex-mathematician out of him, and we dragooned as many as we could of the students and of the younger members of the old-guard faculty.

We were kept busy, and we loved it. My teaching was 18 hours a week for several of the war semesters; the time I taught the "super" ASTP course it was 21. Nevertheless, although my official affiliation with Syracuse lasted only three years, eight of my published papers list Syracuse as my address. Back then the days must have had 36 hours.

My research was almost exclusively in ergodic theory, and a couple of the ergodic papers that came out during my Syracuse years are among the ones I am not ashamed of. In one of them I introduced and studied topologies in the set of all measure-preserving transformations (acting on a fixed measure space) and used such notions to prove a version of Birkhoff's conjecture. The conjecture was that, in some sense, the ergodic case is the general case. Oxtoby and Ulam studied the problem for measure-preserving homeomorphisms, interpreted "general case" in the sense of Baire category, and proved that Birkhoff was right. I studied more transformations, and, therefore, I had to study a new and more general topology, and proved that Birkhoff was right in that sense too.

The story doesn't stop there. In another Syracuse paper I studied the set of mixing transformations—a set much smaller than the set of ergodic ones—and was happy to be able to prove that even the smaller set is large—Birkhoff was more right than he knew. The title of my paper is "In general a measure preserving transformation is mixing"; it appeared in the Annals in 1944. In the Doklady in 1948 Rohlin published a paper titled "A 'general' measure preserving transformation is not mixing"—and he is right. So am I. He was having some innocent fun. The concept "mixing" can be of two kinds, weak and strong; the weak ones, I proved, form a large set, but the strong ones, Rohlin proved, form a small set.

The last Syracuse paper I'll mention is not a research paper but an expository one. The editor of the Monthly in the early 1940's was Lester Ford (the father, not the son). One of his inspired editorial ideas was to run a series of "What is..." papers: "What is an analytic function?",

"What is a stochastic process?", etc. Eager to spread the word about the still new notion that probability was measure theory, and, of course, more than willing to be associated with the great names that were writing "What is...", I wrote and submitted to Ford a paper titled "What is probability?". His answer was quick and negative—sorry, but the "What is..." series is by invitation only. If, however, I insisted, I could re-submit the same paper under another title and take my chance. I did; the new title was "Foundations of probability".

I asked a few people to read the paper before I submitted it; even in the pre-Xerox days of carbon copies that procedure was quite common. Herb Robbins read it and objected to my colloquial abbreviations. I mustn't say, he said, that "it isn't stretching a point too far...", I must say that "it is not stretching...". That upset me; it almost made me angry. I wouldn't (would not) make the change.

Why not? I readily admit—I'd like to be among the first to insist—that expository writing should not deviate from currently accepted standard English; in such writing even puns and other attempts at humor, and, of course, outright vulgarity, are badly out of place. Why? Because they are irrelevant, they are distracting, they interfere with the clear reception of the message. Expository writing must not be sloppy in either content or form and, of course, it must not be misleading; it must be dignified, correct, and clear. Within these guidelines, however, expository writing should be written in a living, colloquial style, it should be evocative in the same sense in which poetry is, and it should not be stuffy, but friendly and informal. The purpose of writing is to communicate, and style is a tool for communication. It should be chosen so as to put the reader at his ease and make the subject seem as easy to him as it already is to the author.

As long as I am talking about communication, I'd like to emphasize one of the phrases I just used. There is a difference between misleading statements and false ones; striving for "the clear reception of the message" you are sometimes allowed to lie a little, but you must never mislead.

One of my favorite examples is what you have to say when you begin to explain the English form of government to an intelligent but ignorant visitor from another planet. If you say "England is a monarchy", you are telling the truth, but you are pointing in the wrong direction, you are misleading. If you say "England is a democracy", you are lying, but, just the same, that's a better first-sentence summary of the facts than the truth.

Here is another example. In 1979, on the Monday before Easter, the question of mail delivery arose during the lunch conversation (at the Union, in Indiana). Richard Timoney, our Irish visitor, asked whether or not Good Friday, which happened to fall on April 13 that year, was a legal holiday. Some said it depended on what state you were in, and some expressed their views about whether it should or shouldn't be a holiday—the conversation rambled in a relaxed, pleasant way. Presently a newcomer joined the group and asked what we were talking about. Richard said: "We are trying to find out whether April 13 is a holiday." At tea that day I described the conversation, and Richard's answer, to Max Zorn, and Max was horrified. "But that's a lie!", he said. Is it clear what Max meant, and what I am trying to say? Yes, it was technically true that we were trying to find out whether April 13 is a holiday, but that is *not* what we were talking about; by mistake, or just for fun, Richard was giving misleading information.

What is important in communication, in lecturing for example, is not what message the speaker sends but what message the listener receives. A part of the art of lecturing is to know when and how to lie. Don't insist on protecting yourself by being cowardly legalistic, but lead the audience to the truth.

All this started when I was describing Robbins's suggested corrections of the "Foundations of probability". The uncorrected paper was accepted for publication, and, a couple of years later, it was awarded the Chauvenet Prize for mathematical exposition. I couldn't help saying (to myself only) "pooh to you, Herb Robbins".

Radiation Laboratory

The scientists' war was being fought on many fronts. The most famous is Los Alamos (called Shangri La, partly as a code and partly for fun), but there were also Oak Ridge (Dogpatch), the Chicago "Metallurgical Laboratory", and many others. In theory even their existence was secret, but in fact "everybody" (which really means a few hundred people in the community of physicists and mathematicians) knew about them. Erdös, who was in some sense an enemy alien, worried a bunch of physicists he sat down to lunch with one day. "The bomb based on atomic fission that you're building—is it ready yet?", he wanted to know.

The mathematicians' secret weapons were the several Applied Mathematics Groups. People like Mac Lane and Kaplansky were at AMG-C (C for Columbia), and Stefan Bergman and Will Feller were at AMG-B (Brown). The latter was run with an iron fist in an iron glove by R. G. D. Richardson, "the Dean". The Dean awarded tiny summer fellowships, designed to retread some of us pures and make us applied; the $300 that I got was enough to bribe me to spend a pleasant summer (1944) in Providence listening to Feller talk about approximation theory.

My main contact with war mathematics started with a telephone call from Bob Thrall late in 1944. Would I come and join his group at the Rad. Lab.? At the *what*? The Radiation Laboratory, in Cambridge, Mass. When? Now, as soon as possible.

I had heard of the Radiation Laboratory; its purpose was to develop radar. The section that Thrall was in charge of studied its airborne possibilities. He didn't tell me what I could and would do, but he used all his powers of persuasion to convince me that I should do it: the work was important, the problems were interesting, Cambridge was full of active mathematicians, the pay was good—I just must come. Some of his reasons tempted me, but I protested at the timing. At the very least I wanted to finish the current term and not just abandon the classes I was teaching; besides I had to ask Ted for a leave of absence, and I needed time to make the usual personal arrangements. Very well—we agreed; holiday or no holiday, hangover or no hangover, I was to show up first thing in the morning on January 1, 1945, at least for an hour or two, to sign in, and thus to make sure that I would be on the Rad. Lab. payroll for the month of January.

Section 91.5 was small when I joined it; it consisted of Bob Thrall, Tom Lawrence (his second in command), one secretary, and three or four young women who did the real work (film measuring and computing). It never grew large, but before I left the "research" staff increased to six and the clerical staff varied between eight and ten.

The first job I was given started as a question from some neighboring group down the hall. It had to do with bombs being dropped from planes; given certain data about times and altitudes, the problem was to determine where the bombs hit the ground. A minute's thought revealed that the solution could be found in the Pythagorean theorem. I wrote down the formula and (in language not yet in common use) I programmed it. That is: on a slip of scratch paper I wrote $a = \sqrt{c^2 - b^2}$, and then on a more formal-looking sheet I wrote "square column 1, square column 2, subtract, take square root". The secretary typed my instructions, and one of the girls went to work: she entered the dozen or so pieces of data in columns 1 and 2, as indicated, punched the calculator in the way the instructions told her to, and then wrote the answers in column 3. The secretary typed the report (half a page), and then stamped it CONFIDENTIAL.

That job was exceptional; mainly I was put in charge of the computers. Our raw data were on film strips. The pictures were taken in a plane equipped with radar, tailgating a target plane. A typical job was to calibrate the radar equipment. That is: the staff would measure the wingspread of the target as it appeared on our screen, and then, using a program based on trigonometry much more profound than the Pythagorean theorem, they would compute how far the target really was from the radar plane. The actual distances could be compared with the ones that the radar equipment gave—and there was your calibration.

Much of what we did was make-work boondoggling. We had some projectors and some calculators, and we had a staff that was used to them; we'd try to rise in internal importance at the Radiation Laboratory by looking for other sections that could use our expertise. We also had the research staff, and we looked for sections that needed mathematical therapy from it. (The Radiation Laboratory had other sections of mathematicians, notably an outstanding one thinking about servomechanisms, but they really had enough to do.)

Some business did come our way, and, as a result, I acquired a small, very small, amount of experience as a consulting applied mathematician. My experience taught me two lessons. First: a large part of the consultant's job is to administer psychotherapy; second, whatever you do, don't *solve* the problem you're asked about.

As for the first: on several occasions I was asked questions that I not only couldn't possibly find an answer to—I couldn't even understand the words in the question. I sat patiently, listening to the client (customer?, patient?), and, when I got really lost, asking for explanations. I never *said* anything, but I did keep *asking*: is this connected with that?, is this component small or large? An hour or so went by, and the client got up to leave, thanking me profusely for my help. He wasn't just being polite; later I heard, from third parties, that my "help" was described in truly glowing terms. My help was psychotherapy: I was willing to listen, to sympathize, and to wait for the questioner to organize his thoughts enough to explain his problem to me.

As for solving the problem you are brought: 99 times out of 100 that's the wrong thing to do for sure. One of my clients had a perfectly definite mathematical question for me, he had prepared for the interview, and boiled all his difficulties down to a theorem about matrices. Now *that* was something I could deal with; it took 10 or 15 minutes to absorb the question, half an hour of fumbling to find the appropriate technique in my toolkit, and there we were. The answer was no, and I gave the client a counterexample involving 3×3 matrices. He looked sad, he said thank you, and he left.

It wasn't till weeks later that I learned how stupid I'd been. I bumped into my victim in the cafeteria, and he was similing happily and reported that the answer to his question was yes after all. Not the matrix question—no, that was simply the wrong question (!)—but what he really wanted to do could be done by using nothing but elementary calculus. Moral: don't answer the client's question, but, instead, help him to formulate it. I suspect that the challenging intellectual difficulty in most problems of genuinely applied applied-mathematics is not to find the technique that provides the answer but to find the question that fits the problem.

In my brief experience as an applied mathematician I never needed high-powered mathematics: a little calculus, a rare bit of linear algebra, and a lot of trigonometry got me a long way.

Some of my friends at the Radiation Laboratory had trouble with classified material; that's how I heard for the first time a story that has reached me many times since then, with different names and places. According to the story a group of Radiation Laboratory mathematicians ran into a mathematical brick wall; they simply couldn't solve the problem they were working on, and they just had to had to know the answer. Yes, there was an expert in Boston, but he was not cleared for secret work; he seemed to be the only one who could help them, but he mustn't be asked. Unless—wait a minute!—couldn't we...?—yes, we could! The work they had already put in wouldn't be wasted; all they had to do, they realized, was to take one of the tortuously reached reformulations—it took months to get it—and ask the expert about *that*. No classified information was visible in the reformulation, and not even the original mathematical question was visible any more—it took a lot of brains, a lot of time, and a lot of work to reach the reformulation, and *nobody* could learn from it what it was that concerned them originally. (An enemy is sometimes eager to know not only what you know but what you want to know.)

Arrangements were made, an appointment was set up, and the expert was presented with the reformulated problem. He listened with interest and intelligence; he took notes; and he said he would think about it. The problem was tough—he didn't get the answer in a day, and he didn't get it in a week. He did get it in 10 days, and he phoned, excited—I've got it. My friends piled into a car and drove to his office, and he plunged right into the explanation. You see, all you have to do—and he did it; he changed variables ingeniously, he transformed the problem, he simplified—and he ended up with his proposed solution. The "solution" was the original problem that my friends encountered; the expert's talent reached the equivalence of the two problems more quickly than they had, and, from his specialized point of view, the equivalence was a beautiful fact—a great step forward—a solution of the problem!

One spring day in 1945, after a little more than four months in Cambridge, I was having lunch in a small restaurant off Harvard Square. The radio was chattering in the background, but scarcely anyone was listening till suddenly, in that curious way that the force of silence can sweep through a crowd, everybody fell quiet. The radio announcer's voice was both sad and harsh: the president just died. Less than a month later Germany surrendered, and VE day (Victory in Europe) was celebrated with joy, noise, and booze—downtown Boston was one gigantic party.

I was sick and tired of 91.5, and I was not at all convinced that I was helping to win the war against Japan. Ted Martin said he would be delighted to have me back in Syracuse in the fall, and in the early summer I initiated the slow process of getting released from the Radiation Laboratory. I was taking my final clearance papers, from one office to the next, one day early in August—it was Monday, August 6—when the news broke of the Hiroshima bomb. The clerks who were stamping my papers were looking at me with wonder and mock suspicion—what secret advance information did I have that made my perfect timing possible?

Referee and review

I didn't have to punch a time clock at the Radiation Laboratory, but it was understood that all of us, scientific staff and clerical, had a really full-time job—48 hours a week. What with the minor chores of housekeeping, and despite eating most meals out, there wasn't much time to do anything other than work, eat, and sleep—research was almost completely out of the question. Even if I had the time, I wouldn't have had the energy and the right mood. I have to have a tranquil Tuesday in order to have any chance at doing research on Wednesday—and there weren't any tranquil Tuesdays.

Still, somehow, in bits and pieces, at nights and on weekends, a little research got done, a couple of little papers got written. One of them was on statistics, sort of; it is one of three (in my whole life) that I published in the Annals of Mathematical Statistics. It was produced self-consciously, deliberately—I had in mind that in some sense it was my duty (at the Radiation Laboratory) to think about the "real world"; and statistics (even if to me that meant a branch of measure theory) is more like the real world than are topologies on function spaces. Perhaps for that reason that statistics paper, the one conceived at the Radiation Laboratory, is the least successful of the three. It is, by the way, a paper without a theorem—or, in any event, without any formally displayed theorems. It is called "Random alms", and it treats the problem of distributing a pound of gold dust at random among a countably infinite set of beggars. Real world.

There was other professional activity I could do by way of a salve to my conscience when the theorems were elusive: more and more refereeing and reviewing jobs were beginning to come my way.

Isn't that curious language? Why should "to referee" mean to judge (in secret, anonymously) a paper or book submitted for publication, while "to review" means to judge (openly, over your name) a paper or book that recently appeared? And, more curious still, how did it come about that the scholarly use (the one I just described) differs from the commercial one? That's how it is; when a publisher asks you (in secret) for an evaluation of a manuscript he is thinking of publishing, he says he wants you to review it. In any event, I was beginning to referee papers and review books, and I became one of the hundreds of collaborators of Mathematical Reviews.

What is the purpose of refereeing? Answer: to help make sure that good stuff gets published and bad stuff does not. What is the purpose of reviews in MR? Answer: to enlarge the fraction of the literature that is accessible to each individual scholar. Both are difficult and responsible jobs, and both can take a lot of time, but when I started doing such things I wasn't in the least intimidated—I was flattered to be asked and threw myself into the work with enthusiasm. By now I have been on all sides of all the problems connected with mathematical publication. I have written papers, I have refereed them, I have reviewed them, and I have made editorial decisions about them. (Whom to ask to be referee? How much to believe and trust the referee? Accept or revise or reject?) I have written books, I have "reviewed" them, I have written reviews of them for the Bulletin and for MR (very different jobs), and I have made editorial decisions about them. (Is the field important? Is the treatment adequate? Will the book sell?) I am no longer eager to be asked, and, when I am, I approach the task with considerably more caution (and trepidation) than I did forty years ago. I am not convinced that I therefore do a better job than I did then; the enthusiasm and the energy of youth might well be worth more than the experience and wisdom (?) of age.

I think I instinctively realized, the very first time I was asked to be referee, that the referee doesn't have to guarantee that the paper is correct, and he certainly doesn't have to rewrite it if he thinks that needs to be done. Hardy said that there are three questions you must ask about a piece of mathematical research offered for publication: is it true?, is it new?, is it interesting? Yes, to be sure. If I am referee, and if I see that the present offering is a special case of something published five years ago, then I so inform the editor and the author involved. But what if the result is just dimly familiar and I cannot annihilate it with a reference? If I see that Theorem 2 is false, and I can produce a counter-example, then I say so—sure. But what if I just don't quite believe Theorem 2 and don't understand the alleged proof? It doesn't matter, I say. The referee neither has to guarantee that the paper is correct nor does he have to prove that it's not. He is called upon, as an expert, to give his views—he should "smell" the paper, and he should recommend acceptance or rejection according as it smells good or bad.

I don't quite like the way I said that. I don't mean that referees should make snap judgments, and I don't mean that they mustn't read the papers they are refereeing. But I do mean that authors should not consider refereeing as an error-catching service; the author is responsible for what is in his paper, no matter what the referee and the editor say. If a paper is not sufficiently attractive and convincing to an expert in the field to smell good to him, then it has failed, at least in part, to do what it set out to do. And most readers of the paper, if it gets published, will not be experts; if the referee had difficulties, they will have more.

Of course referees and editors can make mistakes, and of course they can let personal feelings and other non-professional factors influence their judgments. I have no cure for that—no one has. But careful editors can spot prejudiced referees, and can then turn to others for help, and, in the worst case, a rejected paper can be submitted to another journal. I have faith that by and large the refereeing system works well.

The reviewing system for MR works similarly. The chief purposes of MR, at least for me, are to tell me about things I don't know and to keep me from wasting time trying to learn about things I don't want to know. The first one is obvious: how, other than from MR, could I learn about an interesting paper on unit balls in Banach spaces that appeared in the Atti Lincei? There is no way for an individual to scan all the journals that reach MR, not even if his library did get them all. The negative purpose of MR is almost as useful: if the review tells me that most of that long and interesting-looking paper in Uspehi focuses on von Neumann algebras, and does not apply to the more general C* algebras that I need help with, then I can afford to skip it for now.

When I wrote for MR I tried to remember these purposes. I don't have to guarantee that the paper is correct, and I don't have to point out every blemish on it—what I do have to do is tell my reader what the paper is about. If I'm squeamish about writing "the author proves that...", I can write "the author offers a proof...," or "the author states...", or simply "Theorem:...". If I know for sure that the author's statement is false, or if I happen to know of a reference that the author has overlooked, I may (in fact I must) say so; but my assigned task is not to look for trouble but to report the facts.

There is no harm in trying to make a factual report interesting. I remember one occasion when I tried to add a little seasoning to a review, but I wasn't allowed to. The paper was by Dorothy Maharam, and it was a perfectly sound contribution to abstract measure theory. The domains of the underlying measures were not sets but elements of more general Boolean algebras, and their range consisted not of positive numbers but of certain abstract equivalence classes. My proposed first sentence was: "The author discusses valueless measures in pointless spaces." Ralph Boas, the editor of MR, didn't let me use it; one of his arguments against it was that people who were not familiar with idiomatic usage, people who had to read English with a dictionary, would not understand that I was trying to be funny. I was disappointed—I thought I was being funny.

There is one particular review that I enjoyed when I first read it and have fondly relished ever since. It was written by Clifford Truesdell, and it begins like this. "This paper gives wrong solutions to trivial problems. The basic error, however, is not new." (MR 12, p. 561.)

I'll conclude these reminiscences of reports and reviews by stating two tenets of my professional faith; I acquired them when I started to referee and review and I was strongly confirmed in them when I became editor. One is: be Boolean; and the other is: be prompt.

A referee is not the person finally responsible for the disposition of the paper, but the editor who does have the responsibility cannot be effective in a gray vacuum. Yes, yes, to be sure, the paper has its faults, and it has its good points, and you, as the expert being consulted, see them all. Nothing, you say, is black or white; there are many gray areas. No, I say, a thousand times no. Ultimately, gray or no, with or without suggested revisions, the paper either will or will not be accepted for publication in the journal that is now considering it. The decision will come down to yes or no, plus or minus, black or white—and the main job of the referee is to cast his vote in a firm Boolean fashion—it's 0 or it's 1, and there is no point in fudging in between.

As for promptness, I quote Littlewood's zero-infinity law. Littlewood said that when he got a letter, or a refereeing request, or anything else that required action, he used his self-knowledge to classify it: is this or is this not the sort of thing that I'll eventually do? If it is, then do it *now*; if it is not, then write a postcard or pick up the phone, and say so *now*. Papers to be refereed don't improve by sitting for a few months at the bottom of the pile on your desk, and you don't save time and energy by postponing the day when you refuse the job. You are not only annoying editors and knifing already bleeding authors, but you're wasting your time. Get rid of the sword hanging over your head. Zero or infinity. Do the job now or do it never, and, in either case, say which, *now*.

From Syracuse to Chicago

I spent most of the year 1945–1946 teaching at Syracuse and thinking about the following year. Teaching and research were going satisfactorily, and, in particular, I wrote and published a statistics paper (on unbiased estimation), which was quite frequently referred to later by others working in the field. Early in the year, however, Ted Martin told us that he was leaving Syracuse to become chairman at MIT (a position he held, with great effect on the development of the MIT department, for many years to come), and for me that meant that I didn't want to stay either.

The period immediately after the war was a sellers' market for mathematicians. The economics of the country was in good shape, universities were expanding, and students were knocking at the gates in large numbers, many of them supported by the G.I. bill. As for me, I was no longer a fledgling, I had some teaching experience, and I had published a book and a few papers—there was no need for hundreds of letters of application. Having been a systematic meeting-goer, I knew J. R. Kline slightly (he was secretary of the AMS), and having spent three years at the Institute, I knew the big men there, as well as several other strategically placed people of importance, such as Marshall Stone and Saunders Mac Lane at Harvard. All I needed to do, I decided, was to write a half dozen informal letters saying that I'd like to move, and to keep going to meetings so as to be seen and not forgotten.

I know more about those years now, with hindsight, than I could see clearly as I was living them, but even then it was possible to know enough to make some correct predictions. I predicted that I'd get "several" offers—and I got ten. I wish I could remember where they came from, but all I am sure of is one from Swarthmore and two from Chicago.

I had always admired Swarthmore. I knew and liked the chairman (Arnold Dresden, a cheerful, jolly Dutch Santa Claus), and I was aware of the reputation of the college as one of the best undergraduate institutions in the country. To go there would have meant choosing a teaching career instead of a research one (not exclusively, but in its major emphasis), it would have meant a life very different from the one Chicago offered, and I was tempted. I didn't find the choice easy, and to this day I still wonder: should I have gone to Swarthmore?

The reason for two offers from Chicago was a curious instance of the left hand not talking to the right. The University of Chicago had been a great university since its first day, and its president since 1931, Robert Maynard Hutchins, kept up its high quality at the same time as he started it off on a parallel and novel educational track. He founded what was locally called The College (the capital letters were always audible), and based its teaching on the works of Aristotle, St. Thomas Aquinas, and the rest of The Great Books. The organization of the College (I don't think the use of "the" instead of "The" will make matters less clear) and its relation to the rest of the university were somewhat intricate. Very roughly speaking, the College was the undergraduate school and the Divisions consisted of the graduate departments—there was a Division of Social Sciences, a Division of Humanities, etc.

In mathematics (and in most other subjects) there were two departments: one under the dean of the appropriate Division, and one in the College. Gene Northrop (suave smoothie, bow tie, heavy smoker, former master at Hotchkiss) was chairman of the College mathematics department (consisting of about a half dozen people), and he and a subdean came to Syracuse to interview me. I was eager to make a good impression, of course, I invited them to tea, and I tried to appear smooth too. Later I caught on that the College smooth guys preferred very dry Martinis to Earl Grey's best. No matter—before long I had a telegram inviting me to join the College faculty.

At the same time the mathematics department, or, more formally, the mathematics department of the Division of Physical Sciences, was being reorganized. It had had its great General Analysis days under E. H.

Moore, and later its Calculus of Variations days under G. A. Bliss, but toward the end of the reign of the latter, it had become too specialized and inbred, and its reputation suffered. In 1945 it had a caretaker chairman, E. P. Lane, an elderly differential geometer. Mr. Lane was a Southern gentleman of the old school, his mathematical days were behind him, and he was very conservative. Some of the young turks suggested to him that he hire Sammy Eilenberg who, then in his early thirties, was already established and famous. Mr. Lane's reply was a question: "Who is Eilenberg?", he asked. (Sammy, by the way, went to Indiana instead, but he stayed for one year only, and then moved to Columbia for the next 35 years or so.)

Faculty members at universities demand, or in any event hope for, self-determination for their departments; *they* should decide with whom and under whom they want to work, not those ignorant deans. Yes, that sounds right; it is right most of the time. It's certainly right for departments of high quality. But what do you do when a department goes bad? André Weil suggested that there is a logarithmic law at work: first-rate people attract other first-rate people, but second-rate people tend to hire third-raters, and third-rate people hire fifth-raters. If a dean or a president is genuinely interested in building and maintaining a high-quality university (and some of them are), then he must not grant complete self-determination to a second-rate department; he must, instead, use his administrative powers to intervene and set things right. That's one of the proper functions of deans and presidents, and pity the poor university in which a large proportion of both the faculty and the administration are second-raters; it is doomed to diverge to minus infinity.

Chicago was lucky. Walter Bartky was a reasonably good dean, and Hutchins was a very good president. Hutchins had faults, of course. Many people thought that the College was a scatterbrained, unfeasible, romantic notion of his, and even more thought that he was not a good judge of people. But he was imaginative and energetic, and he often listened to and acted on good advice. He and Bartky intervened in the affairs of the mathematics department; they arranged for Mr. Lane to step down, and they invited Marshall Stone to take his place.

As Stone's nominee, I received an offer from the dean of the Division, as well as from the College, and it took me a while to puzzle out what the various telegrams from Chicago meant. Neither Bartky nor Northrop knew of the telegrams that the other had sent, and when they learned (from me) what happened, they were mildly embarrassed. No harm done, we all ended up friends, and I accepted the job at Chicago — in the Division, not in the College.

PART II
SCHOLAR

CHAPTER 8

A great university

Eckhart Hall

Ever since my graduate school days, I had been thinking about writing the book to end all books about measure theory. Saks was excellent, but it didn't cover enough, and some of what it did cover could be brought up to date. Lebesgue and Borel were good but too old-fashioned. Hahn's and Carathéodory's books on real variables were among the best sources of information, but, for them, the theory of measure and integration was only a small cupful or two in a large bucket. The missionary in me was sure that the world was waiting with bated breath for a modern version of The True Word.

Not everyone agreed. "Don't waste your time," they told me; "Ulam is writing a book on measure theory—oh, yes and so is Robbins." Others were more encouraging. Early in the spring of 1946 (before I got the Chicago offers), Marshall Stone wrote me a friendly letter, suggesting that I submit my book to a new series he was about to edit. At that time the book was only a gleam in my eye and a thin folder of scraps of paper with scrawled ideas on them ("Include probability?", "Must treat Haar measure!"); the encouragement, and, later, the Chicago job being in the bag, were what I needed to get to work. (The Ulam and the Robbins books never did appear.)

The Syracuse spring term ended early in 1946, and I got my last salary check on May 1. In Chicago the Autumn quarter was to begin around October 1, and, consequently, my first check was not due till November. A six-month summer with no income is hard on a young assistant professor, but I thought I might as well put it to good use. I asked the Institute to let me spend the time there, and that worked out beautifully. They gave me an office, and they rented me a small house next to the Institute grounds—and *Measure Theory* began to grow. That's not the end of the story—it was two more years before the writing was finished, and the publication date was a year and a half after that—but I'll tell you how all that went a little later.

In September I moved to Chicago, and the intellectually and mathematically most stimulating and fruitful period of my life began.

The divisional mathematics faculty of about a dozen was housed, together with the College mathematics faculty, in Eckhart Hall, the charming neo-Gothic mathematics building. It was neat and clean, with broad dark corridors and narrow dark offices, tennis courts in front, and a small, cozy, green quadrangle behind—it is one of a cluster of buildings that, according to local wits, Magdalen College (Oxford) copied its architectural design from. If you ever opened your window, in a couple of hours the sill was covered with sticky, grainy soot—the contribution of the steel mills in Hammond and Gary across the lake.

The campus blended smoothly into Hyde Park, a pleasant, old-fashioned residential district—a segregated rectangular island surrounded by a slummy ghetto. The segregation was so extreme that if it hadn't been deplorable it would have been ridiculous. Cottage Grove Avenue was and is the western boundary of Hyde Park, and even a chance visitor to the neighborhood noticed immediately that the pedestrians were white on the near (eastern) side of the street and black on the other. There was nothing except custom to keep anyone from crossing over—but usually no one did. That's how it was in the 1940's; today the areas to the west of Cottage Grove, south of the Midway, or north of 55-th Street are poor neighborhoods, mainly black, but the ridiculously extreme segregation is gone. (The Midway is a straight boulevard, as wide as a generous Chicago city block is long, containing six lanes of traffic, four sidewalks, ample expanses of smooth green lawn, and even a bridle path.) Hyde Park now is like many other middle-class, middle-wealthy neighborhoods: if you can afford it, you can live there. Many of the young faculty, black or white, cannot afford it.

In my day most of the university faculty lived in Hyde Park, and communal life was culturally and socially active. There were bookstores with books in them (not just three-ring binders and the required text for Accounting 101), there were concerts and plays an easy 10-minute walk from home, more than you could possibly have time for, and, if you wanted company, any time of day or night, you were bound to find it at Steinway's Drugs, or at the Tropical Hut on 57-th Street, or at the Hobby House on 53-rd.

The eastern boundary of Hyde Park is the "outer drive"—Lake Shore Drive—next to Lake Michigan. "The Point", a tiny peninsula at the end of 55-th Street, is still a popular swimming and sunning beach. The lake and the Midway are, in effect, fixed natural boundaries, and so is Cottage

Grove, straining under the pressure of the new wings of Billings Hospital (a part of the university) that are constantly being added. The least fixed part of the boundary of Hyde Park is on the north. In my day it seemed to be at about 47-th or 49-th Street—the elegant shops and palatial movies on 51-st Street were well inside; by now it is probably at 55-th Street, within two blocks of the main campus.

The Hyde Park enclave has all the advantages and disadvantages of being a small part of a large city. It is 10 minutes (by the efficient IC electric trains) from the loop, the metropolitan center of several million people. By dint of costly perseverance it has managed to remain apart and immune from the diseases of all the cities of our century. The last time I visited there the shady, quiet streets with charming old brick houses, built to be one-family residences, were heavily patrolled by police cars (four of them cruised past me in 10 minutes), and the emergency phones connected directly to University Security were sprinkled densely around the campus neighborhood. (How do they work? If you think you'll be mugged or raped, you run to the corner, and lift the phone, and... and then what?)

Things were not bad in the immediate postwar period—but they kept getting bad, and then they got worse. I was never attacked or burglarized, but I knew a dozen people who were. There were two or three occasions in my "good old days" when, sitting in my office with the door to the corridor open, I was approached by a panhandler who wandered in off the street and hoped to intimidate someone to make a contribution. The Common Room had to be kept locked, and we were told not to leave our office doors open when we went to the washroom—typewriters had been known to disappear.

Does it seem as if I am describing a state of siege, where the good guys lived behind battlements and ventured forth in bulletproof vests only? It wasn't like that. The city was there, but we didn't live in a constant state of paranoia. All we had to remember was to act sensibly. To take a solitary walk on the dark streets at midnight, or not to lock the car, was not sensible—that was just plain foolishness.

Days of glory

Even though I had lived in Chicago before, I never really knew the university. I lived on what was called the near north side, three miles north, and the university was on the south side, seven miles south of the loop. I had visited the university a few times during my high school days, and later during my graduate school days in Urbana, when I would occasionally attend a spring AMS meeting. The campus and its neighborhood were familiar, yet new; because of my visits I knew my way around, but now I was a part of it, and I was sure I was at the top of the world.

The departmental secretary gave me a key, a master key!, that would open all the mathematics offices in Eckhart Hall. That was a gentlemanly old tradition that added a great deal to the collegial atmosphere in the department. We were all scholars together, members of a family. We respected one another's privacy, but, just as in a family, no one wished to lock the others out. Everybody's books were always accessible to everybody; if I had signed out a volume of a journal from the library, you could check a reference in it any time, with no fuss; a visiting colleague from another university could always be stashed away in somebody's office for a few hours. The master keys were often used and never abused; a great system.

During my first year or two at Chicago another charming old tradition was continued: departmental meetings were held at lunch in the Quadrangle Club on Tuesdays. The business that had to be transacted was sandwiched between the soup and the salad—it was pleasant, friendly, and efficient. As we grew, the Quadrangle Club table became too small and the number of voices that had to be heard too great, but the way we administered ourselves continued to be as nearly perfect as any human government is likely to be. Sure, we were human, we disagreed sometimes, and we had petty arguments (somebody spends too much departmental money on postage, somebody wants somebody else's office), but on balance we genuinely respected one another. We disagreed most when we discussed possible future appointments. One time I wanted Walter Rudin and André Weil wanted Grothendieck, and emotions and voices became elevated. We didn't get either man—and the atmosphere returned to its more nearly normal more nearly calm state.

Teaching assignments in the mathematics department for year $n + 1$ were decided just before Christmas in year n. The whole department met in a small seminar room. André Weil usually brought a stack of reprints and manuscripts to help him over the boring parts of the meeting; Zygmund read the New York Times. The chairman wrote on the board the numbers of the courses that *had* to be offered, and then called on the rest of us to choose from among them and to add whatever fancier courses we wanted to offer. The university bureaucracy put no obstacles in our way—new course names and numbers were made up impromptu. By tradition the "junior officer present" spoke first. Some conflicting desires were bound to be revealed, but, with a little horse trading, they were amicably resolved. Somebody looked after my interests even before I arrived. In my very first term I had the schedule I would have chosen: an introductory calculus course and a graduate course in ergodic theory. That was the normal divisional teaching load: six hours, consisting usually of one elementary course and one advanced one.

Ray Barnard was the second oldest man in the department (after Ernest Lane); he was a General Analysis disciple of E. H. Moore's, and he was not a research mathematician. Another holdover was W. T. Reid (even his wife called him W. T.), a part of the Bliss calculus of variations school; he left a year after I arrived. All my other colleagues, young and not quite so young, were well known and were on their way to becoming even better known figures on the American mathematical scene. Adrian Albert, the algebraic power house, and Lawrence Graves, a real variable analyst, were the senior people; they both belonged to the pre-war old guard. The younger people during my first year included Irving Kaplansky (he was at Chicago for a year before me, and he stayed on for 38 years), Kelley (that's J. L. Kelley, the author of "General Topology'— and nobody except his mother called him anything but Kelley), and O. F. G. Schilling (a non-thin, very clever, but early burned out algebraist with too many initials and a thick German accent—he was called Otto).

Kelley stayed one year only, but other young people came (Irving Segal and Ed Spanier), and, before long, so did four already world famous established mathematicians (Chern, Mac Lane, Weil, and Zygmund); with them in place, Chicago had reached its days of glory.

What makes a great university?

A great university means a great faculty—that's all it means; the condition is necessary and sufficient. Pre-calculus teaching, janitorial and clerical services, and the correspondence school—surely such peripheral things have nothing to do with the greatness of a university? That's right—they don't—but still, somehow, a great university, with a charismatic and knowledgeable president, is likely to do even those things right. I am a great believer in the system of undiluted responsibility; if the mess steward breaks a cup, the captain of the ship is responsible. Very few cups broke at the U. of C.

The bread-and-butter course of the College mathematics department was something like "new math". It concentrated on the fundamentals of mathematics, it was continually being revised and polished, and it was taught by ardent believers. Northrop was the prophet, and Meyer, Putnam, and Wirszup were among the disciples—there were many others, and they kept joining and leaving the department, but these three stayed for many decades. Set theory, Boolean algebras, axiom systems, the definition of real numbers—that's the sort of stuff that Math. 1 consisted of. In principle members of the College were welcome to teach a divisional course from time to time, and vice versa. The exchanges did in fact take place, but not often. The normal division of labor was that the College people handled pre-calculus and the Division did the rest. That's rough—the precise details are complicated, and they are of historical curiosity value only. What's important—what was good—is that the freshmen were not thrown to a bunch of callow graduate students or superannuated hacks. Their teachers were well-trained mathematicians, dedicated to an idea. Sometimes some of them were cynical about the idea and poked fun at it—but they took their students and their teaching seriously. I knew several Math. 1 students who became professional mathematicians, but more important than that, the ones who became linguists or librarians or lawyers knew a lot more about what mathematics and mathematicians were like than the business school students nowadays, forced to take Calculus 111 as a "distribution" requirement, ever have a chance to learn.

Let me describe two other small samples of the atmosphere that helped make the University of Chicago great, samples that have nothing directly to do with mathematics or scholarship.

One time when I went to teach a class I found my students milling about in front—the room was occupied by buckets and wet mops attached to the maintenance workers in charge. I expostulated, but, of course, to no avail: it's hard to win an argument with a mop. The class was called off, I was angry, and I wrote a nasty letter to the head of the physical plant. To my astonishment, he answered immediately, apologized profusely, and promised that in the future he would monitor the cleaning schedules more carefully so as to make sure that classes were not interfered with. Nowhere else have I even heard of such an attitude.

Another time Sz.-Nagy visited Chicago, and a few days after he left I mailed him his honorarium check. Something went wrong—he was on a lecture tour in the U.S., travelling from one spot to another—and his mail never caught up. He phoned me two days before he was scheduled to return to Europe to ask why he never got his check. I pushed the panic button—and with one telephone call to the appropriate and completely cooperative person at the comptroller's office I made Nagy happy. Sure, I was told, why not—we'll write another check and put it in the mail within half an hour, address it to Nagy in New York, special delivery, so that he'll be able to cash it before he leaves the country—sure, don't worry about the first check, when you get it back, just send it to us by campus mail.

What's the explanation for this rare attitude? Private university, as opposed to a public one? Long tradition that scholarly work is what the university is here for? Clearly visible message from the president's office? Whatever it is, the University of Chicago had it.

The research mathematicians and the teachers, the accountants and the janitors—all contributed to the greatness of Chicago, and so, surprisingly, did the extension school, the correspondence courses. The person in charge of that in the mathematics department was Harry Scheidy Everett

H. S. Everett, 1952

(Scheidy to many of us). He was old to my young, he was a grade-point counter, a correspondence teacher—he was many things that the rebellious iconoclast in me wanted to deplore—but he was fiercely loyal to Chicago and its traditions, and he did a great job.

He was friendly to me; he seemed to enjoy teaching me the pre-history and the present ways of Eckhart Hall. Often when we young hotshots made up a new course, he came right along: he designed, offered, and then taught the correspondence school version, with great profit to his pupils. He worked hard at grading his correspondence mathematics courses. When he liked a paper, he'd write words of praise on it and draw a smiling face; when a paper didn't come up to his standards, he'd express sorrow rather than impatience, and, of course, he'd draw the sad circular face instead of the glad one. Iz Singer told me that during his army days he took a course from Scheidy, and Scheidy's friendliness, encouragement, and help were high on his list of reasons for coming to Chicago.

Teaching

True or false: anybody who really knows something can teach it? Conventional wisdom is firmly established on both sides of the question, and so am I.

A part of the difficulty is that the converse is trivially true (to teach something you have to know it), and the temptation to convert the converse is irresistible. But there is more to it than that: if someone knows potential theory as well as Bob Martin did, or, for a more famous example, if someone knows Fourier analysis as well as Norbert Wiener did, why doesn't he tell it to us, why doesn't he teach it to us, how could it possibly happen that he is unable to do so? The brutal fact seems to be that some people just can't explain things, even though, clearly, demonstrably, they know the subject better than anyone else.

I knew Bob Martin, and I knew Norbert Wiener, and I knew many others like them, and I cannot make the paradox go away. I am very much inclined, however, to be in favor of the side that says that one who really knows can teach. I am inclined to suspect that there was a sense in which Wiener didn't "really" understand what he was doing, not as well as almost any writer of a text on harmonic analysis did twenty years later. Wiener viscerally understood the importance of ideals, but (so far as I can remember) he never used the word. He never made the attempt to establish the connection between the powerful analytic techniques he used to draw exciting conclusions and the broad algebraic notions that can put those techniques into a transparent context. He wanted to be understood, and he tried—but he had the wrong idea of what it means to try. He thought it meant to go into detail, to write down the equations between many double integrals many times, proceeding from one to another in small steps. Sometimes, to be sure, he misjudged what "small" means to others, but that's not the main problem; what makes him hard to understand is that he doesn't describe what's "really" going on. Did he "really" know? I say he didn't. I say he was a genius, gifted with deep insight, who could "see" the truth, but he didn't have the large vision that could have led him to "understand", to "know" the truth.

Surely Wiener "understood" trigonometry, and analytic geometry, and elementary calculus—does that imply that he was a good teacher of those subjects? Unfortunately not. If my amateur psychology is correct, he and people like him are as blind to the connecting details of those trivia, and of their importance to most ordinary mortals, and even of their beauty, as a runner is ignorant of the details of the nerves and muscles in his knee. The runner knows something, but a track coach knows it better, in a sense; the track coach "really" understands it, and, unlike the runner, can teach it.

I was led to these contemplations when I started recalling my first year of teaching at Chicago. I was 30 years old by then, and I had already taught at Illinois and at Syracuse (and en passant at Princeton and Harvard), but my attitudes toward teaching and my techniques were not yet in their final form. Chicago students were eager, interested, and talented—not all of them, but enough to make teaching there an intellectual challenge, an exciting and rewarding experience, a joy. I thought I was a pretty good teacher already, but at Chicago I worked hard to become a better one.

The first thing I tried to do in each class was to get to know the students as quickly as possible. I let them sit wherever they wanted to, but I asked them to keep returning to the same seats, so that by use of the seating chart I made I could memorize the one-to-one correspondence between names and faces. Poor artist though I be, I sometimes drew caricatures on the seating chart—long hair, round face, horn-rim glasses; later, when Polaroid cameras came on the scene, I took pictures. If the class was not too large, I asked every student to come to see me in my office for 10 minutes some time during the first two weeks of the term. What we talked about during those 10 minutes didn't matter—what mattered was that at the end of the time I knew *something* about the students (came from New York, learned calculus in high school, wanted to study physics, has difficulty with English), and the students felt that a real live professor was paying some attention to them.

To understand a subject, you must know more than the subject; to teach a course, you must know much more of the subject than you can possibly put into the course. The first time I taught elementary calculus at Chicago, I knew more about calculus than I was expected to teach, but even so I thought it important to prepare for each class, and I still think so. One reason is that students are receptive—no matter how much they resist, they can't help being influenced by the authority of the lecturer—and if you say something wrong, they'll remember it and it will do them damage. I try especially hard never to say anything wrong on purpose (and, even more, never to write it on the board). Crathorne, my Illinois statistics teacher, told me once that, to make the point that "$0^0 = 1$" is a plausible definition, not a theorem, he told a class that he might just as well have defined 0^0 to be 7, and he wrote the equation "$0^0 = 7$" on the board. Years later he got a letter from a student who was in that class: "I have in my notes that $0^0 = 7$, but I forgot how you proved it—could you please remind me?"

Mistakes are something else—of course I make mistakes in my class, and just about equally often I run into a question, in class, that I don't know the answer to. I quickly found out that it's very bad policy to try to bluff your way out; the chances are you'll get into even deeper trouble, and, even if you don't, the students will almost certainly catch on to what you are doing. For mistakes, I have learned to say "Oops, sorry, that was a mistake—I said that wrong—here is what I should have said", and for questions I can't answer "Let me think about that and try to look it up—I'll tell you next time".

Although we try to avoid mistakes and baffling questions, it is well known that they can have great educational value. An omniscient teacher helps perpetuate the myth that mathematics is a rigid body of facts and perfect techniques—a myth that students often find easy to believe but hard to live with. To see an expert make a mistake or admit ignorance, and then battle his way to the truth, can be an eye-opener. Jimmie Savage told me how impressed he was when in class once, at Michigan, he asked

Ben Dushnik a question, and heard the answer: "Savage, I don't know. I never thought about that. But let's see whether we can find out." With that Dushnik proceeded to think out loud at the board, and, in a few minutes, by applying the techniques he had been teaching that term, to reach a conclusion. Jimmie learned a lesson from that experience: mathematical thinking can be used to learn something that you didn't know before.

Between the times of admitting error and bafflement, and most of the time in between those times I hope, I think it's important to be right, to be sure I'm right, and to convey to the student that I am sure—I must have the courage of my convictions. Even questions such as "Is this stuff important? Will that be on the exam? Why don't you just substitute $h = 0$?" deserve crisp answers—I must not waffle and shilly-shally. It's better to be wrong sometimes than to equivocate and instill insecurity.

There is much more to be said about teaching, very much more (although nowhere near as much more as Schools of Education say), but, except for recording here my beliefs about grading, I'll stop for now. Assigning grades to the students in my classes is a part of my job; it is a necessary evil. Grading is bad because students often pay too much attention to it, because it is often regarded as more accurate than it can possibly be, and, incidentally, because it often makes students feel bad. It is, however, necessary because in our present educational and social organization the teacher in a later course must know what the student learned in an earlier one, and a prospective employer wants to know how good the student is likely to be on the job. I can't think of a way of designing an organization of learning and working in which these items of information are not needed.

I do not, however, think that the assignment of informative grades is all that hard. At the end of a course I usually have a pretty clear idea that certain students know the material (A), and certain others don't (F). In between there are those who know some of it but have gaps in their knowledge—possibly big ones (B), and there are those who can use some of it, but don't really understand it (C). Then, of course, there are those who can prove that they have been exposed to it, but for sure don't know enough to go on to a course on a higher level (D). (I found it amusing to grade myself on my knowledge of languages on the basis of the criteria just laid down. Here is what I think I might get: A in English and F in Chinese; Hungarian B, Spanish C, and Russian D. As long as I'm on the subject, I might as well go on: German C^-, and French D^+.) Of course my "pretty clear idea" is subjective, but it's remarkable how nearly unanimous such subjective grade assignments turn out to be: students keep getting the same sort of grades time after time, in different courses from different teachers. I don't agree with those of my colleagues who advocate a more "objective" numerical grading system: Problem 4 is worth 15 points, and you get 3 points if the answer is right, and 2 points off for each of the six most obvious missteps you can take on your way to it. To my mind it is my duty to use my best judgement about how much my students know when I'm finished with them; anything else would be an evasion of my responsibility.

Students and visitors

At Chicago student conversations in the common room, in the Coffee Shop, everywhere on campus, were lively—they were not about basketball and bicycle racing but about St. Thomas and Galileo and Darwin. The genuinely high intellectual atmosphere attracted the potential geniuses, and, at the same time, it attracted the phonies and the screwballs—but the former, and all the rest of us working people, learned to live with the latter, and even to enjoy the spice they added to life.

Life was especially exciting in the Autumn quarter of 1946, the beginning of the first real postwar year. The standard first year graduate course in complex function theory had 95 students in it, and my first-time-ever graduate course in ergodic theory had 38. Among them were Errett Bishop, Henry Dye, Harley Flanders, and Herman Rubin, and at least a half dozen others who grew into respected members of the profession. I gave 27 B's in that course and three A's; the A's went to three of Bishop, Dye, Flanders, and Rubin.

Students like that make teaching a joy. It is unfair to claim credit for teaching them; they are essentially invariant under the educational system and it would be hard to keep them from learning. Here is an example— not a typical one, but one that makes the point. One time I assigned a problem: find, within a collection of infinite series of a given special type, two examples, one that converges and one that diverges. Bishop's homework paper contained a theorem I had never seen before: a necessary and sufficient condition that a series of the type in question converge is that the n-th term tend to 0. The assertion is not hard to prove, but it took real mathematical understanding and courage to think of it. (It became Exercise 13 on p. 19 of my book on measure theory.) I wrote a large A on Bishop's paper, and wishing to congratulate him and to get better acquainted with him I wrote "See me" under the grade. Bishop stopped after class, shy and therefore seemingly surly. "You said I should see you?" It took me a while to get over his barriers, but eventually he became my Ph.D. student, and we became good friends. He was my fourth Ph.D. student chronologically; in quality he was either first or second—history will tell which.

As I was teaching the ergodic course Bishop was in, I started using a grading system that I have found immensely useful ever since. Officially I still gave A's and F's, but in my private grade book I "graded" each student by writing two or three phrases after his name—phrases that would later help me recall the student and the quality of his work. By Bishop's name, for example, I wrote "Messy, intelligent, original; picks hard problems". Some other samples from the same class (names omitted): "Serious, hard worker, no genius, solid"—"Somewhat confused, too glib"—"Unsmart, unwilling to work"—"Talks straight, knows stuff". These notes have worked very well for me. I have used them, for instance, when long after I knew a student I was asked for a letter of recommendation about him; sometimes I included the verbal "grade" in my letter.

Another thing besides outstanding students that contributed to the stimulating atmosphere at Chicago was the policy of permanent visitors—or, better said, the permanent policy of inviting temporary visitors. There were always new faces, both young beginners and established oldsters, new personalities, new ideas to be inspired by. A few among the visitors who stick in my memory: Børge Jessen, Marcel Riesz, and Pierre Samuel—and also Dieudonné, Kodaira, and Littlewood. These were in addition to the promising youngsters who came as strictly temporary instructors, for never more than two years, and in addition to the many who stopped by for a week or for a day.

All the Chicago visitors had greater stature than I, but some were more nearly my contemporaries, and they went out of their way to minimize the effect of the professional distance between us by friendly joshing. Jessen, a spectacularly clear and beautifully organized lecturer, is a case in point. I listened to his course of lectures on almost periodic functions the hour after I myself was teaching a course in the same room. One day a student kept me after my class with a question, and I didn't have time to erase the board. The bell rang, Jessen came in, I sat down, and he erased the board. He noticed, however, that about two thirds of the way down the second panel I had written "almost periodic functions". (As I remember, I was using them as an example of a big Hilbert space.) He rubbed out everything except that, started his lecture, and started filling the board from top left. About 35 minutes later he reached the unerased part and there, sure enough, "almost periodic functions" was exactly the appropriate phrase for his text.

Marcel Riesz was a short man, heavy, a great eater, partyer, and drinker. Gårding said that when he was Riesz's assistant in Sweden, at a time when spirits were rationed there, one of his duties was to transfer his liquor allowance to Riesz. Full of food and liquor and jollity, Marcel Riesz liked to sing (Gypsy folksongs in Hungarian) and to talk about his grandchildren. Every time he mentioned them, he made sure you knew he wasn't ever married. He liked to talk about his older brother Frederic Riesz too. They respected each other, and got along well, I gathered, but sibling rivalry was not totally absent. Frederic was more visibly recognized by the world, and Marcel couldn't hide all his envy. The rivalry continues still: functional analysts and "hard" analysts continue to debate which of the brothers contributed more to mathematics.

I might have thought that I was getting on in the world, but visitors like Riesz quickly set me straight. The day he first appeared in Eckhart Hall, I went to pay my respects. Speaking Hungarian, I introduced myself and welcomed him to Chicago. "Glad to see you, sonny", he said—or, in any event, that's the closest that colloquial American can come to translating the friendly but condescending way an older man is allowed to address his juniors in traditional Hungarian—"and, now that you're here, could you write this down for me, please...?", and he proceeded to dictate a letter. I was promoted from assistant professor to secretary. What could I do? I wrote it down.

CHAPTER 9

The early years

Guggenheim

My second year in Chicago wasn't spent in Chicago. Marshall Stone advised me to apply for a Guggenheim fellowship for the year—and I got it, I got it! After all the many Illinois graduate fellowships I didn't get, not to mention all the prestigious NRC fellowships and named instructorships I didn't get, and skipping over the one-semester fellowship at the Institute that I sort of backed into, I finally got something—I thought of it as the first official recognition I ever received.

My salary at the University of Chicago for 1946–1947 was $5000, and for 1947–1948 it was slated to be the same. The custom of a raise for everybody every year didn't exist yet, not even at places as good as Chicago; it wasn't expected, and inflation didn't make it necessary. The fellowship, however, paid only $3000. (Incidentally, it should be understood that all these numbers were 12-month incomes—NSF summer salaries were still far in the future.) Here, once again, Chicago's first-class way of doing things came to the rescue. The administrative trick was to lower my salary (!) to $2000 per year, and then assign me to work at the Institute for Advanced Study. I proudly told all my friends that I was to be the lowest paid assistant professor in the country that year.

The stated purpose of my Guggenheim application was to write a book on measure theory, and that's exactly what happened. The long summer I spent at the Institute in 1946 gave the book a good start, and it was natural to hope that in a whole year with nothing else to do, under conditions as ideal as the Institute could provide, I could finish the project. This isn't an adventure story, a thriller with ups and downs and cliff-hanging suspense; except for the normal ups and downs of daily life, whose details quickly achieve oblivion, all went smoothly, and I wrote the book.

Life at the Institute was like what I remembered, but more so. I was assigned a tiny (but private) office, very near to heaven: call it attic or call it top floor, it had a sloping roof with a small dormer window. With a desk, a typewriter, and one visitor's chair, it was full of furniture.

Outside my office, everything had grown. In the early forties the School of Mathematics had about 25 "members"; in 1947–1948 there were well over 50. ("Member" is the neutral word for people who aren't either permanent staff, which means professors and secretaries, or temporary staff, which means assistants.) To be sure, the School of Mathematics included the physicists (for instance Paul Dirac and Alfred Schild), but they were way outnumbered by the mathematicians. Einstein was still listed as staff, but he was Professor Emeritus; the other physicists on the staff were Oppenheimer (professor) and Bram Pais (permanent member). Bram didn't stay permanent very long.

There is a silly story about Einstein and me that I like because it shows how famous I had become. I had been introduced to Einstein at two or three public, formal occasions, but he couldn't be said to know me. Ernst Straus was Einstein's assistant that year. Ernst was not a physicist, but he was a clever young man whom the Institute was more than willing to support for a time, and his mother tongue was German. One of the principal duties of Einstein's assistant was to walk home with the old man every day, chatting in German. One day as the walk was beginning, my wife was just going into Fuld Hall, and she and Ernst greeted each other. "Who was that?", Einstein asked Ernst (as Ernst later told us). "Halmos's wife", said Ernst. "Oh", said Einstein, as he thought over that piece of information; "And who is Halmos?" Such is fame.

Speaking of Einstein reminds me of Gödel; they were friends and Gödel frequently took Straus's place and walked Einstein home. Most great men accumulate some kind of legend—stories about them, either true or, if necessary, invented, are told and embellished till they fit the story-teller's idea of the subject's character. Gödel is almost an exception; there are very few stories about him, and several of those are about things he did *not* do. He seemed, for instance, to have almost no social life. On the few occasions when duty forced him to attend a public afternoon tea, he spoke to hardly anyone, and he went to great lengths to avoid touching people or being touched—he pulled his shoulders in and down and weaved around human obstacles like a supercautious broken field runner. He was said to be a perpetually worried hypochondriac with some weird ideas—one was about there being poison in radiator paint, so that to turn on one of the infernal contraptions was to court disaster. The only other thing I remember being told is that he kept trying to use his ultraprecisionistic logical habits in ordinary human affairs. When at the beginning of World War II he had to answer a bureaucratically and un-intelligently designed draft questionnaire, he became confused and sowed even greater confusion. Instead of answering the unanswerable questions with an impatient yes or no, the way most of us did, he would write lengthy and involved essays explaining that if the question meant A, then the answer was X, but if it meant B, then... and so on.

Mathematical life at the Institute was vibrant. At least half of the visitors were, or would soon become, known the world over. Among my best friends were two young men from India (they were young then): Minakshisundaram and Chandrasekharan. (No relation to the astro-physicist Chandrasekhar.) Minakshi died relatively young, some years after he returned to India; Chandra returned, and then left Bombay, where he was at the Tata Institute, and moved to the ETH in Zürich. Just to give an idea of the quality and of the international character of the membership, I mention three names, not quite randomly chosen: Loo-Keng Hua, Dan Mostow, and Paul Turán.

Princeton University still had an important role in the life of the Insti-tute mathematicians. I had two good friends among the students: Norman Hamilton (about 18 or 19 then—still an undergraduate, flunking courses in Latin for non-attendance, and doing brilliantly in graduate courses in mathematics), and Oscar Goldman (later chairman at the Uni-versity of Pennsylvania; at the time a comparatively elderly 21 or 22). Both were frequent guests at the coalminer's house, on the Institute grounds, where I was living. At least that's what I heard: the ramshackle wooden houses that the Institute's housing project then consisted of had been built for a mining town that went broke, and they were floated and towed to Princeton. They were squeezed together, wall to wall; two-storied, most of them, heated by a pot-bellied black coal stove on the ground floor. To get those stoves started was a bit of an art; it took crumpled issues of the New York Times, lots of kindling, experience, and patience. The more modern, architect-designed houses with radiant heat in the ceiling didn't arrive till ten years later.

Seminars, the occasional drive to New York (about an hour through the smelly industrial part of New Jersey), social life, and measure theory—the year was crowded and rich. The book was finished at just about the end of the academic year. There was to be a party in honor of some visitor at my house one afternoon, late in the spring, but (by pre-arrangement) I myself, the host, had to be among the last to arrive. I was pleased and excited when I got home; "I just wrote the last word on measure theory", I proudly announced. "What is the last word on measure theory?", someone asked—and I was stumped—I didn't remem-ber. Nothing would do but that I should rush back to my office and look, and then rush back to tell. The last word in *Measure Theory* turned out to be "X".

Measure Theory

There are at least six different times at which an author can say "I've finished my book". One is when he finishes the first draft, and another when he finishes the "final" draft—the revised, polished version that is sent around to be read and criticized. The third stage is when he finishes the printer's version and mails it to the publisher. The next stage comes after the most frustrating and exasperating work that authors have to do, the reading of galley proofs; how could the copy editor have been so stupid in his directions to the compositor, how could the compositor have been so egregiously inconsistent?! The fifth stage can be a pleasure: the page proofs are clean, the book begins to look like a book, and once that stage is over there is nothing more the author can do. The only time when a book is really finished, however, is the last time, when the first bound copy arrives via express mail from the publisher, and you have something solid in your hand to heft, to admire, to cuddle.

When I left Princeton in 1948 and returned to Chicago, *Measure Theory* was finished in the first sense. I typed it all myself, in my little near-to-heaven Institute office, pounding away hard at a heavy, manual Smith-Corona. (Did electric typewriters exist in 1948? Not on my budget, they didn't.) The pounding was necessary because I wanted five carbon copies. I typed and filled in the original plus the five onionskin copies; then I collated and had the copies bound (for the convenience of the readers) and sent them off to four people. (The original was saved for the printer and the remaining carbon for me.)

For big shots the process was different—but still primitive. I was in Princeton when Hermann Weyl wrote his little "Studies" book on algebraic number theory. The secretary who worked for him told me about the actual mechanics of the writing. Weyl would dictate to her, from rough notes, something like this: "The numbers (8.3), skip a line, constitute an integral basis of, skip nine spaces. By adjunction of, skip a space, the field k, skip a space, changes into a field $\bar{k}$ of degree f over k, skip a space...". She would write all this in shorthand, and type it up; then Weyl himself had to go over the typescript and fill in the formulas.

When my book reached the proofsheet stage, I was back in Chicago. A few students were taking a reading course from me, and what I gave them to read was, of course, the proofsheets. They were bright students and conscientious too. They found the usual number of typographical errors and reversed inequalities; such things are annoying but not catastrophic—they are to be expected. One error that they turned up, however, was a serious one: an exercise blandly stated a false proposition—it was almost meaningless—and challenged the reader to prove it. My excuse was that the exercise made sense, and was true, with respect to the definitions of an early version of the manuscript; when I changed the approach, and changed the definitions, that particular exercise escaped the revision process. The excuse didn't improve the fact: the statement was bad. The book was in page proof by then, and printers throw fits if you add or subtract as much as a line at that stage. If I had just omitted the exercise, the numbering system would have been thrown out of kilter. Solution: count the symbols, spaces, commas, parentheses (there were 535 in all), and then make up an alternative exercise concerning the same narrow part of measure theory that was treated in the preceding few pages and occupying approximately 535 symbols. For writing a piece of mathematics, those were the strangest boundary conditions I ever had to face. (The result, by the way, is Exercise 3 on p. 142 of *Measure Theory*; neither my files nor my memory reveal the falsehood that it replaces.)

Everything about the production of a book takes longer than you think. As an editor I frequently ask an author when his book will be finished; if he says January 1995, I note it in my calendar for July 1995 or, more likely, for January 1996. If the actual date of delivery turns out to be August 1999, I won't be too surprised. It goes the same way with publishers. As an author I have frequently asked how soon I can expect the proof sheets; if I am told February, I note it in my calendar for April, and I'm not too surprised if they arrive in June. *Measure Theory* was no exception. I think I promised to deliver the manuscript in January 1949, but the actual date was almost six months later; I, in turn, was promised proofs "in the early fall", but they actually came in November. "Before Christmas", I was assured; "you'll have the bound copies before Christmas". The printed publication date is 1950; my first copy arrived on January 23.

Master's exams

One of the first tasks Marshall Stone set himself when he became chairman at Chicago was to design a top quality graduate curriculum, and he drafted several of us to help work out the details. The result was called our Master's program; at many other universities it would have been the qualification for Ph.D. candidacy. It was an impressive structure of 12 courses—12 one-quarter courses, that is, or, since the normal academic year consisted of three quarters, the equivalent of four one-year courses. It took most students three years to work through them, but they were not three years of graduate school. At Chicago the definition of "graduate student" was fuzzy; in language that is more widely accepted, the three years were the last two undergraduate ones and the first graduate one.

There were three courses in geometry (including both projective and differential), four loosely called analysis (set theory, general topology,

M. H. Stone, 1973

and real and complex function theory), and five in algebra (from elementary properties of divisibility, à la Birkhoff and Mac Lane, through linear algebra, groups and rings, and ending in a blaze of Galois theory). They were nontrivial courses, meaty courses; anyone who understood them was prepared to enter the Ph.D. program and would have no trouble with missing prerequisites.

There was less of the territorial imperative at Chicago than anywhere else I had been; no one thought it was poaching if an analyst wanted to teach an algebra course. There was quite a bit of swapping around. Kaplansky was proud that he taught all the Master's courses, and I too tried, but didn't finish—I never got around to differential geometry. To be sure, when Kap taught measure-theory, he dropped in to see me several times during the quarter, and when I taught Galois theory, I got stuck and needed to consult him about once a week—but we, and our students, got over the obstacles, and we had nothing to be ashamed of at the end.

When a student finished all the required courses, he presented himself for the Master's Examination. It was both written (three parts: algebra, analysis, and geometry) and oral. The orals were trickily scheduled so that we, the faculty, could circulate among several rooms, the temporary headquarters of the hapless candidates, and spend a half hour or so grilling each one; all it cost us was a couple of afternoons each year. The students were afraid of the examinations, of course, but on the whole their morale was good. "Wholesome respect" might describe their attitude better than fear, and they knew they were fairly treated. Every now and then, to be sure, the student skit at Christmas contained a plaintive line: "Give us Master's exams that our faculty can pass, or give us a faculty that can pass our Master's exams."

Here are a few sample questions from the Master's exams.

Algebra: is there a non-abelian group of order 49? [I confess that I edited that one. The original version was this: prove that a group of order p^2 (p a prime) is always abelian. I strongly believe that an open-ended specific question, ending with a question mark, is more challenging, is more fun, is better training for the "real world" that comes later, and gives more information to the examiner, than a problem that is formulated as a general and brusque imperative. Also, let's admit it, people who set exams can make mistakes, and a "prove that" can be wrong; an "is it" is safer. But nothing is completely safe. In a course once, one of my exam problems was something like this: find the equation of the plane tangent to the surface so-and-so at the point so-and-so. Trouble: I described the surface and the point numerically, explicitly, but I forgot to put the point on the surface. That, and other similar acts of forgetfulness, caused the Christmas-skit character who depicted me one year to be described as the author of a new book: "False theorems for use on tests".]

On with algebra: how many semisimple algebras having fewer than 100 elements are there over the field of integers mod 3? Find the Galois group of $x^3 - 3x + 5$ over the field of rational numbers.

Geometry: find the curvature and torsion of the space curve $x(t) = (t, t^2, t^3)$. True or false: if all points of a surface are umbilical, then the surface is a plane or a sphere? [That too was a "prove that" originally.]

Analysis: if f is an entire function such that $\lim_{z \to \infty} f(z) = \infty$, does the equation $f(z) = 5$ necessarily have a solution? Is there a real-valued, strictly increasing function on the real line whose set of points of discontinuity consists exactly of all irrational numbers? Exhibit a power series convergent in the open unit disc whose sum has a singularity at every point of the perimeter. Is there a nowhere dense perfect set of positive measure in the unit interval whose characteristic function is Riemann integrable?

Judgments

Not everyone who passed the Master's exams became a professional mathematician. Two unusual cases come to mind.

One is not all that unusual: it is just a change of mind. Al Feinstein was an intelligent student of mathematics, but he wasn't sure that he had the call, he wasn't sure that mathematics was what he wanted to spend all his life on. We were friends (it is, alas, easier for a junior assistant professor to have friends among the graduate students than for a graybeard full professor), and as he was preparing for the exams he kept crying on my shoulder (metaphorically only): what should I do?, which way should I go? His decision was to go to medical school, and I have been proud of him ever since: he has become an important figure in the medical world, and he writes papers and books with titles like *Boolean Algebra and Clinical Taxonomy*.

The hero of my other story turned out to be just as successful, but since a part of what I'll say might embarrass him, I'll give him a pseudonym: let's call him Walter. He, too, was a friend, a poker partner, a beer partner, and a frequent dinner guest. He was a member, also, of one class in which I digressed once and talked, for five minutes, not about proving theorems, but about judging students. It's easy, I cockily maintained. Exams and grades are bureaucratic nonsense for the multitudes, I said. It's easy to decide whether a student has what it takes to become a mathematician. Give me an hour alone with him in a relaxed atmosphere, and just from the way he talks about mathematics, from the way he asks questions, from his attitude to language, I can tell whether he's got it or not. (I was impressed once when a cop, acting at the time as examiner for driver's tests, told me that he can tell whether someone will pass the test just from the way he gets in the car and closes the door.)

My glib bluff was called. Walter came up to me after class and asked if I meant it, if I would judge *him* that way. I had, of course, never actually done what I said I could do, and I was a lot more certain in theory than in practice. But, what the hell, I said, sure Walter, let's spend an hour and see what happens. So we did. We sat out in the quadrangle behind Eckhart Hall and we chatted. I asked questions, and Walter asked questions—I was tense, and he was anxious—but from the outside I'm sure we looked relaxed. When the hour was up, I said: Walter, do something else. You are intelligent, and you are quick; you can reason, and you have good intuitive insights; but you are not a mathematician, and you'll never be one. Of course I may be wrong—don't take my word as that of an oracle—but you asked, and that's my answer.

Walter believed me—he took my oracular word, and he left mathematics. He went to Wisconsin and he became a graduate student in somethingology. Three years later he got his Ph.D., and he's never looked back since. He has written a huge stack of articles, he has turned out many Ph.D. students, he became a respected full professor at a first-class university, and he is known as one of the best mathematical [!] somethingologists in the country. We are Christmas-card friends still, and on the rare occasion when we can be together, we reminisce happily.

That was over thirty years ago, and I still think that you can pick a winner early, that mathematicians tend to run true to form. It is the traditional wisdom of a frequently repeated movie plot that a piano teacher can listen to a neophyte play one Beethoven sonata and immediately predict the future—this one will never be more than an amateur, but that one has a great career ahead. I'm not enough of a musician to know whether that's true. What I am saying, however, is that the mathematical analogue is true for sure.

In musical composition and in others of the non-ephemeral arts—such as painting, sculpture, literature, and mathematics—such a test is considered less necessary than it is for performers. The work is there and it can and should speak for itself; it will live or die according to history's judgment of its merit. The insecure beginner doesn't want to wait that long—he may seek expert opinion before time has had time to tell. I firmly believe that any mathematician with just an ordinary amount of insight and experience, and, of course, the courage of his convictions, can predict as well as the legendary piano teacher. I believe, moreover, that it is the mathematician's professional duty to make such predictions, even unasked. It is not kind to allow a youngster to flounder for years before foundering for good; it is kinder to be harsh.

One rule I follow, with exceptions, but only a few, is that when a student comes and asks "should I be a mathematician?", the answer should be no. You have to want to be a mathematician to be one; if you have to ask, you shouldn't even ask.

Another reason for my predictive courage is the number of graduate students I have seen waste four years of their lives. At some universities it takes that long, believe it or not, before a weak-kneed departmental examination policy ("we must be fair, we must give them a chance to show what they can do") allows students to be told what was obvious after the first month of their performance in Math 401.

Is the movies' piano teacher ever wrong? Can we not make a mistake in evaluating a student's mathematical talent? The answer is surely yes: the piano teacher is bound to be wrong sometimes, and so are all of us. But two comments are relevant. One: the probability of error is very small; most teachers make the same predictions and most of their predictions turn out to be right. Two: an error is not a catastrophe. The error of failing a good student costs hurt feelings, anger, disappointment, but that's all—he can and will go on, somewhere, sometime, somehow, and his life will not be ruined. I firmly believe that you can't keep a good man down.

We are all being judged all the time. The two most important judgments in the professional life of a mathematician come early in his graduate career, and, if he gets a Ph.D. and then a job, early in his employment career. Am I good enough to keep the job? Will I be promoted? Will I get tenure? Once again false kindness frequently influences the way departmental committees come to decisions. At many universities appointment to an assistant professorship is, in effect, a six-year contract—it takes that long before the chairman dares to tell no-longer-quite-so-young assistant professors what was obvious within a month of their arrival on campus. I believe in one-year appointments—yes, one year. Yes, I know, the second-year reappointment procedure has to be started within the first or second month of the first year—but, I say, that's time enough to evaluate a junior colleague's probable contribution to the department. If we fire a good one, the loss is ours. That will happen once in ten times, or once in twenty, and the maintenance of high scholarly standards in the university is worth the risk—and, remember, we haven't committed murder. You can't keep a good man down. Not for long.

There are many who disagree with me about these things, but I feel certain that they are just not facing the facts. Did I say it all badly? Am I taking an extreme position? Time will tell—or has it already told?

Jimmie Savage

I had heard about Jimmie Savage years before we met; when he was a graduate student at Michigan he had the reputation of being an abrasive genius who always sat in the front row and asked irritating questions. He wasn't at all like the picture that suggests.

He was a big man, not fat but above average in height, solidly built, usually rumpled and dishevelled, with extremely thick glasses that didn't do him much good. He had nystagmus (uncontrolled almost perpetual motion of the eyeballs) and he was myopic enough to be legally blind by some definitions. He read by holding a book within two inches of his left eye and peering at it sideways—and he read a lot. He was broadly cultured in the humanities, in the arts, and in the sciences—he understood people and he liked most of them. He was so used to living with incomplete vision that he was not always wise about it—when I met him at the Institute in 1941 (he was 24 then), his friends had not yet succeeded in dissuading him from riding a bicycle around town. As for the front row: he had to sit there, he couldn't see anywhere else. He always had a powerful monocular magnifier handy, a mini-telescope, but even so he would frequently have to leave his seat and go to within a couple of feet of the blackboard to peer at it. His questions—oh, he asked them all right—might have irritated some lecturers, but they were always incisive and got to the heart of the subject.

When you talked with Jimmie he listened to you—he really paid attention to what you said and what you wanted to say. We were together at the University of Chicago for 14 years and we talked often. Jimmie was slightly younger than I, but when I needed a wise advisor, more often than not I chose him.

He didn't like everybody, and he could be quite cold to people he didn't like, but his tastes in people and in pastimes had a broad range. One of his friends was Milton Friedman, famous for his right-wing economics, and others were on the radical left. In mathematics he could hold his own in pure differential geometry (the subject of his thesis) as well as in very concrete special applications of statistical techniques to medicine. He was fun to be with, but he could be serious, formal, and even stuffy when he got onto his professional religion, the Bayesian approach to statistics.

Our 14 years together at Chicago came to much less than that really, because Jimmie was on leave for several of them, and so was I, and we were frequently out of phase. The first time we were out of phase was my Guggenheim year at the Institute, during which Jimmie stayed in Chicago. We corresponded a bit, and among the letters I managed to save was one strictly on business. The subject is the sales of my book, *Finite-Dimensional Vector Spaces*. Love of linear algebra is one of the things that Jimmie and I had in common, but he was ambivalent about it. Later in life, in the course of an autobiographical talk, he mentioned the attitude that he used to have during his junior and senior years at the University of Michigan. "The stumbling block to becoming a mathematician appears in retrospect to have been a modest, but certainly important, topic in mathematics, commonly called linear algebra. Nothing then written about it got through to me, and little does today. I have reread the books that puzzled me then, and I don't see how anyone can read them."

While I was at the Institute, in the spring of 1948, the first printing of FDVS was sold out, and the Princeton University Press produced the second. Something went wrong, however, with the distribution procedure, and bookstores weren't getting the copies they had been promised. I knew that at least two courses in Chicago wanted to use the book, so I took advantage of being in Princeton by strolling over to the Press to ask what could be done. I found everybody helpful—they couldn't immediately cure the distribution disease, but they could let me have 100 copies if that would help. Yes, it would; I shipped them to Chicago, and suggested that they be made available to the students who needed them through the mathematics department. A week later I received Jimmie's business letter.

"Dear Paul, The Marshall [that's Stone, of course, the chairman] strictly forbade the sale of your book through the mathematics office, an act of great wisdom as the unfolding narrative will shortly reveal.

"Since my statistics course was in desperate need of copies, I undertook (with no one's permission) to sell some myself. Schilling helped me, selling 19 copies to date.

"Sales opened Monday morning and at this moment [Thursday] the 100 copies are to be accounted for thus:

1	given to Schilling,
19	sold by „
6	held by „
67	sold by me,
7	held by me
100.	

"Yesterday the University Bookstore called up and threatened to throw me in jail for violating a statute of the trustees prohibiting competition with the University Bookstore. I evaded incarceration by selling the Bookstore 20 copies at $2.50, which they will retail at $3.00. This move was not altogether selfish, since veterans in Schilling's course would rather let Uncle Sam buy them a book for $3.00 than buy it themselves for $2.50. Having shouldered some responsibility, I would be pleased to hear whether things have been handled as you would have wished. [They had.]

"P.S. Will the book be on the open market soon?"

Speaking of linear algebra, here is a puzzle that Jimmie included in one of his letters. "D'ja know", he wrote, "that if $S = A + B$ is invertible, then

$$AS^{-1}B = BS^{-1}A?"$$

I enjoyed working it out even though, actually, it is not linear algebra, but pure ring theory. A one-line proof can be obtained as follows: replace the A in $AS^{-1}B$ by $S - B$ and, at the same time, replace the B by $S - A$.

Jimmie had what he called "a long-standing neurosis about Pólya–Szegö" (the most famous and long-lived problem book in analysis). Even when he was working on his first (and major) book in Paris, he was spending evenings on that neurosis. "Pólya–Szegö humiliates me", he wrote. "I never really know what's going on, but I can now work quite a few of the problems and seem to learn thereby some things of general interest."

That was one of his pastimes; another (later) one was the wonder-diet, pemmican. According to legend, pemmican is the homogenized carcass of a buffalo; you grind it all up, horns and hooves, meat and fat, tail and tripe, and then press it into cakes. Not everyone agrees that this is the right recipe. The Indians are rumored to have mixed in some leaves and herbs, and at one time a commercial version was marketed by one of the major meat packers. In any event, the result is touted as the complete convenience food; easy to carry, it provides everything you need to survive as you're crossing Death Valley or rounding Cape Horn. Half for fun and half as a scientific experiment, Jimmie lived on pemmican and water for three months. Result: he was hungry for three months, he lost 17 pounds, and he gorged himself on ice cream when he stopped.

Jimmie and I collaborated on a paper that became quite well known in statistical circles. The collaboration started with a casual question that Jimmie asked about conditional probabilities. My heart was very much in measure theory those days (as well as in *Measure Theory*), and measure theory is the only way that the subtleties of conditional probabilities can be made clear and rigorous. Jimmie taught me about sufficient statistics and I taught him about the Radon–Nikodým theorem; the relation between those two subjects is what we ended up writing our paper about.

He will not be remembered for contributions to pure mathematics, but there is no doubt in my mind that Jimmie was a mathematician. Others tell me that he was a great statistician; I testify that he was the most mathematically perceptive of all the many mathematical statisticians I have known. He knew mathematics, he understood it, he had a feeling for it, and he saw the connections among its parts like a pro.

He was human, and therefore he was not perfect, but the only failings of his that I know of were minor foibles. He kept smoking although he kept wanting not to. He didn't like music; he called himself tone deaf. Although he was witty, he was in some ways humorless. I was surprised once when, having overheard me make a date for a poker game, he chided me: "You mean you make an appointment to waste time?" As he grew in stature and importance, he grew in self-importance too; the Jimmie of 1940 would probably have thought that the great Savage of 1970 took himself a bit too seriously.

Jimmie died when he was 54. There was a memorial service for him at Yale, and I was deeply touched by the words of Allen Wallis (a long-time Chicago colleague, statistical collaborator, and personal friend).

"It was beyond my capacity", Wallis said on that occasion, "to portray his charm, his wit, his zeal, his curiosity, his intelligence, his eloquence, his enthusiasm, his generosity, his intensity, his subtlety, his complexity, his simplicity, his loyalty, or his colorfulness. He had great joy in life; and he brought great joy to many lives. He was a deep friend, a true one, and a strong one. He was an authentic genius and a towering personality."

Students and courses

Bishop and his classmates were not the only superstudents around — a quick glance through my grade records (yes, I kept them) shows that in all my many years at the University of Chicago I met four or five outstanding students each year. By "outstanding" I mean students who went on to become established mathematicians, many of them with names that are recognized and honored in the entire mathematical world. Here is a selected list of a half dozen of them: Hy Bass, Paul Cohen, Moe Hirsch, Dick Kadison, Iz Singer, and Eli Stein — each one unquestionably among the foremost mathematicians that America has produced.

Not all Chicago students were good — some were nutty and some were there just to be able to say that they had been there. There was "Nature boy" who had some talent, perhaps, but was totally disorganized and had some odd personal habits (whence the nickname — not only his feet were bare).

Then there was the one I'll call Wilhelmina, a wealthy, middle-aged ex-drop-out, who tried to learn about vectors from me. In an effort to make an emphatic point, in one lecture I listed a half dozen things that vectors were not. A vector is not, I said, an *n*-tuple of numbers, and it is not an arrow in the plane, and it is not... several other possibilities that I described. All a vector is, I went on to say, is an element of a vector space (and a vector space, of course, is what I just defined). Wilhelmina didn't like that. She looked up the definition of vector in the Encyclopedia Britannica, and then went complaining to the appropriate associate dean of the division. I was narrow, I was prejudiced, I was using classroom authority to teach a private point of view. The dean was too experienced to be impressed, but he had to ask me to drop in and tell him what was going on. He was satisfied with my explanation, and I continued teaching my perverse private point of view. One question on the final exam was the old chestnut that I mentioned once before: is the set of complex numbers a vector space over the field of real numbers? Wilhelmina's answer (in toto): "That's a trick question, and if there's one thing I learned in this course, it's not to answer trick questions."

Between the first-rate people I have already named and Wilhelmina there were many other first-rate students, and others who were only very good (at most other universities they would have been called excellent), and still others who were intelligent visitors to the world of mathematics from other worlds such as linguistics, physics, engineering, and philosophy. There were bad students too, and unintelligent ones—but very few.

Chicago has no engineering school, but one of my calculus students from my early years there turned into an engineer—he worked for Western Electric for a long time and we stayed friends over the poker table. What ever happened to Bob Maguire, and how did he manage to convert a University of Chicago education into an engineering one?

In my first year or two most of the names on my official class sheets had asterisks by them. That meant they were veterans, on the G.I. bill, and operationally it told me not to give them R's. That was an odd Chicago grade, standing for "registered", and I never did discover what purpose it served. It had something to do with residence requirements and with the financial arrangements between students and the university. If you had an R in a course, there was a sense in which you had credit for it—the grade was your receipt. It was a non-prejudicial grade, and, in particular, it did a student no harm, but Uncle Sam would have none of it—veterans had to have *real* grades.

There would be no point in copying my grade books here—although some of them may make interesting reading—but there are bits and pieces that I can't keep to myself. They concern people who have, since then, come to play non-trivial roles in the mathematical scene. In one course, for instance, I classified students very roughly as O.K., poor, and lazy. Bishop was in that course; he was "O.K. +". There was only one student whose performance couldn't be covered by the notation; Murray Gerstenhaber's classification was " + + + ". He became an algebraist with other interests; in his middle age, as an established scholar, he took time out to earn a law degree and was admitted to the bar. I know three or four other mathematicians who have done that, and I am in total sympathy with them—I am a little surprised that I never did it myself.

Ray Kunze, who became an outstanding harmonic analyst, took Math. 251 from me (the first course in algebra, at the Birkhoff–Mac Lane level). In the next course, Math. 252, I had Mary Bishop, Errett's sister, who became a recognized analyst in her own right (a student of Zygmund's), and who met her future husband, Guido Weiss, in Math. 252. (She died tragically, too young.)

Math. 253, in the spring of 1950, was spectacular. The registration was down to 40, from 50 in 251, but the 40 were not a subset of the 50. Schedules and sections frequently changed between quarters, and you were bound to lose a few and pick up a few new ones each time. Among the 40 were Dick Block, Anil Chowdhury, Bert Kostant, Paul Mostert, Ed Nelson, and Vera Stepen. Do you know most of those names? You probably do even if you think you don't.

Dick Block is an algebraist of renown (and I had the nerve to teach him about vector spaces), and Bert Kostant is an outstanding geometer, if that's the word for the hard Lie mathematics that he does. Paul Mostert is a semigroupie, and many years later he played a brief but big role in my life. At the same time as I was being offered a cushy chair at the University of Massachusetts, Paul had a nibble for a deanship there. We kept in close touch during those negotiations—but they ended with neither of us going. Neither of us ever quite said that if the other one didn't accept, then he wouldn't, but we were both inclined in that direction.

Ed Nelson, an analyst with leanings toward physics, and, more recently, a convert to non-standard number systems, showed up at Chicago straight from an excellent high school education abroad (Italy, I believe), and made full use of the Chicago system of placement exams. According to the legend that sticks in my memory, he went through "The College" without ever going to class; he took all the exams in the first two weeks and was admitted to the Division as a full-fledged third-year undergraduate. As for Chowdhury and Stepen, they changed their names. The former is now Anil Nerode, a recursive logician, and the latter is Vera Pless, an expert mathematical computer scientist. She too met her husband-to-be in my class. Irwin Pless got one of the six A's in Math. 253; Kostant and Nelson were two others of the six.

Ernie Michael was a Chicago student in those days. He became a general topologist specializing in selection theorems (when can you choose a continuous point mapping in a set mapping?), but as a graduate student he was interested in many other things. He even participated in a couple of my courses, and we became friends outside class. Being one of Ernie's friends meant, for several of us, that we had to buy his lunches on

Saturdays. He was an observant Jew and, as such, he couldn't carry money on the Sabbath; he would deposit money with us in advance, or conscientiously pay us back the following week, but when we went through the cafeteria line, he wouldn't pay. Another feature of his observing the holy day was that when he came to visit Saturday afternoon, he couldn't push the buzzer button on the ground floor of the apartment building—that would be doing "work" on the Sabbath. In good weather there was no problem—he would just stand in the court below my window and shout; I'd wave to him and push the door release button, so that he could come up. In the winter the windows were closed and shouting was useless. An alternative that frequently did the trick was that Ernie would make a snowball and, aiming carefully, throw it up at my third floor window. That did not come under the definition of work.

One of my students in an advanced calculus class in 1951 was Larry Wos, and he kept me on my toes. Since then Larry has made fundamental, prize-winning contributions to automata theory, despite the handicap of being blind since early childhood. Lecturing to a class with a blind student in it takes extra care. You cannot say "... and now take *this* and substitute it into *that*..." as you wave your chalk hand and your eraser hand all over the board; you have to say "... and now take this expression for y that we have just obtained and substitute it into that earlier formula that connects y and z...". I tried to adapt, and Larry made it easy. He was always alert, he listened hard, he took Braille notes, he frequently volunteered answers to my questions, and he asked good ones of his own. He came to consult me during office hours from time to time, and he was a pleasure to teach. I found it fascinating to observe how he could "see" mathematical concepts. Once, for instance, he said "... but it's all right if the function is monotone...", and, at the same time, he moved his right hand up to the right in a curve concave downward. I gave six A's in that class of 40; Larry got one, and Guido Weiss another.

It's fun to look back on all these beginners who turned out to be strong finishers, and I could go on this way for a long time, but I'll stop for now by mentioning my last measure theory course before the year in Uruguay. Like many other Chicago classes then, it was good-sized (over 40), and it had several subsequently famous mathematicians in it. Four among them were Walter Feit (of Feit–Thompson simple group fame), John Isbell (who, God forgive him, became a categorist), Karel de Leeuw (an excellent harmonic analyst, brutally murdered in his middle age by a psychotic student), and Sterling Berberian. For his friends of those days Berberian's name is Sam, and even today he still remembers to sign his letters to me that way. Sam's interests and mine are not identical, but they overlap quite a bit; like me, he too wrote one book on measure theory and one on Hilbert space, and he went on to write, among other books, one I never got around to writing on functional analysis.

There was a "night shift" and a "weekend shift" at Eckhart Hall; the building was always alive. Nostalgia?—yes, maybe, but with top-quality students in such breath-taking quantity, there is something to be nostalgic about. Personal attitudes aside, the mathematics department at the University of Chicago in the late 40's and 50's was either the best in the world or close enough that I can be forgiven for regarding it so.

The beginning of Hilbert space

In the late 1940's I began to act on one of my beliefs: to stay young, you have to change fields every five years. Looking back on it I can now see a couple of aspects of that glib commandment that weren't always obvious. One: I didn't first discover it and then act on it, but, instead, noting that I did in fact seem to change directions every so often, I made a virtue out of a fact and formulated it as a piece of wisdom. Two: it works. A creative thinker is alive only so long as he grows; you have to keep learning new things to understand the old. You don't really have to change fields—but you must stoke the furnace, branch out, make a strenuous effort to keep from being locked in.

As my own focus on measure theory began to waver, I published a couple of comments on other people's measure theory. One was on Liapounov's theorem (to the effect that the ranges of well-behaved vector-valued measures are closed convex sets). Kai Rander Buch published a paper on closedness, and that paper made me angry: it struck me as wordy and pretentious and unnecessarily complicated. Surely one can do better than that, I said; I thought about the question, saw a way of doing much better, and dashed off a note to the Bulletin of the AMS. My proof was a lot slicker than Kai Rander Buch's and a lot shorter, but his was right, and, to my mortification, mine turned out to be wrong. Both Jessen and Dieudonné wrote and told me that my Lemma 5, the crucial lemma, was false. A pity; it was such a nice lemma. It says that the span of two compact topologies is compact (span, supremum, generated topology)—a statement for which it's not only easy to find counterexamples but it's hard to find any non-trivial instances where it is true. Being caught stumbling in public was all the motivation I needed to sit down and think matters through more deeply and more effectively. My second note came out a year after the first (1948), and it was twice as long (six pages), but it was elegant and correct, and has been quoted quite a bit since then. It is all superseded by now; in 1966 Lindenstrauss came out with the slickest proof to end all proofs (J. of Math. and Mech.).

In 1949 I published another little note precipitated by an emotional reaction. The irritant in that case was a paper by Shin-Ichi Izumi proving a non-theorem. The subject is ergodic theory; the statement is that under certain rather restrictive conditions on a measure-preserving transformation T it can be concluded that series such as $\sum_{n=1}^{\infty} \frac{1}{n} f(T^n x)$ converge almost everywhere. Izumi's proof is rather complicated and, so far as I could tell, absolutely right. There is trouble, however; the conditions are so restrictive that the only transformation that satisfies them is the identity transformation acting on a space consisting of exactly one point.

With that, my career as a research measure and ergodic theorist came to an end. In 1948 I gave an invited address to the AMS, and in 1960 I followed it with another; the written versions of those addresses, as well as a couple of subsequent publications, are expository in nature, but they help perpetuate the myth that I am an expert measure theorist. People who learned measure theory from *Measure Theory* a quarter of a century ago still write to me from time to time, or have their students write, to ask about the state of current research in the subject. I wish I knew it.

In 1950 my study of Hilbert space began to pay off. In my research diary I asked myself a question (I no longer remember why), which turned into a gold mine. The question was "which matrices can appear as the northwest corner of a normal matrix?". The answer is easy by now (all of them); there is something called dilation theory that disposes of the matter in a couple of sentences. From dilations it was a short step to extensions (the more natural concept anyway), and lo! I had discovered subnormal operators. In the course of the next two years I took the first few steps toward building their theory. If anything I have ever done is remembered a hundred years from now, subnormal operators are likely to be on the list.

My friend Leopoldo Nachbin was editor of an obscure journal called Summa Brasiliensis Mathematicae (it's still obscure—in fact I don't even know whether it's still alive), and about the time I was ready to publish my first paper on subnormal operators he asked me to send him something. I did, and there it is.

The same year that the Summa Brasil. paper appeared, I took my next big Hilbert space step by offering a course on spectral theory. The main focus of the course was multiplicity theory, whose study was begun by Hilbert's student Hellinger. Like many other Germans of his generation, Hellinger became a refugee in Hitler's time. He was a professor at the Illinois Institute of Technology—which was called "minus T" by some mathematicians ("IIT"—get it?)—and I knew him slightly. What he did was to generalize to infinite-dimensional spaces the concept of the multiplicity of an eigenvalue of a Hermitian or normal matrix. The generalization is not trivial. Since normal operators on infinite-dimensional spaces can fail to have any eigenvalues at all, the main difficulty is to figure out how to attach multiplicities to things that aren't there.

The answer turned out to be measure-theoretic. Measure theory is a generalized kind of counting; you "count" the number of elements in a "set" that doesn't have any (because each element, being a singleton of measure zero, becomes identified with the empty set). That kind of generalized counting is exactly what's needed in infinite-dimensional multiplicity theory. Stone's book (1932) contains a systematic but complicated exposition of Hellinger's work, and Nagy's monograph (1942) also touches on the subject. In the middle 1940's another German refugee, Plessner in the USSR, collaborated with a young Russian (who later abandoned ergodic theory and Hilbert space and became a famous topologist), namely Vladimir Rohlin, and produced an improved approach. The concepts were very much in the air, and I was eager to learn what Plessner and Rohlin had done. The best way to learn something is to teach it; I stuck my neck out and announced that I would teach a course whose purpose was to draw a straight line between the axioms of Hilbert space and the theory of multiplicity.

What I did next is something I recommend very highly to anyone who wants to teach a big chunk of material in a small chunk of time and who doesn't mind working like a slave. I sat down and wrote out the whole course: prefatory comments, list and discussion of prerequisites, definitions, theorems, proofs—everything. The result was not a book—it was not intended to be—it was a set of lecture notes. I was ruthless in my effort to keep it short: I treated the central special case only (bounded Hermitian operators), I wrote in a telegraphic style, and I omitted all the connective tissue, the expository chit-chat that turns a coldblooded Satz–Beweis type of report into a readable and enjoyable presentation. The notes were mimeographed and placed in the students' hands on the first day of class. My slavery, however, was less than half finished then.

I told the class that I expected them to read the notes (70 pages in all) at a steady rate of six or seven pages a week for each of the 10 or 11 weeks of the quarter; the definitions, theorems, and proofs in the notes would not be discussed in the lectures at all. What I did talk about when we met was the side issues: the history of the subject, its connections with and analogies to other parts of mathematics, its applications, exercises that the students were expected to work to develop their technique, and whenever possible, the research problems that still remained open in the field.

It worked. I was enthusiastic, the students were cooperative, and we covered the ground in gigantic steps. The notes and the lectures had mistakes, and the students enjoyed finding them; morale was high.

When the course was officially over, the biggest single part of the job I had been planning to do all along was still ahead of me: to write the book that contained it all. With the notes and the lectures and the discussions behind me, the labor was a pleasure, much easier than it could possibly have been without them, even for such a little book as my *Introduction to Hilbert Space . . .* (114 pages).

My publisher in those days was Van Nostrand, and I stayed with them till they went out of the advanced mathematics business twenty years later, but in connection with the Hilbert space book they did something that upset me. When the manuscript was nearly finished, I wrote them and asked if they wanted to consider its publication. I didn't expect them to buy a pig in the poke; my specific question was whether, judgments of quality aside, the book was too short to be viable. The answer came by return mail, emphatically: no, definitely not, by all means please let us see it, brevity is no obstacle. Very well—a few weeks later I submitted the manuscript and sat back to wait. The answer came soon enough (perhaps even sooner than the time it would have taken to send the book out for a detailed evaluation): no, thanks—we're sure that what you wrote is just fine, but it's too short to fit into our program. Wasn't I justified in being at least a little peeved?

Aaron Galuten was a graduate student at Columbia during the war years. In the course of his studies he and some of his friends found that they needed a certain book—one recently published in Germany—that was not easily available in this country. A few copies of anything could always be obtained (one from Mathematical Reviews, one or two imported through neutral nations, some brought in by travellers), but wide distribution was difficult or impossible. Galuten's solution was to found an ad hoc company, and, via the usual war-time arrangement with the Alien Property Custodian, arrange to re-publish the book in question. The technique worked, and one thing led to another—the Chelsea Publishing Company became permanent. I sent my short book to Galuten, and he published it, and we've both been glad ever since. His company (begun in the back room of a relative's fur shop) grew and prospered; by now, 40 years later, it has a list of two hundred titles or more, many of them originally foreign and, before Chelsea rescued them from oblivion, out of print. My Hilbert space book is on the list (and so are a couple of others added later, *Ergodic Theory* and *Algebraic Logic*). By now, after more than 30 years, only one or two hundred copies of it are sold each year, but it is still in print, and the customers are satisfied. So am I. If I were still a smoker, my royalty income from Chelsea would not be enough to keep me in cigarettes.

Ph.D. students

I taught many graduate courses at Chicago and through them came to know a good many graduate students. It was quite natural that some of them should approach me to be their Ph.D. supervisor. The first batch consisted of Dan Orloff, Arlen Brown, Joe Bram, and Errett Bishop.

Dan Orloff was in the first course I taught at Chicago, one on ergodic theory. He was a good student, a mature person, a veteran, and a friendly guy—we frequently had coffees together and lunches. A year later, when I was at the Institute, he wrote me a letter: would I take him on as a Ph.D. student? Flattered, I said yes. Dan never knew, but what I answered was word for word what Joe Doob said to me: "I think we'll be able to find something for you to do." We did, and it worked out well; Dan became my first Ph.D. student. He didn't stay in the academic world—he went to work for an airplane manufacturing company. I ran into him about three years after he got his degree. We got talking about salaries, and he was horrified when he learned what mine was ($7500). His was twice that.

Salaries remind me of ranks. One pleasant spring afternoon in 1949 Marshall Stone knocked on my office door. I was in stocking feet, horizontal on the couch (an essential piece of office equipment), but my after-lunch nap was over, and I shouted "come in" goodnaturedly. Marshall came in, commented on my informal attire, and told me the good news: I was to be promoted to associate professor in the fall. It was my third year on the Chicago faculty, my eleventh after the Ph.D. My salary was to rise from $5500 to $6000. The new rank automatically carried tenure with it; the faculty called it "permanent tenure" but the administration insisted on using the phrase "indefinite tenure". Another seven years were to pass before I was made full professor. That promotion came a little slower than average (Mac Lane was chairman by then and he was able to contain his enthusiasm for my sterling qualities), but not much slower. I was 40, and had had my Ph.D. for 18 years.

Arlen Brown practically grew up under my eye. He is one of the few who really went through the college and learned mathematics in the order in which the system meant it to be learned. Judging from him, the system was good. I met him in a relatively elementary calculus course, and then in a fancy special one, all about infinite series. Although we are colleagues now, equals, and the age difference between us seems to have converged to zero, he told me not long ago that those early courses still come to his mind from time to time and affect how he thinks of me. Once I fulminated against undetermined δ proofs. I said that to say "choose x so that $|x - x_0| < \delta$, where δ will be determined later" is lazy thinking that can lead to circular reasoning and other errors. It is the prover's responsibility, I argued, to choose δ when it must be chosen, and thereby to reassure the provee that the order of choice of the parameters is as it ought to be. I am no longer all that emotional on the subject, but Arlen says that even now, when on rare occasions he does present an undetermined δ proof, his conscience hurts a little—no, no, I mustn't do that, Teacher said it wasn't right.

Arlen started seriously working with me for the Ph.D. in about 1949 or 1950, but when I took a year's leave in 1951, he shifted and finished with Kaplansky. A little later Buzz Galler started with Stone but, for similar reasons, finished with me; between the two of them I want credit for at least one full Ph.D. student. Arlen and I stayed in close personal and professional touch for a long time. He went to Texas, to the Rice Institute (as it was then called), but two years later, when I was at Michigan, I was able to wangle a job for him there, and we were together again. Later still he moved to Indiana, and I was at Hawaii, and he wangled a job for me at Indiana; turnabout is fair play.

Buzz Galler and I, on the other hand, didn't manage to stay in touch; he had a lot to do with computers, and also with administration, and that meant that we moved in very different circles. I remember Buzz fondly, and at least one reason for that is that he is one of the very small number of logic students that worked with me. (The number is three if you count bodies, or one plus two halves if you count fractions.) And the way he got that way is a story that pleases me: during the years that I was bitten by the logic bug, one mathematician who became at least temporarily influenced by my ideas was my revered senior, Marshall Stone. Marshall discussed my work in a course (or two?) and began doing some thinking and writing about it; when Galler wanted to work with him, Marshall suggested that the work be on algebraic logic. In a sense therefore my father figure, Stone, is also partly my professional son, and my (half) student Galler is to an incalculable fractional extent both my son and my grandson.

Joe Bram was the most quiet student in some of my courses, and he was almost as invisible as he was inaudible. He sat near the back of the room, he answered (correctly) when a question was addressed directly to him, he never volunteered, and he got almost perfect grades on examinations. His thesis is the best I ever had anything to do with. I was interested in subnormal operators at the time, and Joe gave the subject a powerful push. His only paper on the subject (or any subject?) is still very useful and widely quoted.

Errett Bishop was the Chicago student I was proudest of. He was always a strong-minded man, and I cannot claim to have influenced him much or taught him—he made himself. He chose his own topic, he wrote a good thesis, he then found that most of it had been anticipated (by Naimark, no less—the main result was the extension theorem for positive operator-valued measures), so he scrapped it and went on to write an even better thesis. When we discovered the Naimark scoop I told him that I was ready to accept the thesis anyway. I knew it was his original work, and, as such, it accomplished what a thesis was supposed to accomplish: it gave him experience in doing research-level mathematical thinking and writing. Bishop would have none of that: he wanted to do something that no one had beat him to. He did so, and having done it he went on to become an even more powerful mathematician. He discovered many of the fundamental concepts about function algebras and the relations among those concepts. Then, almost discontinuously, he "got religion", went into constructive mathematics, wrote the book that

E. Bishop, 1980

made the phrase famous, and started the sect of which he was the leading but somewhat reluctant guru till the day he died. Functional analysis misses him, and so does constructive mathematics, and so, most of all, do we, his friends.

The Cambridge Congress

International Congresses of Mathematicians used to occur on leap years. The last pre-war one was in 1936; the one scheduled for 1940 in the U.S. was cancelled, or, in effect, postponed to 1950; since that time the International Congresses have been scheduled halfway between leap years.

In the late 1940's I began to have more and more to do with the AMS. I went to many meetings, and I worked with compulsive conscientiousness at the low-level committee chores that I was assigned from time to time. In 1949 I found myself part of a small but effective opposition to the establishment, but that only made my ties with the Society stronger than ever. The central issue was the forthcoming Congress.

When Congress arrangements were being discussed, my friends and I discussed them vociferously enough to be noted by the powers. They were worried about the effect of the noises we were making. At the summer meeting in Boulder, Lefschetz sought me out and, making an unaccustomed effort to be diplomatic, told me to cool it—we were, he said, not helping the cause we wanted to help. Maybe he was right, and maybe he was not. I am inclined to think that our noises caused the powers to be aware of our views, to think about them, and to take steps without which the Congress might have been very different.

The source of the trouble was political; the motto of the discontents was "A free Congress or none!" The U.S. was between the euphoria of postwar patriotism (we won!) and the hysteria of McCarthyism (they're winning!). I am sure that the mathematical powers, such as Lefschetz, wanted a free Congress, but the U.S. State Department, scared possibly of the House of Representatives, or of public opinion, or of the Red Menace, was making difficulties that were not easy for the AMS to overcome. Laurent Schwartz was to be invited as a distinguished lecturer, and Jacques Hadamard was to be honorary president of the Congress. Schwartz was a known Trotskyite activist and Hadamard (aged 85) an anti-Nazi liberal Jew. The problem was that they and several others equally dangerous were threatened with not being granted the U.S. visas they needed to attend the Cambridge Congress.

In the end, whether because of the young rebels or in spite of them, it all came out right. Schwartz and Hadamard were allowed in, and, so far as I know, no one was kept out.

Schwartz became a frequent visitor to the U.S. in later years, and for a long time he kept encountering the same difficulties and delays each time that he applied for a visa. He was in great demand as a visiting professor at American universities, but high-level administrative influence and pressure was needed to make it possible for him to accept the invitations. The last time I saw Schwartz was in Berkeley in the 1970's, and he told me that by then his very undesirability was making it easy for him to come to the States. The way it worked was this: since he had been so "dangerous" once, he was thoroughly investigated, all his activities and associations were described in his file in detail, and the functionaries at the U.S. consulate in Paris were thoroughly familiar with the file. A new person, a young leftist, an unknown danger had to be looked into very carefully, but Schwartz, that familiar, elderly, pleasant, and well-known applicant— why, we know all there is to be known about him, and in granting him a visa we're just following established precedent.

Attendance at the 1950 Congress was between 2000 and 3000; since then Congresses have grown, and numbers like 6000 and 8000 have been seen and foreseen. The Cambridge Congress was almost too big; the later ones were no longer "almost" but definitely so. In Cambridge, if you saw someone you wanted to talk to, you were well advised to grab him right then—the chances that you'd run into him again were nil. I enjoyed meeting Schwartz and Hadamard, and others who had been only glamorous names till then, but I am beginning to be skeptical of the mathematical value of Congresses.

My own role at the Cambridge Congress was small but positive. The organizing committee appointed a panel on measure theory consisting of Dieudonné, Maharam, Oxtoby, Ulam, and me as spokesman. "Spokesman" meant that I was in charge of preparing the written report and delivering the oral one. I busied about the task, but it was a frustrating experience that did not end in a blaze of glory. Maharam and Oxtoby cooperated just fine. Both contributed the material I asked for, well thought through and carefully prepared; all I had to do was homogenize the terminology and the notation. Dieudonné was polite but impatient; he had other and better things to do, I was allowed to infer, and what he sent still needed quite a bit of work to make it publishable. Ulam didn't answer my first n letters (I forget the value of n, but it was non-trivial), and answered letter number $n + 1$ with a postcard. He didn't have anything to send me, but he'd be passing through Chicago next Thursday, and he would gladly give me whatever advice I needed—perhaps we could have lunch? I was sad about that—it meant, in the end, that the efforts of the rest of us had to remain incomplete and had therefore been a waste of time. I did the best I could with the lecture and with a noncommittal one-page abstract. The incomplete manuscript is still in my files—it is an up-to-date partial report on the state of measure theory, thirty years out of date.

Follow the sun

I have taught at many universities, and in that sense I probably have more teaching experience than most of my colleagues. A disconcerting disadvantage of having known many students over many years at many places is that times, towns, and stories tend to blend into one another. I am not likely to forget student S, but was he at U university in year Y, or at U' in Y', and was he or was he not in the same class with S'? The question is not too important—but it could be embarrassing (it's happened a few times) to ask S whether he is still in touch with good old S' and then to learn that they have never heard of each other.

In any case: why has my teaching experience been so geographically scattered? One reason is an apparently congenital and perpetual feeling of restlessness; another is that I am, in a mild way, a sun worshipper, and Chicago's helpfully flexible attitude toward leaves of absence made it easy to follow the sun; and a third is the quarter system.

The year really had four quarters in Chicago. The Autumn, Winter, and Spring quarters were the main part of the academic year, but the Summer quarter was not just summer school, an add-on, but a fully functional part of the operation. In addition to the earnest school teachers who kept coming back to work for their Master's degrees, many graduate students stayed through the summer. Why did they? Partly because they were advised to, and partly to get their degrees more quickly. I was one of several who advised them to stay. I firmly believe that if a mathematician stops doing mathematics for a time interval of length t, it'll take him time t to get into the swing again. A full summer's holiday means, according to that belief, not a three-month setback in mathematics but a six-month one. (A shorter holiday was built into the quarter system: the month of September didn't exist.)

As a result of the seriousness of the summer operation, a decent fraction of the faculty had to be present, and that's where my sun-worshipping came in handy. The people with children in school couldn't get away for any full quarter except in the summer. Not having any children, I could (and I was eager to) take my quarter out of residence during the nasty winter or during that depressing, dark, rainy, slushy period that Chicago calls spring. That made me popular with my colleagues at scheduling time, and it made it possible for me to accept short-term invitations to visit universities with better climates, such as Tulane and the University of Washington. The visit could take place during the host university's regular academic year, and everybody was a winner.

Sometimes a whole year's leave could be arranged: I spent one of my Chicago years at the University of Montevideo and, later, one of my Michigan years at the University of Miami. That made for a lot of sunshine, and many different grades of students. From the point of view of student quality I order my employers as follows: Chicago, Michigan, Illinois, Tulane, Indiana, Santa Barbara, Syracuse, Miami, and Hawaii. In Montevideo and Seattle my contacts were too limited to draw reasonable inferences from: I was in touch with only three or four graduate students, and, of course, the faculty.

Every time I moved to a semitropical environment, it took only a few months (or less) before I discovered, again, that there must be something else in life besides the sun—and each time I returned to the Middle West. When I say that I "moved" to a warm environment, I don't just mean that I sublet my house and rented another. When I went to Hawaii from Michigan, and later when I went to Santa Barbara from Indiana, I meant to stay—sold the house, gave away the winter clothes, boxed *all* the books, loaded the furniture and the piano on the moving van, bought a new house, and settled in. Hawaii lasted one year and Santa Barbara two—then, sell the house, box the books, load the van, and buy a house.

What was wrong with the sunny places? In theory, nothing. In fact, however, for me, they had nothing else but sun. I don't think that a lower quality of mathematics departments is a necessary consequence of better weather, but it almost seems that way. Berkeley is a counterexample, and La Jolla is probably another (I know it from a couple of one-day visits only). What I missed in the warm places was not only a lively mathematical atmosphere, but some other important features of culture. Notable example: music. In Chicago and in Ann Arbor, and, above all, in Bloomington, the musical life is rich. A lot of good music is easily accessible almost every day. (Correction: in Bloomington it is literally every day.) At the sunny places I have been it is rare, and expensive, and far away.

What brought all this to mind was the Montevideo trip, and I'll tell about that now. It was not an adventure, but it was an experience. It put me in close touch with a culture as different from Chicago's as Chicago's was from that of Budapest. I learned a new language, I met new kinds of people, and, somehow, I feel that my soul became richer. That increase in wealth would be hard to account for in day-to-day ledgers—all they would show is the occasional small insight or startling view balanced by the frequent petty annoyances, and the fascinating new attitudes, behaviors, and customs versus unexpected embarrassments, delays, and discomforts. It all started when...

CHAPTER 10

Montevideo

Where to go?

It all started when my Uruguayan friend Rafael Laguardia approached me at the Cambridge Congress (I knew him quite well several years before at Brown and at Princeton) and asked whom I would recommend, and specifically what statistician I would recommend, as visiting professor at Montevideo for a year. Without a moment's hesitation I said "me". He took my answer for the half joke that it was, laughed good-naturedly, and asked whether I considered myself a statistician. I allowed as how not exactly, but I had published two or three papers on statistics, did have some idea of what probability was about, and, besides, I was really interested in going to Uruguay. This last qualification, which, it appears, I was the only one in the United States to possess, was what impressed him the most. He liked me pretty well and wanted to have me visit—the trouble was that he could persuade his deans to allocate money for a visiting professorship only under the guise of doing something practical. I gave him the names of a couple of young statisticians as a back-up, and we parted with the idea that maybe, who knows. . . .

Academic people in general and mathematicians in particular are a peripatetic lot. Given half a chance, they love nothing better than to pack a suitcase and visit friends and colleagues in Pakistan or Tasmania. Shortly after the end of the second world war they were given considerably more than half a chance. Fulbright grants and Guggenheim fellowships were to be had practically for the asking, and there were many other gravy trains to be dipped into. The situation took on the proportions of an epidemic—one after another of my friends took off for

Europe, South America, India, and other exotic places, and one after
another came back and told wondrous tales of exciting times. I felt I
had to get in on all this. Where to go? Under the Fulbright program I
could have tried for France, or England, or India, or Burma, or Egypt—
culture, romance, and exotic adventure beckoned. How to choose?

I began by ruling out the places that had nothing but their exotic
character to recommend them. I wanted to continue to do mathematics—I just wanted to do it in a different place and with different people
for a while. The other considerations that I kept in mind were climate,
language, politics, and novelty.

On strictly scientific grounds my choice should have been England
or France—it was clear that I would find interesting colleagues to learn
from and interested students to teach in either one, that I would find
well established academic traditions, and that, in general, I could come
closest to duplicating the scientifically well nigh ideal conditions that
I had become used to. For these very reasons I decided against Europe.
I thought I could profitably spend a year away from scientific stimulation, developing my own thoughts. The difference between Chicago and,
say, Oxford appeared to be of an order of magnitude comparable to the
difference between one midwestern state university and another.

Climate also contributed to the decision: the climate of France has
nothing special to recommend it, and that of England is well known to
be foul. Language pointed the same way. French I had never learned
and what little I knew made me think that it was very hard, and
English, on the other hand, was too easy—I wanted to acquire a new
language and to put myself in a position where I'd be forced to acquire
it. As for politics, I had nothing against the political systems of either
England or France, but I didn't particularly trust their stability. At any
moment a war could have broken out, some people thought, and I
didn't want to be stranded in Europe. And, to cap it all, Europe didn't
offer much novelty for me—after all, I was born in Europe, and even
though I never lived in either England or France I had briefly travelled
through them and thought I knew the atmosphere and the culture well
enough. It could well be that all this reasoning was merely rationaliza-
tion—that for various reasons that I didn't admit even to myself I had pre-
judged the issue, decided against Europe, and then proceeded to hunt for
reasons for the decision. Be that as it may, I succeeded in fooling myself,
and decided to concentrate my efforts on trying to go to South America.

Mathematical considerations immediately ruled out all but three
countries—except in Argentina, Brazil, and Uruguay there were virtually
no mathematicians to be found. Mexico and Cuba didn't count, of
course: they are not South America, and, besides, it would have seemed
like cheating. They were so near the States and so much like them, that
it would hardly have been like a trip to go there. Well then—Argentina,
Brazil, or Uruguay?

Mathematically, so far as I could judge from a distance, there was not much to choose among A, B, or U. All were about 50 years behind the States, all were trying to build mathematical traditions, and all had already made some progress. Since Argentina meant Buenos Aires, Brazil meant Rio or São Paolo, and Uruguay meant Montevideo, all subtropical, the climate of all of them was attractive. I could speak not one word of either Spanish or Portuguese, but they had the reputation of being easy and I thought I could pick up either one well enough to get along. Any of A, B, or U offered plenty of novelty— I knew nothing about Spanish culture, I had never been in South America, and I was eager to learn. The decision would therefore have to be made on the grounds of politics and, most important of all, practicability. It's all very well for me to decide that I want to go to Brazil, say— but how do I get there? Politics and practicability—what to do?

Politically the alphabetical order, A, B, U, happened to coincide with the order of my increasing preference—I liked Argentina the least and Uruguay the most. About Brazil I was sort of neutral. My dislike of the Perón regime was strong enough that I might have decided against going to Argentina even if I had had the chance—but I decided to leave all doors open at the beginning.

The first steps took place at the Cambridge Congress: chats with two Argentine mathematicians (González Domínguez and Rey Pastor), two Brazilian mathematicians (Candido Lima da Silva Dias and Leopoldo Nachbin), and my Uruguayan friend (Rafael Laguardia). GD and RP I had not known before. Even though I could name Stone as a reference, the conversations were somewhat formal and not very helpful— yes, yes, we'd love to have you, but we're awfully broke nowadays. (And so they were.) To make things slightly worse the conversations took place in broken English (GD) and even more fractured German (RP). The explanation of the German is that when I approached RP, he said: "Español? Italiano? Deutsch? Français? English —no!" When I agreed to use my poor German, I faced the difficulty of singing my own praises in that tongue—it turned out that he had never heard of me. Where am I, again? And what's my name? And what do I work in? What was that I said I wrote a book about? What? Oh.

The conversations with Candido and Leopoldo went better, but no more effectively. Candido said yes, sure, he'd see what he could do. Leopoldo shook his head sadly and said he couldn't do anything—he himself was in a bad political situation at the time, everybody was broke, and he was doubtful in the extreme.

All this, and of course Laguardia, pointed to Uruguay—and something else pointed there too. I had been talking about my hopes and plans for about a year before the fall of 1950 (when all these conversations took place). Many of my friends expressed great interest in Uruguay and volunteered facts about the country. They had learned these facts because of a semi-serious (or, more accurately, one-per-cent-serious) fear that the time would come when it would be wise to have a

refuge on tap. The foreign policy of the U.S. could not be called ideal—
it looked like we were heading for a stupid and destructive war as fast
as we could head—freedom of speech and freedom of thought still existed, but were decreasing—the witch hunt was on—the atom bombs
were on their way—and both the foreigners, who had already had to
emigrate from somewhere once, and the native Americans were looking
around. I was no more fond of having atom bombs dropped on
me than the next guy—and I became interested in investigating Uruguay
as a possible haven. Everything I was able to gather from a superficial
study—from cocktail party talk with friends—looked good. The sanest
country south of the border—perhaps in the world. Democratic government—no bloodshed revolutions in 50 years—low crime rate—no un-
employment—no oppressive rich class and no oppressed paupers—
liberal immigration policy—sound currency... a heavenly haven?

Saturation in Spanish

When Laguardia and I talked about the possibility of my Montevideo
visit, we discussed the language problem also. He said that it would be
better if I learned Spanish—it would be much easier to communicate
with the students that way. I found the statement strange: I should have
thought that it would be indispensable that I know Spanish, and,
indeed, I learned later that that was true. Very few university students
could speak or understand English; for most of them a lecture in
English would have been as mysterious as a lecture in Spanish for
a class at the University of California at San Diego. Laguardia was
probably just being subtle and polite.

Accordingly early in January, even before receiving the official offer,
I set about learning Spanish; I had nine months to prepare for
living and lecturing in a language that at the time I knew absolutely
none of.

The saturation method was put to work. I began gently with a paperback called *Spanish Through Pictures*; I found it amusing and helpful. Even before finishing it I started on Hugo's teach-yourself course,
and, in addition, I went to a Berlitz school for a total of 20 hours. That
had a scary part. The Berlitz idea is never to translate—just start right
in speaking. After a couple of hours of saying things like "this is a
pencil", my teacher and I started exchanging personal information, and
presently I learned that he was a recent arrival from Peru and couldn't
speak a word of English. Once I got over being scared, however, I felt
a touch of pride.

After finishing Hugo, I bought what turned out to be an excellent
old-fashioned grammar (Ramsey), and several novels. The novels were

mostly translations such as *The Prince and the Pauper* by Mark Twain, and *The Mysterious Island* by Jules Verne. In the early spring of 1951 I bought and systematically worked through the Linguaphone records. In the latter part of the spring and throughout the summer I worked with private tutors.

One of my tutors was Lucio Chiaraviglio, a young Italian mathematics student who had lived almost all his life in Argentina. I made a deal with him: I'd teach him some of the mathematics he wanted to know if he'd let me do it in Spanish and if he'd correct my errors as we went along. That worked out fine—we both profited from our collaboration.

Another tutor was Homero Castillo (from Colombia?), a graduate student of Spanish literature at the University of Chicago. He was pleasant to work with, and our acquaintance led to a curious turnabout of the student-teacher relation. A couple of years after he was my teacher, and after I had gone to Uruguay and returned to Chicago, a professor in the Spanish department phoned me: could I help him out? His student, Homero Castillo, had satisfied all the requirements but one for the Ph.D. degree, the missing one being the final exam. It was the summer quarter, most people in his department were away, and he couldn't raise the legally required quorum. A member of the examining committee had to be on the graduate faculty of the university, which I was, but it didn't matter which department he was in. Since I knew Homero and knew Spanish (!), could I, would I, help make a quorum, help administer the exam, and make it possible for Homero to get the degree that his job offer for the very next term depended on? Sure, why not—it sounded like a lark. I went and quorumed; tried to look intelligent during the literary technicalities (in Spanish) that the real members of the examining committee asked about; and even managed to ask a question or two myself. Having learned Spanish from Homero, here I was acting as his teacher, examining him in preparation for certifying that he had the qualifications needed to be a teacher of Spanish.

I took every step of my Spanish saturation with characteristically compulsive thoroughness. I studied the picture book at a slow but steady rate of five pages per day. I copied out every word of it, and, in a parallel column, I wrote the translations. I quizzed myself, from English to Spanish and back, till I had them perfect. In Hugo, and in Ramsey, I did every exercise, five times altogether, till I could do them perfectly. I followed all the instructions of the Linguaphone course.

As the official negotiations made it seem more and more nearly certain that I'd actually go to Uruguay, I stepped up the amount of time I put in on Spanish. It was never less than two hours a day, seven days a week; the average over nine months was probably three hours a day; toward the end I was Spanishing four hours each day.

The results were satisfactory. Not perfect—but adequate. At my best I spoke Spanish about as well as the average European visiting professor at Chicago spoke English. The average, I say, but nowhere near the best: not as well as Jessen or Gårding, but infinitely better than, say, poor old Stefan Bergman. My Spanish was perhaps comparable to Laguardia's or Marcel Riesz's English—I might have been slightly better, but not much.

I could deliver a mathematical lecture fairly well, particularly if I prepared for it carefully. The grammar and the pronunciation didn't come out perfect, but no utterly ridiculous errors appeared. Impromptu lectures were slightly worse—but not enough so, for example, to warrant the effort to prepare for each class presentation with the same care as for an invited address at a formal conference.

An hour after landing in Montevideo I found myself in a hotel room, and the telephone rang. That presents one of the hardest linguistic problems to cope with—I almost panicked—but the saturation method prepared me to cope with it. In social conversations, in the months that followed, I could follow the drift, understanding something like 85 per cent of the words and almost all the sense, and I could make my contribution. I could not make it as well as in English, and jokes tended to come out quite painfully, but I managed to keep from being a dead weight at a party.

Shopping, ordering in a restaurant, and asking for directions were not too hard; the trouble was with understanding the clerk, the waiter, and the kind passerby who gave the directions. That varied tremendously. In some cases there was no trouble whatever, but in some the other guy might as well have been talking Chinese with dumplings in his mouth. After six months in Uruguay I still had no idea what one of the maids at the pension was saying: was she making a comment on the weather, asking permission to clean the room, or telling me that the hot water heater was bust again? The rule observed by many foreigners in many languages applied to me: the easiest people to understand were other foreigners, and the next easiest were educated natives. I never reached the point where it was fun to talk Spanish; I was always under tension, and, after a couple of hours, ready to relapse to English.

All that was a long time ago—over thirty years—and by now, of course, my Spanish is not even as good as it was then, but a surprising amount of it stayed with me. I give most of the credit to the saturation method—do everything, do it all at once, and do it every minute you can possibly spare—the best way of learning there is.

Room and board

The trip to Montevideo was good and bad, as such things usually are. First class on the S.S. Uruguay for 18 days on the South Atlantic made a splendid holiday: gorgeous swimming pool, luxurious public rooms, tiny but efficiently laid out staterooms, deck chairs in the sun, snacks served almost every minute, and elegant appetizers with every drink at the bar. Not all the company on a cruise ship like that would be among my early choices; the cruise director's efforts at first-name jollity were wearisome in 30 seconds and became worse as time went on; and the attitude of most of the staff was blatant tip-grabbing, smileless service for a ransom. Some good, some bad. The schedule called for a one-day stop at Rio de Janeiro, where, by prearrangement, I gave a colloquium talk and began to learn about the way mathematics was organized in South America.

In Montevideo we were met at the dock by three people from the Instituto (Laguardia, Massera, and Forteza, the latter being a junior would-be mathematician), by three relatives (cousins of my mother who had emigrated there), and by two representatives of the embassy (Francis Herron, the Public Affairs Officer, from whom I had been instructed to ask permission for everything from breathing deeply to cashing checks, and his assistant, Miss Bransford). Hands were shaken, insincerities were pronounced, and our life in Montevideo began.

It did not begin auspiciously. The time was late September—the tail end of winter—it was nasty, dark, rainy, chilly (about 40°F). Mrs. Laguardia took us house hunting, or rather, as everybody advised, pensión hunting. We studied newspaper ads, and in the course of the next two days we looked at about a half dozen places. Some of them were frightful. In our first sample the odor of stale grease hovered about the dark and dingy stair hall where one entered, the woodwork was not only dusty but greasy, the rooms had no windows, and the landlady and the four year old nena who followed her around sucking on a pacifier hadn't had a bath that month. All the pensiones we saw were like that, except one. That one we agreed to take on the spot, and the next day we moved in.

From the outside the house looked cool and clean, and the inside was cool and clean. When we first arrived and rang the bell, the sun was shining and as we were waiting we could peek into the front hall. It was large and airy, with very little furniture—two easy chairs, a coffee table with a vase of fresh flowers, and a couple of small landscapes on the walls—more like a show room than one for living. The landlady, Mrs., or rather Señora, or best yet Frau Mayer was a shortish, plumpish, cheerful woman of 40. To our untutored ears her Spanish sounded perfect—only later did we discover that her pronunciation, her diction, and her grammar were all echt from Vienna. She was a relaxed, pleasant person, willing to go far out of her way to please.

Our room was small (12×12, looking a little larger because of the contribution of the third dimension, which was also 12 feet). The furnishings were sparse; they included a big wardrobe (built-in closets are an American invention) and a comfortable looking (but deceptive) easy chair with one leg shorter than the other three; all the furniture was a highly polished maple yellow. Adjoining the room and also part of our private domain was a small alcove, as wide as the room but only four feet deep. It was separated from the room itself by ceiling-high French doors, which, naturally, we kept open in all but the coldest weather. The alcove was completely filled by a tiny pigeonhole desk (my study), a narrow two-shelf bookcase, and a table the size of a bridge table flanked by two straight chairs. From about three feet off the floor all the way to the high ceiling and along the entire width of the alcove there were windows looking out on the street.

The rent that we paid Frau Mayer included pensión completa, i.e., food. That meant, specifically, breakfast (at about 7:30 or 8:00), dinner (at 12:30 more or less), and supper (at 8:00 plus or minus a half hour). This meal schedule was a concession to the fact that the landlady was not Uruguayan and neither were most of the guests. The more usual meal hours in Uruguay were about 1:30 or 2:00 for the heavy midday dinner and about 9:00 or 10:00 for the almost equally heavy, and sometimes even heavier, supper. In between, around 6:00 p.m., the native Uruguayan has tea, or some reasonable facsimile thereof.

Our midday dinner was served in the public dining room of Frau Mayer's empire. Every individual pensionista had his private little table; we were the only couple, but there was another family table for an elderly lady and her two middle-aged sons. We could arrive at dinner at any minute that pleased us, within reasonable limits, and we were then waited on by one of the maids, at whatever rate pleased us. Except for a vaguely muttered greeting of an embarrassed sort when someone entered ("buenas...", "guten Tag"), and equally vague mutterings when people left ("buen provecho", "Mahlzeit"), there was virtually no conversation. When, rarely, one pensionista spoke to another in the dining room, all the others glared at him, astonished, and made him wish he hadn't. There were, of course, no introductions; only by accident did we find out the names of some of our fellow boarders.

In the back of the pensión was a little garden, with a table, several folding chairs including one of the deck chair type, and what seemed to be dozens of cats of all sizes and ages. (On one occasion I counted a mangy old dog, five cats, and seven kittens.) In good weather, and we had lots of good weather, one could sit out in the garden, read, play with the cats, talk to a neighbor, knit, type, or do whatever else it occurred to one to do. Frau Mayer was likely to bring one a cup of coffee if one stayed long enough to be noticed by her—sometimes even a glass of beer.

This more or less describes the atmosphere of the pensión; with a few words about languages, I leave the subject. One of the reasons that made us want to live in a pensión was the opportunity it would give us to meet people and practice Spanish. As it turned out, Spanish was, to be sure, the common language of all the pensionistas, and in cases of emergency they did speak Spanish with one another, but most of the time they preferred to speak German (the overwhelming majority were Viennese), Czech, Hungarian, French, or English. The only ones to whom we had to speak Spanish were the maids.

In a discussion of room and board the quality and quantity of food should at least be mentioned; in Uruguay the words were mediocre and stupendous.

Breakfast is a very light meal, consisting usually of rolls and coffee, or, more precisely, café con leche. Usually coffee means a tiny demitasse of very black, very hot, very strong, and very bitter substance; the breakfast café con leche is served in an ordinary size cup and consists of the same substance diluted fifty-fifty with hot milk. You may order café con leche at tea time too; if you order it after a meal, you are American.

It is at his midday dinner, to which he sits down at about 1 or 2, that the average Uruguayan really starts absorbing calories. A typical dinner consists of soup, fiambre, fish, entree with trimmings, dessert, and coffee. The soup is a generous bowlful, accompanied by a French roll or two. The word "fiambre" means something like cold cut, but it is not necessarily interpreted literally: it can be a slab of tongue, a tomato stuffed with hard boiled eggs, or even a plate of hot spaghetti with cheese. The fish course is not obligatory but customary; in all cases when I ate out and didn't order myself (as at a banquet or in a private home) the fish course was present. Then you settle down to the entree, which is generally beef or mutton. Steak is what's most likely to be on your plate, accompanied by fried potatoes, a vegetable or two, and a salad. The meat is likely to be at least an inch thick and larger than your hand; don't be surprised if you can't cover it with both hands. If you survived the meal so far, you face dessert: most probably a slab of cake, about the size of four packages of cigarettes, with thick orange icing, and chunks of candied pineapple, drowned in sugared whipped cream. All this time you have been helping yourself generously to the appropriate liquids: mineral water or beer on week days, and wine on formal occasions. After the dessert you sip your coffee and light up your cigarette; your work is over for now.

The next time you eat will be around 6 p.m. No matter how much of a home body you are, your tea is likely to be in a public place, in a confitería. The simplest order in a confitería is té completo. What you then get depends on where you are; you must either know the meaning of the magic phrase at the particular confitería you selected or take a chance. What you are likely to get is a pot of tea with some slices of toast, butter and jam, several sweet cookies, or, if you are lucky, a twin of the cake you had for dinner, and two or three small sandwiches. A sandwich is a thin slice of cheese or else a thin slice of ham (there are no other kinds—cheese or ham is built into the definition), decked by two very thin slices of white bread.

If instead of tea you ask for a drink—a Martini, say—you're in for a surprise. In the first place the amount of liquid you get is about half or at most two thirds of what I was used to in a Martini; in the second place, no matter how much you insist that you want it dry, it'll be 90 per cent vermouth. But, in the third place, you'll get a meal, a free lunch thrown in extra. The waiter will start unloading from his tray onto your table a succession of little saucers (about the size that pats of butter are sometimes served on). Each little saucer contains some hors d'œuvre substance: cubelets of cheese, potato salad, peanuts, tiny biscuits, potato chips, etc. All this—and it adds up to a good-sized amount of food—is part of the definition of a Martini. If you order a second Martini, you get a second free lunch, and if you order a third one, you get a third one.

By this time you have consumed about twice the number of calories needed to maintain you in good health each day, but you are nowhere near finished. At 9 or 10 you must sit down to eat again. I won't describe what you'll get for supper; except for minor variations, both the contents and the size of your evening meal will reproduce your mid-day experience. Between your meals you support the tremendous number of ice cream and hot dog (franfrute) vendors that roam the streets, you drink more Coca Cola per capita than your U.S. counterpart, and—surprise!—you have liver trouble. Everybody in the country complains of liver trouble.

The variety of the food is not great, and neither is its quality. Steak and cake are alternated monotonously, and, surprisingly for a country that exports meat, the quality of the steak is not the best. I had some very good steak in Uruguay, but none that I couldn't duplicate in any decent American restaurant, and what I most frequently got was tough. All the roast beef I ever encountered was chewy and stringy. Chicken and pork exist, but they are delicacies; they are expensive and rare enough to be saved for Sunday dinner. Other fowl (duck, goose, turkey, squab) I hardly ever saw. There are no oysters; lobsters and shrimps are very expensive and rare events.

I think I can conscientiously advise you: if you want to eat well, don't go to Uruguay. I have fairly described the middle class average, and, I am afraid, very little that is above average can be found. I didn't do a lot of exploring, looking for those darling little Italian restaurants where they whip up the most amazing ravioli in the world, but with what I saw, together with what the natives told me (if you want to eat well, go to Buenos Aires), I think I am safe in concluding that they don't exist.

Weather and climate

Winter in Uruguay usually begins around the middle of May and lasts till the middle of September. According to reports, the winter of 1951 was unusually mild, with the temperature hovering near 70° (Fahrenheit). We arrived on September 24, and for the next three weeks we saw weather that was much worse than it had been all winter. It rained almost every day, and it kept chilly even while it was sunny. The thermometer hovered around 45°, frequently reaching 40° and rarely 55°. When it was 45° in the streets of Montevideo, it was 45° inside our house; we were miserable.

In our room there was a three-section radiator, a size that might be considered sufficient to heat a small bathroom. Since the room had about two or three times the floor space and one and a half times the height of a small bathroom, and since the radiator practically never functioned, things were rugged. The reason the radiator didn't function was, of course, that the heating season was over. One turns the heat off on September 15, and cold weather is not enough reason to turn it back on again.

My office at the Instituto was, in all dimensions, larger than the room we lived in—larger, that is, than the room plus alcove. The windows of the office faced the sea—beautiful view—from where the cold winds blew. The building was new (in fact unfinished), but badly constructed; when a wind decided to blow, really to blow, the windows were not enough to keep the papers on my desk from fluttering. There was no central heating. They intended to install central heating, but the money ran out before the building was finished. The problem was solved by not even installing the ducts for central heating; if and when the building finally gets finished, considerable portions of the walls will have to be torn open to make the installation possible—or else tack the ducts on the outside. (So far as I know the central heating is still not there.) The situation was considered bad enough that an electric heater was supplied to each office. If I held my feet carefully and rigidly near the heater (but not too near), my feet got warm (but not too hot). My fingers were cold no matter what I did, as was the rest of me. In the classrooms there were no electric heaters—and not even functioning outlets into which a borrowed one could be plugged. During the first few weeks overcoats were worn in the classrooms—not only by me, but by all the natives as well.

It was the same everywhere. We were cold at home, cold in the movies, cold in the restaurants, and cold in the private houses we visited. The only place we were not cold was on the streets; if one is suitably dressed and walks briskly, an outside temperature of 45° is rather pleasant.

The chilly, dark, damp, and nasty weather lasted a little more than three weeks; by the middle of October things began to improve. For the next six weeks or so we were still chilly, but no longer miserable. At tea time it was possible to sit at the sidewalk tables, if one sat in the sun. At home I began to find it possible to work without being bundled up in my heaviest clothes and without a woolen scarf wrapped around my neck. During this time the daily maximum was around 55° or 60°.

When summer arrived, it arrived with a bang and it stayed. Around the first of December the sun went to work, and the city changed. Bath houses and franfrute stands were put up on the beach; the tiny electric heater that failed so valiantly to help out the non-functioning radiator in our room was put away; the weather was gorgeous. It rained only rarely, and it was never cold. The temperature was almost always between 70° and 80° and frequently a sea breeze made you feel comfortable though slightly sticky. In Chicago my friends were shovelling six foot high snow drifts.

How to get a chair

I mentioned before that on my one-day stop in Rio I began to learn the difficulties of the mathematical organization of South America. The main difficulty was not that the number of mathematicians was below the critical mass needed for effective mutual stimulation. That would have been bad enough, but, in addition to that, the financial, intellectual, and governmental support, respect, and recognition that mathematics received were too low; the number of jobs was even smaller than the already too small number of mathematicians.

In Rio, for instance, Leopoldo Nachbin was struggling to make a living, although he was far and away the best mathematician within a thousand miles. When he wasn't proving theorems, he worked part time for the Brazilian Center for Physical Research and he taught a couple of courses at the University. Then came a ray of hope: the university authorities announced that a chair of mathematics was vacant and called for all candidates to come forth and present themselves. The chair had in fact been unfilled for some time—the duties associated with it were being carried out in an unofficial and admittedly temporary manner by a man who was chronologically senior to Leopoldo but mathematically not worth half of him.

Two candidates for the chair presented themselves—the only two possible ones—namely Leopoldo and the temporary holder. Since it was evident to everyone, including T, the temporary holder, that T's qualifications were as nothing in comparison with Leopoldo's, behind-the-scenes, parliamentary machinations were set in motion. Here is what some of them were.

Leopoldo was applying for a professorship. One cannot hold, and therefore one cannot apply for, such a position unless one has a Ph.D. from the Philosophy faculty. However, one cannot, or in any event according to T one should not, have such a degree, unless one also has the Brazilian equivalent of a master's degree from the Philosophy faculty. Leopoldo received his master's degree from the Engineering faculty. Conclusion: Leopoldo's Ph.D. should be revoked and, consequently, his right to apply for and hold a professorship should be taken away from him. These things T said officially and publicly.

As a consequence of these charges it was necessary to appoint a fact-finding committee whose duty it was to rule on Leopoldo's eligibility for candidacy. A curious aspect of the situation was that not even T hoped that the committee would rule against Leopoldo. The point, however, was that at the moment Leopoldo was riding high, his fame and his power were growing, and if the final committee to decide between the two candidates had been appointed right away, there would have been a very great probability of Leopoldo's victory, too great for T's comfort. The point of T's move was to play a delaying game, hoping to postpone the final decision for as long as possible. In the meantime T could go on temporarily filling the chair and trying to build up support. If, by some chance, the fact-finding committee ruled against Leopoldo, the game would have been won by T. If they ruled for Leopoldo, there would still have been a chance for T with the final selection committee—although, of course, in that case the exoneration itself would have been a positive factor in Leopoldo's favor.

This kind of official tomfoolery takes time—it took months, years, and in some respects decades before all the returns came in. In the end virtue triumphed. The first committee found in favor of Leopoldo, unanimously. Later, the selection committee awarded him the chair, and, in fact, I was in Rio again thirty years later, participating in the pomp and ceremony of Leopoldo's retirement from it.

Humanities and sciences

Nothing like the horror story of Leopoldo Nachbin's professorship could have happened in Uruguay when I was there—because nothing like a chair of mathematics existed yet in the entire country.

The University of Montevideo is very different from anything that exists in the U.S.; the system is more like what I was told about the

European one. The university is organized into nine or ten faculties—
roughly speaking a "faculty" in this sense is what a "division" is
at Chicago. There are, for instance, faculties of medicine, of engineering, of humanities and sciences, of architecture, of chemistry, and of
law. The faculties are completely independent of one another, and de-
pendent on the government for their budgets only; they are proud to be
autonomous little entities. Their independence is really complete: no
such thing as a joint appointment is ever dreamed of, and the buildings
(of the separate faculties) are separated from one another by literally
several miles. A student of economics would no more dream of taking
a course in mathematics from Laguardia than an average economics
student at Northwestern would of coming to the south side of Chicago
on Mondays, Wednesdays, and Fridays at 9:30 a.m. to take Math. 101.
A good aspect of the absence of joint appointment *de jure* is that
de facto they occasionally exist: it is perfectly possible for one person
to obtain two distinct appointments.

The quality of the faculties varies a great deal. I was told that medicine and engineering are the best, and there seems to be unanimity on
declaring Humanities and Sciences the worst. The ideal of H. and S. was
in a way similar to that of the College at the University of Chicago: it
was to be a faculty not interested in such crude, materialistic things as
degrees and professional training—a faculty where professors and students would teach and take courses and investigate mysteries for the
sheer love of doing so. That ideal didn't seem to work out. As far as
mathematics is concerned, for instance, the faculty modified its attitude
and established a degree you can work for—the degree (licenciatura)
is roughly the same as a master's degree in a midwestern state university. It is a bit of a mystery in itself what the degree is good for. It does
not aid one in getting teaching positions in the high schools and junior
colleges—for that one must have a different kind of training. It is
equally irrelevant to becoming an engineer; it stands in a vacuum all by
itself.

The faculty of H. and S. is organized around a large number of little
chairs (one might almost call them stools); each chair corresponds in
reality to one course. For teaching that course, if you are fortunate
enough to win the competition leading to the award of the chair, you
get about 20 or 25 per cent of a middle class income for life. (Originally, to be sure, you are appointed for one year only; the second
appointment, if you have proved yourself not unworthy of it, is for
five years, and all succeeding reappointments are for five years. It is
virtually unheard of to get the first five-year appointment and not
to get all others—reappointment is *de facto* automatic.) It isn't much,
but it helps.

A psychologist, Rimoldi, looked us up when he learned we were from
Chicago, and became a friend; he got his Ph.D. from Thurston, and he

enjoyed talking, more than slightly homesick, about the good old days in Mandel Hall and Steinway's drugstore. He was on the faculty of H. and S.; he was, in fact, the only full-time member of that faculty. The institution of full-time professor was virtually unknown in Uruguay, and what was known was mistrusted. There are, of course, historical reasons for this. Universities were established rather late, and they began with such relatively practical subjects as medicine and law. How are you going to get someone to teach in a recently established university? The founding fathers' answer, in medicine for instance, was to ask a successful practicing physician (with a degree from Italy) to take time out for two or three hours each week and teach a course in anatomy. The university, being poor, couldn't pay much; the money the doctor earned by his lectures was probably an infinitesimal fraction of his income. But being invited to be a university professor was a great honor—and, what's more, if you put it on your calling card, it attracts patients.

All the bad aspects of the system jelled and became permanent—and in my day you were extremely unlikely to find a university professor in Uruguay. You were likely to find an engineer who was also in the faculty of engineering, or a newspaperman who taught a course in Spanish literature, or an architect who taught descriptive geometry.

Rimoldi told me that his working conditions were intolerable. He needed students and assistants, he needed laboratory space and equipment, and he needed office space and equipment. He had no complaints about the human material: he could find plenty of intelligent and enthusiastic people. The rest was hopeless. They not only wouldn't give him the lab and the rats—that sort of thing is what all laboratory scientists are used to fighting for. The trouble was that he was having difficulty even getting pencil and paper, and an item like an electric heater for his secretary's office (a necessity in that land of frozen toes) required what seemed like an act of congress.

Here is one more anecdote about the faculty of H. and S. that Rimoldi told me. The faculty advertised that it had a vacant chair in comparative anatomy and that all candidates should come forth. Several did. Their credentials were examined, all the usual formalities were complied with (there were quite a few), and eventually the successful candidate was selected. He was notified by letter and by an official appointment form, and he was told that he could begin to consider himself a member of the faculty. When, a little time later, classes were just about to begin, and the brand new faculty member had not yet been told anything specific about his duties, he went around to the administrative authorities and asked what he was expected to do, and when, and how he was to collect his pay, and when. Oh! Well, you see, there has been an unfortunate misunderstanding. We cannot think how it could have happened. But the fact is that this chair that you were appointed to—well, you see, we just found out that there is no such chair. It doesn't exist on the books, there is no money for it in the budget, and it is all a mistake. We do hope you don't mind.

Faculty of engineering

The organization of each faculty is pretty much the same as that of the others; I'll describe what I learned of the faculty of engineering.

The administrative power of the faculty is in its executive council of 11 members. Four of the members are elected by the faculty, four others are elected by the professional engineers of the country (each one of whom is presumably a graduate of the faculty), and two more are elected by the students (but are not themselves students—usually they too are engineers). The ten elected members in turn elect a dean—if they happen to elect one of themselves, then a supplementary election is held to fill the vacancy thus created. The deanship is a great honor and is usually held by a commercially successful engineer. The council over which the dean presides runs the faculty: all important steps (and, in particular, all appointments) are made by the council. Example: while I was there, an assistantship for one of the mathematics courses became vacant. Since Laguardia was not on the council, he had absolutely no say as to which of the two competing candidates should be chosen. The choice was between Villegas, a part time member of the Instituto de Matemática and an earnest student, and Petracca, a high school teacher with no scientific merit. The job was given to Petracca on the grounds of seniority.

Since the professors (and in particular those representing the faculty on the council) were in almost all cases practicing engineers, and since they earned most of their income outside the university (being on the faculty is a great commercial asset), there was virtually no academic esprit de corps. No one on the council, for instance, had the least interest in getting a salary raise for professors—it wouldn't matter a hoot to them if their classes were worth an extra five or ten per cent a month.

The engineers on the council (all the councils in the history of the faculty) are very much influenced in their academic decisions by the fluctuations of the market for engineers. If there is a shortage, they put on the pressure to ease up on the courses and speed up the diploma mill; if engineers are a drug on the market, they tighten up the requirements. The fact that the dean is not an academic person, and has therefore no feeling for the continuity of the university, and is both ignorant of and not affected by what his predecessors have done, has as a consequence an interesting almost-periodicity. The dean's job lasts for four years. Once every four years, therefore, the new broom reorganizes the curriculum. By 1951, in Laguardia's observed twenty years, the successive deans have tracked back on themselves four times.

Students exercise a rigid control over some of the day to day procedures—for instance exams. All exam questions, with the answers completely written out, must be delivered by the professor to the dean's office before the exam is given to the students. If the students subsequently decide that the exam was too hard, "not in the book", or for any other reason unfair, they have the right to complain to the council, which then calls the professor in question on the carpet. That happens frequently.

Among professors there is virtually no swapping around of courses. Laguardia was surprised and envious when I told him about the set-up in Chicago. Geometry, for instance, was taught to the engineers by an engineer who learned it when he was a student and knew nothing else of mathematics. Much as Laguardia would have liked to teach the course for a change, he could not do so; its permanently assigned teacher neither would nor could teach analysis. (Analysis, by the way, meant advanced calculus.)

The red tape is frightful. If, for instance, you want to have a short holiday (less than or equal to three days), the head of your department (in my case Laguardia) is entitled to grant it to you, but you have to report it to the central administration, and they in turn record it and charge it against your vacation. If you want to go away for a number of days between 4 and 30, a formal petition to the dean is required; even if your petition is granted, the chances are that your pay will be given a leave of absence too. If the number of days that you wish to spend on your vacation is 31 or more, the executive council has to meet and ponder the matter, and your payless status becomes a statutory certainty. The amount of paid vacation you are entitled to is exactly 20 days a year (not 20 working days—20 days altogether), to be taken at a time approved by the dean.

There is one thing in this connection that I didn't make clear. Since in my day there were virtually no full-time professors, it might sound paradoxical to speak of vacation. The point, however, is that there are inside the faculty of engineering a certain number of Institutes—they are the closest analogues of departments. The work of the Institutes has nothing to do with the educational program of the faculty—it is more like a research program. In mathematics, for instance, there was an Institute, and Laguardia was a member of it. Laguardia also taught engineering mathematics. The two jobs were separate and he had to arrange to do them at different times; if, for instance, he worked at the Institute in the afternoon, then his engineering classes had to be in the morning, and vice versa. The vacations I've been referring to are Institute vacations, not faculty vacations.

Instituto de Matemática

The high sounding name, Instituto de Matemática y Estadística, describes the mathematical research branch of the faculty of engineering; its very existence is a great achievement of its founder, Laguardia. Originally the faculty of engineering had no department of mathematics—it had only three professors of mathematics who taught the standard courses that engineers think they need. Laguardia conceived the institute and by fiat brought it into existence, at least on paper. It stayed on paper for almost ten years. Officially no such thing existed, there was no allotment for it on the faculty budget, and all that it really meant was that Laguardia and Massera tried to inculcate some mathematics into a few promising youngsters and started to accumulate a few books. Neither Laguardia nor Massera were paid a penny for their troubles. They earned their half salary as professors in the faculty of engineering, they eked it out as professors in the faculty of humanities, and they donated their services to the institute. Eventually, in the late 1940's, Laguardia talked the council of the engineering faculty into recognizing the institute and giving it a line in the budget.

Physically the institute consisted of about a half dozen rooms. One of them was the director's office; that's where Laguardia was perpetually writing reports (overdue) and planning budgets (for the year that began two months before). Another room was where Massera worked at mathematics, keeping rigorously to the prescribed timetable, and (even during coffee breaks) talking less about politics than any of the others.

A third room was the señorita's (in the nine months I was there I never heard anybody call the secretary by name), and a fourth belonged to the computers, which in this case meant two young men with moustaches in charge of two desk calculators. One of the computers left for a better job soon after I arrived. Since, as a way of saving money, no state position could be filled for six months after being vacated, during most of my stay there was only one computer. Ruibal was a pleasant, baby-faced young man, a student of engineering of course, who had some small curiosity about mathematics; when he needed a part time job, it was natural for him to look for one in the Instituto. To be a computer was a reasonable job for a budding engineer–mathematician; unfortunately there was no computing to do. Laguardia frequently told the story of the time (but even he always referred to the same, single occasion) when the computers did some computing for one of the other institutes, one of the practical ones. Unfortunately the results of the computation were obtained by the originating institute also, by a simple experiment—and except for that brief flash of glory the computers never had anything to do. Ruibal was set to work filing library cards, writing call numbers in library books, and, when the señorita was overloaded, helping to type. It's not what he came to do, but it was a job.

One of the rooms was my office, and the one next to it was the institute's library and, simultaneously, the office of some of its junior (student) members. It was not surprising, in view of the institute's youth and poverty, that the library was not good; it was surprising that it was as good as it was. Even before the institute came into being some of the classical books of mathematics and a few of the old and respectable journals (such as Crelle, Liouville, and the American Journal) were already in the library of the faculty of engineering. The newer books, and subscriptions to the modern journals, came directly to the institute. The book shelves in the institute library covered all of a largish wall, and the books in the shelves filled about half of them. Bourbaki, the AMS Colloquium books, the Princeton series, and, generally, items that are frequently seen in a mathematician's personal library, were there. The obvious American journals came regularly, but there were serious gaps. There were virtually no back files, the Pacific Journal (then very new) and the Journal of Math. and Physics (quite old) were never heard of till I spread the word. Serious bibliographic research was impossible.

Institute people

The total number of people whom I saw at the institute more than once during my stay was 11—that includes professors and students, but it excludes the occasional distinguished one-day visitor, the engineering student using the library as a place to study, the secretary, the panadero who came each day with a basket slung around his shoulders selling pretzels and rolls for our coffee break, and the peón (not an insulting term) who waxed the floors. The 11 included three students whom Laguardia found somewhere near the bottom of the barrel—once he succeeded in getting a small amount of scholarship money on the budget, he had to give it away whether he found anyone deserving of it or not. The 11 included also Forteza and Infantozzi, who were pleasant, intelligent, and cultured people, but who were not (and had never set out to be) research mathematicians. In the U.S. they would rate as outstanding teachers at the junior college level, with the additional unusual feature that they had not completely stopped reading mathematics, and, from time to time, would even try some minor research about things like rings satisfying a curious identity or topological spaces in which a singleton was dense. Forteza was the ever hopeful amateur. He would read the first three pages of Hasse on number theory one day, the first three pages of Hilbert on geometry the next day, and the first three pages of Hobson on real variables the day after that.

That's 5 of the 11. Two others were Alfredo Jones (pronounced Choness), and Cesáreo Villegas, with whom I had mainly friendly contact, but very little of the professional kind. There was Laguardia, of course, whom I definitely regarded as a friend, but with whom my professional contacts were almost exclusively administrative.

Then, last but certainly not least, come Massera, Lumer, and Schäffer. Massera is the same age as I (35 in 1951); he was already an established and respected mathematician. Lumer and Schäffer were some 15 years younger, ambitious and talented students.

J. Massera, 1960

Massera was a stocky chap, below average height, with dark but graying crew-cut hair, and a quiet, almost placid behavior. He was warm and friendly to me. While he made no secret of being an active member of the local communist party, he managed to keep his political life carefully separated from his professional one. I saw him every day during the official hours of the institute (which during the winter were from 1:30 to 6:30 in the afternoon). He stayed in his office more than most others (all offices could be spied on from all others thanks to an ingenious system of windows designed for that purpose), and he appeared to be busy doing what mathematicians do. We discussed the weather over cups of tea, he told me the contents of his last year's course on Hilbert space, and he tried to interest me in his current differential equations research problem. He frequently became interested in the questions the rest of us were asking one another, and, after disappearing into his office to think about them, would pop out an hour or so later with the answers. He typed his own papers—long ones; he read the current journals; he conducted a seminar; and he guided several students in their seminar-related studies.

His ideology had at least an indirect effect on the daily life of the institute nevertheless. Because of his ideology he asked for a month's leave of absence to attend a political congress in the USSR. The request came at an inconvenient time, in the middle of a term; Laguardia had to find someone to take Massera's classes, and he had to get along without Massera's help with the routine administrative chores. The worst part was that such a leave of absence heavily and publicly underscored the existence of a dyed-in-the-wool communist in the Instituto. While communists were not as much feared and hated in Uruguay as in the States, they weren't exactly popular, especially not with the higher echelon of political officials (such as the ones in the Ministry of Education, whence all blessings flow). And, of course, communists were not popular in the U.S. Embassy, and the Rockefeller Foundation, and other North American sources of the staff of life. Laguardia was not happy.

Gunter Lumer was born in Germany, received some of his early education in France (where he was Guy Lumer), and went to university in Uruguay (which was the permanent haven his parents found away from Hitler). He spoke both German and French like a native, and Spanish almost as well. He is a short man, with a hooked nose, a wide grin, and a vivacious manner; he is always in motion. His size used to vary from normal to nearly obese so often that he needed two complete wardrobes, the fat one and the thin one.

His positive personality was visible even in his mathematical attitudes, even when he was a young student. What he was interested in at the moment was the most important problem in the world, and what's more he had just solved it—well, almost solved it—there is, you see, just this minor point that remains—is there a curve in the plane such that... ?—and, great generalization, is there a surface in space such that... ? When I met him he was (without being aware of it) a spiritual descendant of R. L. Moore. I took it to be one of my jobs to show him a window to the rest of the world of mathematics. I opened the window, I pointed, I argued—he closed his eyes, he resisted, he argued—and it all ended well. He was energetic and talented enough that it's not at all clear that I really did him any good; I am sure he would have become a mathematician no matter what I did.

Juan Schäffer's story is similar: born in Austria, he lived in France for a while, spoke German, French, Spanish, Italian, and English, all very fast, all with complete colloquial mastery. He was, even as a student, spectacularly bright and quick—when I asked a question, he would catch on when I was half through and the answer would start pouring out in staccato Spanish.

Lumer made a lot of small mistakes, kept correcting them, and converged to the truth by successive approximations. Schäffer was quick enough to catch his own mistakes after he half said them, and to correct them while I was still bewildered. Lumer's stuff would come out structurally organized but chaotic in its details; Schäffer's would be a snow job—each crystal was impeccably clean but the mass could be incomprehensible in its impressive totality.

That's how 11 people struck me then, way back then in 1951. Laguardia has died since. Massera became a communist member of parliament and, when the government changed from a Swiss-type of committee-presidency to repressive extreme rightism, he was jailed and tortured, and kept in jail for eight years. Jones is a respected algebraist in Brazil, and Villegas a respected statistician in Canada. Lumer is doing hard Hardy spaces in Belgium, and Schäffer (partly still under the impetus of Massera) has done a lot of work on differential equations in Pittsburgh. Mathematics in Uruguay is dead now, as dead as it was before Laguardia; perhaps it will come alive again some day.

Teaching in Montevideo

The institute's schedule changed with the seasons. We had to be there for five hours every working day: 1:30 to 6:30 p.m. in the winter and 7:30 a.m. to 12:30 p.m. in the summer. The theory of the summer schedule was that since during summer afternoons the weather is tropical, one must do the day's work in the morning. Both the premise and the conclusion are false.

The weather, though warm, is not tropical: it is never as hot as Chicago or New York can be. As for doing one's work in the morning: the economic organization of the country made that impossible in my time (and things got worse since then, not better). To make an even approximately decent living, almost everyone had to have two jobs. I mentioned already that a professor of engineering is an engineer whose professorship amounts to moonlighting, and the same is true for professors of medicine and architecture and agronomy. The system extends, with very rare exceptions, to professors of mathematics, philosophy, astronomy, Latin, and economics: outside of office hours they are likely to be newspapermen or insurance agents. It's the same in the real world: the morning bus driver is the afternoon waiter and vice versa. That it might be more efficient and less tiring for these two people to swap half their jobs with each other is yet to be thought of. Inside the university there are too many students to make it possible to educate them all during a half day: summer or winter there are both morning and afternoon classes.

To make the national schizophrenia possible, many important institutions (such as banks and government offices) are open for half of the day only. The bank that was open between 2 and 6 in the winter conducts its business between 8 and 12 in the summer. The bus driver changes to driving in the afternoon and waiting on tables in the morning; the morning classes of the university are changed to the afternoon, and the afternoon classes to the morning.

I had to be at the institute five hours a day, but what I did was almost entirely up to me. I would prepare my notes (in Spanish) for my lectures (in Spanish), drink coffee, swap problems with Lumer and Schäffer, browse in the library, drink coffee, open my mail, read the New Yorker, try to prove a theorem, read a statistics book, drink coffee. The important thing was to be there. One morning when the panadero didn't arrive, I dared to suggest that we go down the street to have our "coffee and", but my suggestion was treated as a joke in bad taste; that just wasn't done. When a couple of years later an Uruguayan friend came to Chicago and did not find me in my office Tuesday morning at 10:30—I was at home thinking about algebraic logic—he was horrified—how can I get away with absenting myself from my duties? When he telephoned me next Sunday at 2:00 p.m. and learned from my wife that I was not at home—I was in my office doing some editorial work—his confusion was complete—why am I in my office when I don't have to be?

My duties at the institute were affected, but only peripherally, by the student strike. "Huelga", meaning "strike", is the first Spanish word I added to my vocabulary when I arrived in Uruguay. Strikes were almost as popular as saints' days, and much more effective at cancelling work days: though there were fewer strikes, they were less easy to ignore and they lasted longer. The bus strike was called off soon after I arrived, but a week later it was on again. The gas strike made me shave and shower in cold water for the first three weeks; the student strike took five weeks out of the winter term.

It was all about politics, of course. Parliament was debating a proposed constitutional reform, and some people feared that the new proposals were more likely to make for governmental interference than the status quo ante. Countermeasure: the entire student body of the country (about 50,000 strong) called a strike: laboratories and libraries closed, classes and exams were called off. Since the work of Laguardia's institute was superimposed on the university structure, rather than a part of it, our activity during the strike weeks was greater than usual—people had more time to come.

The strike had some effect; some of the university's recommendations were incorporated in the new constitution. The financial autonomy clause, however, which the university badly wanted, did not make it.

What financial autonomy would have meant was that parliament would stop rigidly dictating how the university's budget was to be divided among salaries, buildings, books, and everything else that a university needs to spend money on. It would have given the university authorities greater fluidity of decision—so that, for instance, they could invite a visiting professor such as me without what was almost literally an act of congress. To show its disapproval of parliament's rejection of its suggestions, the university council voted that this time the professors would go on strike.

Most of my "teaching" was directed at the research-minded members of the institute—which meant Lumer and Schäffer, with Laguardia and Massera politely adding to the audience, and some of the others, such as Forteza and Villegas, participating some of the time. The audience was intelligent but unfamiliar with the language and attitude of modern mathematics; almost without realizing it I kept having to lower the level of my lectures as the term proceeded. I tried to begin with rings of operators (that's what von Neumann algebras were called then), continued with subnormal operators, and concluded with an introductory course on topological algebra (an alternative name of functional analysis). That took 29 lectures; the next term I revealed some of the secrets of Boolean algebras in 16 lectures.

Because a part of my reason for being in Uruguay was to introduce the country to probability and statistics, Laguardia sent out a circular letter inviting some forty students, the ones who he thought could conceivably be interested, to come and meet with me and discuss a possible approach. Six showed up, and we discussed. Laguardia did most of the talking, and he tried very hard to overcome their apathy. Here I am, I am expert; don't they want to take advantage of my presence, profit by the opportunity, and squeeze in another course on top of their rigidly prescribed engineering curriculum? The answer seemed to be a listless "no". They were still taking their courses, the academic calendar was pushed back two months because of the strikes, pretty soon the exam period would begin and they would be even busier—maybe later. When Laguardia finally said, well, all right, we'll begin such a course in February, they seemed relieved; that was nice and far away. Two of them subsequently came to talk to me privately and half apologized for their lack of interest—there just wasn't enough time, they said. They asked my advice about what to read, and after finding out what they were prepared to read, I made a couple of recommendations. We parted friends.

The probability course actually materialized; it was a short (16 lectures) but standard elementary course, with five regular attendants. The total amount of teaching I did in Uruguay in an academic year came to 70 hours—about the same as I normally did in one quarter at Chicago.

Research in Uruguay

My own research kept chugging along. I thought some about subnormal operators and managed to make some progress; the spectral inclusion theorem belongs to this period, and so does my first paper on commutators.

The spectral inclusion theorem is one of my chronic worries. It says that if you pass from a subnormal operator to its natural enlargement, the minimal normal extension, then the spectrum goes in the "wrong" direction; instead of growing, it shrinks. What worries me is that there is a similar theorem in another context (if you pass from a Toeplitz operator to its natural enlargement, the corresponding Laurent operator, then too the spectrum shrinks), and I cannot believe that the resemblance between the two statements is accidental. I suggested to a Ph.D. candidate once that he try to discover a general context that includes both spectral inclusion theorems, but, after thinking about it for a couple of weeks, he offered a theological proof that there is no such general context and went on to do something else. I have heard of one or two further mumbles on the subject since then (yes, there probably is a relation between the two results, and it has to do with the representation theory of C* algebras—no, there is no relation between the two results), but I remain dissatisfied. I keep seeing similarities between subnormal theory and Toeplitz theory, and I would like to know why. It was E. H. Moore, I think, who said that if two subjects exhibit such a close analogy, then they just *must* have a common generalization that explains it, and I don't think we really understand either subject till we understand the generalization.

Commutators of operators are algebraically interesting in their own right, and quantum mechanics has made them even more so. The Heisenberg uncertainty principle can be formulated as the statement that the commutator of two specific (unbounded) operators is the identity ($PQ - QP = 1$). In the late 1940's two different and interesting proofs appeared (one by Wintner and one by Wielandt) showing that bounded operators cannot satisfy the Heisenberg equation. Very well then, I said to myself; if $PQ - QP$ cannot be 1, what can it be? I kept returning to the subject and getting results that were surprising but fell short of being conclusive. (Sample: every operator is the sum of two commutators.) The beautiful conclusive answer was found by Arlen Brown and Carl Pearcy in 1965; their penetrating analysis is quite complicated, and no one has been able to simplify it yet. Their result is that scalar multiples of the identity operator are essentially the only ones that are not commutators—and the word "essentially" is used here in its technical sense ("modulo compact operators").

The final piece of Uruguay research I'll mention here has to do with square roots. Every complex number has a square root; the number 0 is slightly troublesome, to be sure, and you have to be careful about not slipping from one sheet of the Riemann surface to the other, but existence is always guaranteed. For the most natural "hypercomplex" numbers—square matrices with complex entries—the facts are different, but not too frightening. If a matrix is not even partially zero, if, that is, it is non-singular, then it has a square root. Natural conjecture: the same is true for bounded operators on Hilbert space. Surprising discovery (made in collaboration with Lumer and Schäffer): the conjecture is false; invertible operators can fail to have square roots. The trouble is that even if 0 does not belong to the spectrum of an operator, it can be surrounded by that spectrum (the way that a circle "surrounds" its center), and that can be enough to kill square roots. (Remember that it is dangerous to surround 0 on the Riemann surface of $\sqrt{z}$.)

The biggest step that Uruguay took toward becoming a part of the research world of mathematics the year I was there was, however, not a small set of papers credited to the University of Montevideo but the conference that Laguardia and I arranged. We started it off in October by a visit to the local Unesco office, presided over by señor Establier. He was a Spaniard who had lived for many years in France, a dapper round-faced man with a hairline moustache, a professional internationalist who used to work for the League of Nations, and one whose Madrid Spanish sounded much more like the books say it should than what you could usually hear in Montevideo. We couldn't have come at a better time. The end of the year was near, and Establier still had some money in the till—he snapped up the conference idea hook, line, and sinker. Yes, by all means let's have a conference; yes, Unesco will fund it; now here's what we'll do. Having observed Uruguayan administrators for several weeks, I found it a pleasure to watch Establier operate. He started talking about names, dates, and places, writing notes as he went; he understood all the difficulties and knew how to get around them; he knew how to get things done.

One: the conference must take place before the end of the year—say, for instance, around the middle of December. Two: the topic of the conference must be as general as possible, so as to attract the maximum number of people and not to offend anyone. Three: we are to make out the list of people to be invited; Establier reserved the right to add others, but not to subtract any. Four: to maximize the efficiency of the gathering, it should not take place in the city of Montevideo but somewhere about 50 miles away, so that the members should not be tempted to spend their time visiting long lost friends and taking in a show. Five: Establier will pay all travel expenses, but we are to look for other sources of money for food and lodging—except that he, through his local contacts with municipal and national government, will help us, and in fact he'll practically guarantee to find what we need.

Establier even went so far as to suggest a topic for the symposium: "Mathematics, classical and modern". Laguardia and I winced—that sounded near the absurd side—and Establier was immediately conciliatory and diplomatic. He mentioned the topic, he said, only to show us the extent of generality that we had better have. How about "Unsolved problems in mathematics"?, I suggested. It's a topic as general as you can get, it'll take the participants less time to prepare, and it looks like a good way to help tell some of our South American colleagues what current mathematics is all about. Laguardia and Establier agreed and the title was tentatively adopted.

The main decisions to make concerned, of course, the list of participants: whom should we invite and from where? Establier's answer was crisp and clear: invite anyone, he said, who can truthfully be called a mathematician and who is at present living in Latin America. He wanted delegates from Argentina and Brazil, but also from Chile, Peru, Bolivia, and Paraguay, as well as Cuba and Mexico. We quickly agreed on a dozen names (including, for instance, Santaló, Beppo Levi, Monteiro, Ricabarra, Catunda, and Murnaghan), and then Establier and Laguardia settled down to a friendly and noisy disagreement session about another thirty names, most of which I had never heard before. From what I could gather, Establier was inclined more toward famous big shots than toward active mathematicians—but that was to be expected.

As the weeks rolled rapidly by, names were added and plans were sharpened. Cotlar and Calderón, and Thullen (famous among mathematicians for his work on several complex variables, he was currently an actuary in Paraguay) and Godofredo García might all be able to come—fine, the more the merrier. González Domínguez suggested that a similar conference be held in South America every year, and that part of the agenda of this one should be the discussion of plans for the next one—fine, Unesco seemed willing to pick up the tab regularly.

Establier's office did all the hard administrative work—arrangements for transportation, beds, food, lecture rooms, chalk, microphones—and did it very well. The location helped—it was Punta del Este, a luxurious resort on the coast, where we stayed in comfortable semi-private bungalows (separate bedrooms, with two people sharing a bath and a sitting room), and there was nothing to do except enjoy the beach and talk about mathematics.

We were scheduled to leave Montevideo for Punta del Este by bus at *exactly* 4:00 p.m.; Establier in his official letter underlined the word. The bus left at 5:40. (I went in private and regal comfort in Laguardia's car.) The minister of public instruction was invited to the conference—that was traditional for such affairs. He must be invited and he must accept, but he almost never comes. He didn't come. Since that meant one less formal speech to listen to, no one minded.

Establier's official crew did come—himself, two subadministrators, and three secretaries—and attended every single minute of every meeting. The rest of us, the mathematicians, thought that was funny—and sad—and pitiful. Those poor administrators! They came, no doubt, on Establier's orders; but why did he come? The only reason we could think of was that he didn't trust us—if he wasn't there to watch us, we'd probably play hooky on the beach. Since the "crime", his crime, his sitting there and being somewhat of a wet blanket, carried its own punishment with it, none of us was really resentful, but there was some mild clucking of tongues.

Several scheduled speakers didn't come. Murnaghan sent a paper (in English—Frucht did a very creditable job of looking it over for five minutes and then giving a résumé in Spanish). Cotlar's paper (in Spanish) was read, word for word, symbol for symbol, by Pi Calleja, who apparently hadn't had even five minutes to prepare, and wasn't familiar with the subject—the result was painful. Besides, the paper was too long for the half hour that it was supposed to fit into, so that only the first, introductory, half was read—the meaty results that Cotlar presumably put at the climax were, so to speak, accepted by acclamation. Monteiro neither came nor sent anything, Massera was still in Moscow, and Ricabarra had Argentinian passport trouble. Of the half dozen Brazilians invited only Nachbin came—the others seemed to be the victims of administrative bungling. The gossip factory was working overtime: the Brazilian invitations never arrived?; they arrived, but too late?; they were on time, but were so badly worded that the recipients didn't take them to be invitations? You expect this sort of thing whenever you arrange a conference; the Punta del Este chaos was of the same kind as elsewhere, but there was more of it.

The Argentine delegation was there in force, and they told us about their library problems. They hadn't had any money to buy foreign journals for the three years preceding; their only contact with current work was Mathematical Reviews. During the few hours they spent in Montevideo, on their way to Punta del Este, they pounced on our journals at the Instituto and spent their time taking notes and copying references. Some of the German ones among them were still fighting the war (six years after it ended); their battlefield was MR versus Zentralblatt. The words were the German party line: Zbl is truly international, since it publishes in four languages, not just one; for the Zbl all Russian material is available and many of its reviewers can read Russian, which is not true of MR; the Zbl is up to date; and the US rejection of the 1939 Zbl proposal to collaborate in the publication of a review journal was unscholarly and inimical. (World War II started in 1939.)

The business of taking notes added some interesting seasoning to the proceedings. Our three junior members (Forteza, Lumer, and Schäffer) were charged with taking notes and so were Establier's three secretaries;

and, in addition, a wire recorder (before the days of tape) was running all the time. Nobody was quite sure what was to be done with all the material so collected. In any event, the wire recorder playback was inaudible: there was only one low-power microphone for a room full of 25 people, who muttered, interrupted one another, and scraped their chairs on the floor. The professional secretaries had no scientific experience—the foreign words, technical terms, proper names, and mathematical symbols that were flying through the air served to confuse them thoroughly. They tried nobly, and stayed up nights typing up rough drafts which they then asked the speakers to correct—but the drafts were hopelessly uncorrectable.

One amusing mistake they made, which was easy to correct, occurred in their transcription of one of my comments. I referred to a vector x and its sequence of images under the iterates of an operator A, and I said, of course "x, Ax, A^2x, etc.". In colloquial Spanish mathematical language A^2 is pronounced not as "A cuadrado" (A squared) but as "A dos" (A two). What I said, therefore, was "equis, A equis, A dos equis, etcétera". Every mathematician "hears" the commas in the right places —they come where he expects them. The poor secretaries didn't; they reassociated my syllables and typed "xa, $xa2$, x etc.".

There was also a jurisdictional dispute between our boys, whose notes though not a word-for-word transcription were true in spirit and made sense, and the Unesco brigade. The latter not only refused to use the boys' version in the published proceedings; they refused even to compare it with their own and strike a happy medium.

Much went wrong, some of it was ludicrous and exasperating, but the whole was a success. The conference did take place, people did meet and talk and learn, and it was a good thing.

Spy, junior grade

Because of my semi-official capacity I had more to do with the U.S. Embassy officials in Montevideo, just a little more, than the average American tourist. With a small number of exceptions I found them efficient and friendly. The finance office filled out all the necessary forms for me—all I had to do was to tell the nice lady at the counter how much I spent, and she had my expense money ready for me the next day—and on one or two occasions when a question I asked of the cultural attaché had to be relayed to Washington, he did so quickly, and the answer came quickly.

The first time that I reported to the Embassy I was taken on a handshaking circuit; one of the hands belonged to Robert Gahagan, by

appearance a tall, vacant-faced farm boy. We met again later, at a cocktail party, and we chatted a few minutes. Could I come to see him at
the Embassy sometime soon?, he asked. I said yes, of course, but I was
puzzled; he seemed to want to exude a mysterious cloak-and-dagger air.
When I went to see him, he kept on being mysterious for a while. Did I
know about this here mathematical conference that was coming off
soon? Did I know about it?!—why, hell, I practically started it myself.
Aha! Well, well. And who is coming and where all are they coming from?
Oh—lots of places—all of South America—mostly Argentina and Brazil,
and, of course, Uruguay, but some from as far away as Mexico and
Cuba. Uh-hm; and Peru? Yes, sure—Peru too. Who is coming from
Peru? Only one guy that I know of—the only mathematician Peru
seems to have—someone called Godofredo García.

Now we seemed to have arrived where Gahagan wanted to be. That's
what all the cloak-and-dagger was about; it seemed, according to
Gahagan, that old Don Godofredo was a dangerous communist. Is he
likely to bring anyone with him? *Exactly* when and where is he arriving
and exactly how long is he staying? Could he use our colloquium as a
sounding board for political propaganda? Whose idea was it to invite
him? Who made out the invitation list? Who can attend? I gave short
and truthful answers to all his questions, and left, as puzzled as when I
came.

Laguardia was not just puzzled when I told him all this, as I immediately did—he was amazed. (I saw no reason to treat the matter as confidential; I was certainly not asked to do so.) He knew Don Godofredo
quite well and never suspected him of being a communist. Besides,
García is old, quite wealthy, and a big shot in a fascist country (rector,
four-term dean, etc.)—it really wasn't very probable that he could be
an agent of the Kremlin.

I reported my conversation with Gahagan to Marshall Stone also;
here is his answer.

> "If Don Godofredo is a communist, we might as well give up the struggle
> right now! When I last saw him, in 1948, he was overjoyed at the con-
> servative revolution which put Odría in power and was spending half his
> time at the Presidential Palace palling around with the General.... Of
> course if anyone is in the business of suspecting Fascists, he might let
> his eye light on Don Godofredo.... If you get a chance to lay it on the
> line for the snoopers, I hope you'll do so."

A couple of other similar things happened. Once, for instance, I was
chatting with the cultural attaché, Allen, at the Embassy. He asked me
a few questions about the academic set-up; I was surprised at how little
he knew. He also asked me to do a little spy work, junior grade. It
seems that there was a student organization, Federación de Estudiantes

Universitarios del Uruguay, that is politically suspect. (I had never heard of it before Allen's description.) It was, in Allen's phrase, a "third-position outfit"—the expression was meant to indicate that they are just as anti-yankee as they are anti-commy. To illustrate for me the flavor of the phrase, Allen described Peronism as third-positionism too. Well then, there was recently an Inter-American conference and a few of the FEUU people went to Rio to attend it; mostly the FEUU seemed to consist of anarchists. Allen would have just loved to see the by-laws of the organization, and the names of the people who went to Rio—what could I do? I was too flabbergasted to say anything—I muttered politely that if the subject arose naturally in a conversation, I'd try to find out something, but I refused to promise anything.

After leaving Allen's office, I got angry. I didn't consider spy work my job—if that's what the State Department wanted me to do, they should have said so before they gave me the grant. I came to Uruguay to get away from the hysteria of the witch hunt, not to add to it.

On another occasion Allen claimed to have been informed that Laguardia was a communist, and what did I think about that? I said that I thought it was the damnedest nonsense I ever heard.

Small memories

"We're invited for 6:00, and it'll take at least a half hour to drive there", Laguardia said; "I'll pick you up at 5:30." He picked us up at 6:30. On another occasion he taught us the expression "hora Uruguaya" (Uruguayan time, as opposed to "hora Americana")— meaning the appointment that's always missed, the clock that's always slow, the delay that's chronic and predictable. "But I'm not like that", he assured us; "having lived in North America, I learned the advantages of being on time. When I make a date, I try always to be punctual —only major last-minute developments ever make me late." Like most of the world, Laguardia didn't see himself as others saw him—I had many appointments with him, and every one of them was preceded by a major last-minute development. After a few controlled experiments I devised an algorithm. "I'll pick you up at 5:30" meant, I decided, "start to shower, shave, and dress at 5:30—I'll pick you up somewhere between 6:00 and 6:30."

It was a Columbus day custom for the Uruguayan engineers to get together at a banquet, listen to a few after dinner speeches, welcome the recent licensees to their ranks, and then go home and sleep it off. Laguardia wangled an invitation for me, but he couldn't pick me up, so he gave me instructions. The dinner was scheduled for 1:00 p.m., but the preliminary vermouth would appear earlier, and he wanted to introduce me to a few people and allow time for at least a brief chat. "Just take a taxi", he said, "and time it so as to get there at 12:30; that's the customary time." I carefully timed my arrival for 12:40, but that was not late enough. That early, it turned out, only about 20 people were there, standing around, smoking and chatting. Sixty places were set, and the guests assigned to them all arrived eventually. Laguardia (who told me he'd meet me there at 12:30) arrived at 1:00, and, as people kept drifting in, we kept smoking and chatting. At exactly 2:00 o'clock we sat down to dinner and started to eat—we got up from the table at 4:00. Everything was pleasant—the food was worthy of the name of a banquet, the service was fast and polite, the wine glasses were filled with agreeable frequency, and the speeches were not unduly long. When it was all over, I went home to sleep it off, and I broke my almost invariable habit of tea in the middle of the afternoon—I hardly woke up in time for supper.

Much of social life in Uruguay revolved around the family. Family Sunday dinners and Christmas dinners, weddings and christenings, and family excursions to the forest or the river were the rule—evening parties or dinners for mere friends or visiting foreigners were the rare exception. Most of the time that I was "entertained" the occasion was a Sunday afternoon drive, or tea in a restaurant, or an evening movie followed by a drink and a snack in a restaurant.

Birth and death are the same everywhere, but the way in which humanity copes with the smaller aspects of day to day life—such as traffic and money and art and industry—vary tremendously with time and place.

In Uruguay there was a national statute ordering drivers to honk their automobile horns before every crossing. "How else is anyone to know you're coming?" There were no traffic lights, and traffic policemen were posted at only three or four of the major intersections downtown. What made me extra nervous was that cars were permitted to pass streetcars on either side, and that streetcars and buses would stop to pick up women only—for men they would slow down, enough (?) that they could be boarded on the run. But, to mention another side of the coin, it was quite safe to walk in town at 3:00 a.m.—thieves, robbers, and muggers either didn't exist, or, in any event, were a negligible problem.

Checking accounts existed, but they were a rarity. Salaries were paid in cash; you went to the appropriate window once a month and you had a large stack of wrinkled pesos counted out to you. Credit, and more generally trust in one's fellow human beings, was in short supply.

I went to see the movie *Hamlet*, with Lawrence Olivier in the title role. It was accompanied by Spanish subtitles, and, so as not to distract the audience with irrelevant noise, the sound was turned down almost all the way—Olivier was inaudible.

There was an all-pervasive shabbiness—the country gave the impression of being run down and coming out at the elbows. There were no dirty or dangerous slums—and there was nothing first class. The buildings with marble walls had paint peeling off the ceiling. In the restaurant that served crêpe suzettes, the waiter needed a shave and his white coat was soiled. Bathrooms usually had a faint odor of sewage hovering about them. The luxurious beaches were blessed with God's sunshine, but weeds were growing in the sand. The finely wrought leather briefcase had a defective buckle (proudly marked "Industria Uruguaya") that made it unusable, and the flimsy clasp on the native jewelry either couldn't be opened or else opened without encouragement when you were walking in the grass.

I went to Uruguay with great expectations, some of which were fulfilled and some not. Petty daily annoyances often overshadow the adventure and novelty and beauty and memorability of life, whether at home or abroad. Since I have the tendency to enjoy remembering the past, anticipating the future, and complaining of the present between them, I didn't enjoy Uruguay as much as I could have. By now I think fondly of it and wish it well; it gave me something and I left part of me there; I am glad I went, and I am glad I am not there now.

CHAPTER 11

The fabulous fifties

Back home

The 1950's were the most exciting and glamorous and productive years in the history of Chicago (where "Chicago" means, of course, the mathematics department of the University). That's a subjective view, but many share it, and it hasn't been shot down yet. The research atmosphere of Eckhart Hall, with its first class faculty and, if anything, better than first class student body, and its contrast with the languorous mediocrity of Montevideo were invigorating. I dived into it full of enthusiasm.

I started by teaching a course on topological groups, again. The last time I did that was the year before Uruguay, in the Autumn quarter of 1950. I have been conducting a continuing flirtation with the subject ever since Pontrjagin's book appeared, I am fond of it still, and I really loved the 1950 course. The class was largish: its slightly fluctuating membership ended up with 34 grades, including three R's and one bona fide, earned F. It had many excellent students, including, for instance, Mauricio Peixoto from Brazil and Andrew Wallace from Scotland; I based it mainly on one of the best books available at the time (André Weil's *L'intégration dans les groupes...*); and, even if I say so who shouldn't, I did a pretty good job of it. I worked especially hard preparing the (take-home) final exam. Since a look at it might convey the flavor of the course better than any sequence of adjectives, here it is—have a look. Can you pass it? I think I still can, but I'd have to rethink some of the questions before I could be sure.

TOPOLOGICAL GROUPS   EXAMINATION

In each of the following assertions, the word *group* means *topological group*. Discover in as many cases as you can whether or not the assertion is true and prove your answer. The difficulty of doing this for the n-th assertion is more or less an increasing function of n. The complete solution of seven problems will be considered passable work; ten is good; thirteen is excellent. Geniuses are expected to solve all fifteen. No solutions will be accepted after 11:59 a.m., Saturday, December 2, 1950.

1. A one to one, continuous homomorphism onto a compact group is open.

2. The group product of two compact sets of (Haar) measure zero, in a compact group, is a set of measure zero.

3. If to every element (except the identity) of a group there corresponds at least one character which does not take the value 1 at that element, then the group is locally compact.

4. If a locally compact, abelian group has no non trivial, closed subgroups, then it is finite.

5. A compact group is finite or uncountable.

6. Every connected, locally compact, non trivial group has a non trivial, closed, invariant subgroup.

7. If every element of a compact, abelian group is of finite order, then the orders of the elements are bounded.

8. A locally compact group is normal.

9. If every element (except the identity) of a compact group is of order two, then the group is the (topological) Cartesian product of groups of order two.

10. Every non trivial, abelian group has at least one non trivial character.

11. If every element (except the identity) of a non trivial group is of order two, then the group is not connected.

12. A measurable homomorphism from one locally compact group into another is continuous.

13. Hilbert space may be retopologized so that it becomes a compact group.

14. Every compact group has arbitrarily small, closed, invariant subgroups such that the quotient group is separable.

15. Every totally disconnected, compact group is the inverse limit of finite groups.

The 1952 course was similar but different. The class was smaller (nine students, including Jack Feldman and Ed Nelson), it was pitched more at the research level (and I gave no judgmental grades—everybody got a P), and it relied less on the classic texts than on some of the classic papers that came out in the 1940's (such as Raikov's "Harmonic analysis on commutative groups...", Cartan and Godement's "Théorie de la dualité et analyse harmonique...", and Segal's "The group algebra of a locally compact group").

The students I saw around Eckhart Hall in the post-Uruguay years, not only in my course on topological groups, but in other courses, some of them quite elementary, were a pleasure: it was a pleasure to see them be enthusiastic, to see them grow, and, before long, to welcome them to the ranks of the best active mathematicians in the country. I met Harold Widom in those days, and Ray Smullyan; at about the same time Moe Schreiber and I were beginning to talk about his thesis. I had Don Ornstein in several courses and he began to learn ergodic theory from me (but wrote his thesis with Kaplansky on algebra). He did pretty well in general topology, and very well in algebra (the course on Galois theory); in complex function theory he wasn't so good. Didn't care?

Speaking of ergodic theory: I met Larry Wallen in an ergodic summer course (1955), and I am proud to say that I taught measure theory to Arunas Liulevicius, in the usual course, and John Thompson, in a special reading course. John had already drawn the straight line between himself and group theory, and he paid attention to other things only temporarily, when he had to, or cursorily, as momentary relaxation. (In the latter manner he found the simple, elegant proof that the unilateral shift has no square root—his proof made it possible to forget the tricky one that Juan Schäffer and I worked out.) I taught baby algebra to Steve Wainger (who became a sharp analyst), set theory to Jimmy Milgram (a leading topologist), and topology to Rob Kirby and Bob Solovay (one of them did well, and the other just so-so).

It was good to be home.

Is formal logic mathematics?

In the 1950's I set out to revolutionize logic. I didn't succeed (not yet!), but it wasn't for want of trying.

Logic has always fascinated me. In the common-sense sense of the word logic was easy; I had no inclination to commit the obvious errors of reasoning, and long chains of deductions were not intimidating. When I was told about syllogisms, I said "sure, obviously", and didn't feel that I learned anything. The first genuine stir in my heart toward logic came from Russell's popular books, and, just a few weeks later, from Russell and Whitehead: the manipulations of symbolic logic were positively attractive. (I liked even the name. "Symbolic logic" still sounds to me like what I want to do, as opposed to "formal logic". What, I wonder, is the history of those appellations? At a guess, the old-fashioned word wanted to call attention to the use of algebraic symbols in place of the scholastic verbosity of the middle ages, and the more modern term was introduced to emphasize the study of form instead of content. My emotional reaction is that I like symbols, and, in addition, I approve of them as a tool for discussing the content—the "laws of thought"—, whereas "formal" suggests a march-in-step, regimented approach that I don't like. One more thing: "symbolic" suggests other things, "higher" things that the symbols stand for, but "formal" sounds like the opposite of informal—stiff and stuffy instead of relaxed and pleasant. All this may be a posteriori—but it's the best I can do by way of explaining my preference.)

I liked symbolic logic the same way as I liked algebra: simplifying an alphabet-soup-bowl full of boldface p's and q's and or's and not's was as much fun, and as profitable, as completing the square to solve a quadratic equation or using Cramer's rule to solve a pair of linear ones. As I went on, however, and learned more and more of formal logic, I liked it less and less. My reaction was purely subjective: there was nothing wrong with the subject, I just didn't like it.

What didn't I like? It's hard to say. It has to do with the language of the subject and the attitude of its devotees, and I think that in an unexamined and unformulated way most mathematicians feel the same discomfort. Each time I try to explain my feelings to a logician, however, he in turn tries to show me that they are not based on facts. Logicians proceed exactly the same way as other mathematicians, he will say: they formulate precise hypotheses, they make rigorous deductions, and they end up with theorems. The only difference, he will maintain, is the subject: instead of topological notions such as continuity, connectedness, homeomorphism, and exotic spheres, they discuss logical notions such as recursion, consistency, decidability, and non-standard models.

No. My logician friend misses the point. Algebraic topology is intellectually and spiritually nearer to logic than to organic chemistry, say, or to economic history—that much is true. It makes sense to house the logicians of the university in the department of mathematics—sure. But the differences between logic and the classical algebra-analysis-geometry kind of "core mathematics" are at least as great as the analogies on which the logician is basing his argument. ("Core mathematics", by the way, is a recently invented weasel phrase whose users think that by frowning on the expression "pure mathematics" they can make the pure-applied chasm go away.)

He argues well, that logician, and he has a lot going for him. Certainly every mathematician finds the famous accomplishments of Gödel and Cohen, about the necessary existence of undecidable propositions and about the consistency status of the axiom of choice and the continuum hypothesis, both fascinating and admirable. On a less spectacular level, logicians have shown us that there are unsuspected relations between the Peano axioms and Ramsey's theorem, and both of those subjects are indubitably parts of "core mathematics". And, finally, logicians point proudly to a handful of mathematical theorems whose first proofs used the techniques of formal logic; the one nearest to my work is the Bernstein–Robinson result concerning invariant subspaces of polynomially compact operators.

All that is true, but it is not enough to alleviate the mathematician's discomfort. Why not? A partial answer is that, however admirable some of the accomplishments of the logicians may be, the mathematician doesn't need them, cannot use them, in his daily work. The logic proofs of mathematics theorems, all of them so far as I know, are dispensable: they can be (and they have been) replaced by proofs using the language and techniques of ordinary mathematics. The Bernstein–Robinson proof uses non-standard models of higher order predicate languages, and when Abby (Abraham Robinson's nickname) sent me his preprint I really had to sweat to pinpoint and translate its mathematical insight. Yes, I sweated, but, yes, it was a mathematical insight and it could be translated. The paper didn't convince me (or anyone else?) that non-standard models should forthwith be put into every mathematician's toolkit; it showed only that Bernstein and Robinson were clever mathematicians who solved a difficult problem, using a language that they spoke fluently. If they had done it in Telegu instead, I would have found their paper even more difficult to decode, but the extra difficulty would have been one of degree, not of kind.

As long as I am on the subject, this is the time to put into words what I suspect is the role of non-standard analysis in mathematics. It is a touchy issue: for some converts (such as Pete Loeb and Ed Nelson), it's a religion, and they get touchy if someone hints that it has imperfections. For some others, who are against it (for instance Errett Bishop), it's an equally emotional issue—they regard it as devil worship. To most mathematicians it is a slightly worrisome puzzle: is there really something there?—do we have to learn it? My suspicion is that, yes, there is something there—it is a language that is, for those who can use it without stuttering, a convenient tool in studies of compactness. But, if I am right, that's all it is: a language, not an idea, and a rather sharply focused one at that. Here is a somewhat unfair analogy: Dedekind cuts. It's unfair because it's even more narrowly focused, but perhaps it will suggest what I mean. No, we don't have to learn it (Dedekind cuts or non-standard analysis); it's a special tool, too special, and other tools can do everything it does. It's all a matter of taste.

As for the spectacular theorems—Gödel and Cohen—we admire them, but they haven't changed our work, our philosophy, our life. If someone should succeed in proving that the Riemann hypothesis is undecidable, we'd be shocked—as shocked as we were when the parallel postulate was proved undecidable—but we'd recover. We'd probably go on to discover and study non-Riemannian number theories, and live happily ever after. In the meantime, when the subject of the next colloquium speaker is announced as an application of the second order predicate calculus, we go to listen—politely, respectfully, but not eagerly, not hopefully. We don't think we're likely to learn something; at best we'll be entertained, at worst worried. When the logician says "it's all the same, it's all mathematics", we feel insecure and humble—we don't know how to refute him—but, at the same time, we feel sure that what we are about to hear is not, cannot be, the same as a talk on Brauer groups or Baire functions.

Both the logician and, say, the harmonic analyst, look for a certain kind of structure, but their kinds of structures are psychologically different. The mathematician wants to know, must know, the connections of his subject with other parts of mathematics. The intuitive similarities between quotient groups and quotient spaces are, at the very least, valuable signposts, the constituents of "mathematical maturity", wisdom, and experience. A microscopic examination of such similarities might lead to category theory, a subject that is viewed by some with the same kind of suspicion as logic, but not to the same extent.

Ah, there's a clue: the microscopic examination. The logician's attention to the nuts and bolts of mathematics, to the symbols and words (0 and + and "or" and "and"), to their order ($\forall\exists$ or $\exists\forall$), and to their grouping (parentheses) can strike the mathematician as pettifogging—not wrong, precise enough, but a misdirected emphasis. Here is an analogy that may be fair: the logician's activity strikes the modern analyst the way epsilons and deltas might have struck Fourier.

It's been said that the biggest single step in the logic of the last 200 years was the precise explication of the concept of proof; as a tentative parallel let me advance the thesis that the biggest single step in the analysis of the last 200 years was the precise explication of the concept of continuity. But oh, no, says our hidebound mathematician; the two situations are not alike at all. Epsilons and deltas were a step forward, they made it possible to avoid many errors, and they opened new vistas— they allowed us to call many more functions continuous than we had dreamt of before. But as for formal proofs, in the sense of certain admissible finite sequences of well-formed formulas—what, besides technicalities within proof theory itself, have they done for us? A clever graduate student could teach Fourier something new, but surely no one claims that he could teach Archimedes to reason better.

Here is another analogy. A Shakespeare scholar might want to know about Shakespeare's art, characterization, depth of insight, understanding of history, and technique of dramatic construction; or he might want to study Shakespeare's vocabulary, grammar, and use of similes and metaphors. Both scholars would proceed, let us presume, in a correct academic manner, using the available literature, and not jumping to unjustified conclusions—but, I suspect, in private they would think of one another with less than total charity. The litterateur is vague and unscholarly, says one, while the grammarian is an uncultured fussbudget, says the other.

Are they both wrong? The litterateur might grudgingly admit that his colleague's observations about the probable authorship and date of a disputed sonnet are of interest, and, of course, the word-counter knows that if it weren't for the poetic merit of Shakespeare's work, no one would want to know the computerized statistics of Shakespeare's vocabulary. Are they both right?

I didn't say all this to convince anyone of anything—I said it in an attempt to describe the reaction, as I see it, of many mathematicians, myself included, to formal logic. It is not a question of right or wrong. It is a matter of native competence, professional upbringing, and mathematical taste; the fact is that, for whatever reason, a certain negative reaction does exist.

Boolean logic

An exposition of what logicians call the propositional calculus can annoy and mystify mathematicians. It looks like a lot of fuss about the use of parentheses, it puts much emphasis on the alphabet, and it gives detailed consideration to "variables" (which do not in any sense vary). Despite (or because of?) all the pedantic machinery, it is hard to see what genuine content the subject has. Insofar as it talks about implications between propositions, everything it says looks obvious. Does it really have any mathematical value?

Yes, it does. If you keep rooting around in the library, trying to learn about the propositional calculus, you will find more and more references to Boolean algebra. That's a pretty subject, one that has deep, hard, and still unsolved problems, but, at the level that is pertinent to logic, it is easy—almost trivial. The most helpful books that I found were Hilbert–Ackermann, Paul Rosenbloom, and Kleene's first book on metamathematics, the big one; and bit by bit the light dawned. Question: what is the propositional calculus? Answer: the theory of free Boolean algebras with a countably infinite set of generators.

Imagine that you are a mathematician who just doesn't happen to know what a group is, and that, consequently, you go to an expert and ask for a definition. Imagine next that the expert, instead of giving you the usual axioms, starts talking about finite sequences of letters, the operation of concatenation that makes new sequences out of old, and equivalence relations that lump many sequences together. If both the expert and you are sufficiently patient, you might presently learn what a free group is, and then, via generators and relations, what a group of any kind is. That's what the propositional calculus does for Boolean algebras: it laboriously constructs the free ones, and then sometimes, via so-called non-logical axioms, it shows how to get all the other ones as well.

To impose a set of relations on (the elements of) a group is in effect the same as to specify a normal subgroup and divide out by it. The same thing is true in Boolean theory: to superimpose a set of non-logical axioms on a Boolean algebra is in effect the same as to specify a Boolean ideal and divide out by it. (Algebraists are used to ideals, but logicians are more fond of the dual concept of filter. In logic, filters, which arise in the algebraic study of provability, are indeed more natural than ideals, which have to do with refutability.) Group theory may be considered to be the study of ordered pairs (F, N), where F is a free group and N is a normal subgroup of F; similarly a general (not necessarily "pure") propositional logic may be considered to be an ordered pair (B, I), where B is a free Boolean algebra and I is an ideal in B.

The parallelisms, analogies, between groups and Boolean algebras, and also between the customary approach to the propositional calculus and free Boolean algebras, go much further and deeper than the preceding comments. Consider for instance "the deduction theorem", which says that an implication $p \Rightarrow q$ can be deduced from a set of axioms if and only if the conclusion q can be deduced from the enlarged set obtained by adjoining p to the axioms. Dualized algebraic translation: an inequality $q \leqq p$ is true modulo a Boolean ideal if and only if q belongs to the ideal generated by p together with the given ideal. Another example: a logic (B, I) is "consistent" if I is "small" (meaning that it is a proper ideal, not as large as B—not everything is refutable), and it is "complete" if I is "large" (meaning that it is maximal—everything not refutable is provable).

The illustrative concepts mentioned in these examples are expressed in terms of the structure theory of the underlying algebraic system; logicians call them "syntactic". There are also "semantic" concepts, which have to do with representation theory. "Truth tables", for instance, are nothing but the clumsiest imaginable way of describing homomorphisms into the two-element Boolean algebra, and that's what representation theory is always like: choose the easiest algebras of the kind under study, and consider homomorphisms from arbitrary algebras into the easy ones. (A standard, non-trivial example is the theory of locally compact abelian groups, where the role of the "easiest" ones is played by just one group, the circle.) The algebraic analogue of the logical concept of "semantic completeness" (everything not refutable is satisfiable) is semisimplicity (every group element different from the identity is mapped onto a number different from 1 by some character).

I found many of these beautiful, exciting, and illuminating analogies, sometimes only implicitly, in the books I was devouring, I guessed some of them, and a few of them I had to discover myself, stumbling painfully from one blind alley to the next. I was looking for a dictionary, I believed that one must exist, a dictionary that would translate the vague words that logicians used to the precise language of mathematics. Yes, vague, or, at best, precise in a forced, ad hoc manner. When I asked a logician what a variable was, I was told that it was just a "letter" or a "symbol". Those words do not belong to the vocabulary of mathematics; I found the explanation that used them unhelpful—vague. When I asked what "interpretation" meant, I was answered in bewildering detail (set, correspondence, substitution, satisfied formulas). In comparison with the truth that I learned later (homomorphism), the answer seemed to me unhelpful—forced, ad hoc. It was a thrill to learn the truth—to begin to see that formal logic might be just a flat photograph of some solid mathematics—it was a thrill and a challenge.

The road to polyadic algebras

The challenge was to discover the solid mathematics of which the predicate calculus and its special "applied" versions are photographs—to find out what plays for propositional functions the role that Boolean algebras play for propositions.

If propositions, whatever they may be, are most cleanly studied via Boolean algebras—if, to put it more radically, what a proposition "really" is is just an element of a Boolean algebra—then, presumably, the even more mysterious propositional functions are functions with values in a Boolean algebra. If "Socrates is mortal" is a proposition (or, more carefully, a sentence representing a proposition—an element of the equivalence class that a proposition really is), then a propositional function (of x, say) must be something like "x is mortal". But the degree of complication that that indicates is not yet great enough. Indeed: if "Socrates died before Plato" is a proposition, then "x_1 died before x_2", considered as a function of two arguments, should also be regarded as a propositional function. In mathematical contexts functions of three, four, or, for that matter, any finite number of arguments are needed—the number of available arguments must be potentially infinite. (I must keep remembering to avoid the word "variable" here—that way confusion lies.)

Very well then: the prototypical propositional function is a function from a domain X (integers, Greek philosophers, whatever) to a Boolean algebra B. No, that's not general enough: it should have been from X^2 to B (for functions of two arguments), or from X^3 to B, or, to be prepared for anything, from X^I to B, where I is an arbitrary index set, preferably infinite.

What in this model is a "variable"? It is not an element of X, or X^2, or X^I: that's a constant (such as 7, or Socrates). Logicians in their texts have been telling the rest of us what a variable is for quite some time, but they didn't have the extensional mathematical courage that Russell had when he defined 2 to be the class of all pairs. They used a linguistic analogue instead: a variable, they said, is like a pronoun. In the classic proverb "for all h, if h hesitates, then h is lost" (sometimes more pithily expressed as "he who hesitates is lost"), the "variable" h that enters is a pronoun; its role in the pithy expression is played by "he".

Suppose, for an example, that X is a set of Greek philosophers, and that p is the function from X^2 to an appropriate Boolean algebra B, defined so that $p(x_1, x_2)$ has the value "x_1 died before x_2". The use of symbols such as x_1 and x_2 here is a mathematical convention, but most mathematicians would be very hard put indeed to say just what x_1 and x_2 are. (They might say "variables", but they would soon admit that they don't know what that word means.) An instruction such as: "evaluate p at the ordered pair (Socrates, Plato)" makes unambiguous sense and uses mathematically defined or definable terms only. The instruction could have been expressed this way: "evaluate p at the pair whose first coordinate is Socrates and second Plato". Equivalently: "put coordinate number 1 equal to Socrates and 2 to Plato". Again, telegraphically but clearly: "put 1 equal to Socrates and 2 to Plato". The symbols "1" and "2" here have a clear and fixed meaning; they are the elements of the index set I ($=\{1, 2\}$) that is being used. They are the "names" of the variables—why not be courageous and say they *are* the variables? And that's what, in effect, logicians do. When they say "consider a set I of variables", they mean "consider a set I whose elements will be used as indices to distinguish coordinates in the Cartesian product X^I from one another".

Even after we agree that to study propositional functions we should have a set I of variables, a domain·X, and a value-algebra B, and that the objects of study are functions from X^I to B, we must still decide which operations on such functions are deserving of axiomatic abstraction. Repeated looks at the logical literature, and introspective mathematical daydreaming, suggest two important kinds of operations: substitutions, which change "$x_1 \leqq x_2$", say, into "$x_2 \leqq x_1$" or even into "$x_1 \leqq x_1$", and quantifications, which change "$x_1 \leqq x_2$" into, for instance, "$x_1 \leqq x_2$ for some x_2". They are equally important, but not equally difficult; the algebraic and geometric properties of quantification are for most people less well known, and therefore harder.

Suppose that $I = \{1, 2, 3\}$, X is the set of real numbers, and B is the Boolean algebra of subsets E of X^3 (or, equivalently, the Boolean algebra of propositions of the form "$(x_1, x_2, x_3) \in E$"). Suppose, to be specific, that E is the closed unit ball $\{(x_1, x_2, x_3): x_1{}^2 + x_2{}^2 + x_3{}^2 \leqq 1\}$, and that p is the corresponding propositional function (so that $p(x)$ is, for each x in X^3, the assertion "$x \in E$", true for some x's and false for others). Then "$p(x)$ for some x_3", a possible abbreviation for which is $\exists_3 p$, is the propositional function q of two arguments such that $q(x_1, x_2)$ is, for each (x_1, x_2) in X^2, the assertion "$(x_1, x_2, x_3) \in E$ for some x_3". Geometrically p is (corresponds to) the unit ball; what, geometrically, is

q? Answer: the infinite cylinder whose central axis is the x_3-axis and whose cross section in the (x_1, x_2) plane is the unit disc.

Motivated by examples such as this, Tarski and his school, in the early 1950's, began to study the algebra of quantification, and introduced structures that they called cylindric algebras. Similar work was being done in Poland at the same time, some in contact with Tarski and some independently; notable representatives of the Polish school at that time were Rasiowa and Sikorski. Motivated, possibly, by the desire to put forth a perfect finished product, Tarski and his collaborators published almost nothing about cylindric algebras for many years. If you wanted to learn what they had accomplished, you could read the two one-paragraph abstracts in the 1952 Bulletin or you could go on a pilgrimage to Berkeley. Sooner or later I did both.

The heuristic meditations of some of the preceding paragraphs (the ones about propositional functions and the operations of importance on them) are typical of a certain kind of pre-research thinking. That's the kind of thinking that Pontrjagin (probably) did when he examined finite abelian groups, the circle group, the line, tori, vector groups, the Cantor set, and the groups obtainable from them by the formation of direct sums, subgroups, and quotient groups—the kind of thinking that led him to the "right" abstract concept, namely locally compact abelian groups. Aha! (he might have said)—that's it—that must be the right general context—now let's see whether a duality theorem can indeed be formulated and proved in that context.

Inspired by my desire to make honest algebra out of logic, and by what I could learn of the Tarski secrets, I began in 1953, and continued for about six or seven years, a study of polyadic algebras. (Besides, it was time to change fields again, and six or seven years later it was time to go on to the next change.) The name was suggested by the presence of many ("poly") quantifier operators (such as $\exists_1$, $\exists_2$, and $\exists_3$); the important special case in which there is only one such operator, monadic algebras, is the algebraic approach to Aristotelian syllogistics, and the important degenerate case in which no such operator occurs is the familiar Boolean approach to the formal propositional calculus. The program was ambitious. I wanted to make algebra out of *all* logic, and, in particular, I wanted to understand the algebraic meaning of the two famous Gödel results, the completeness theorem and the incompleteness theorem. Did I succeed?

All logic and all mathematics

No, I didn't succeed in making algebra out of all logic; yes, I did succeed in getting some insight into how that could be done. I still think it could and should be done, and I hope it will be done, but not by me, not any more; a new idea is needed. It was work, hard work, as far as research goes almost full time work for six years of my life, and, as far as research goes, quite possibly the best work of my life.

The algebraic axioms for existential quantification are simple, but it took me months to absorb them into my bloodstream, to understand them intuitively and emotionally as well as merely technically. I wrote them on a small card, which I then carried in my wallet and pulled out frequently to look at during those many minutes that we all waste every day—waiting for my lunch date to arrive five minutes late, or waiting with my mouth blocked open for the dentist to come back. Is there anyone who still doesn't know what those axioms are? Very well: an existential quantifier is a mapping $\exists$ of a Boolean algebra into itself, such that

$$\exists 0 = 0,$$

$$p \leq \exists p,$$

and

$$\exists(p \wedge \exists q) = \exists p \wedge \exists q$$

whenever p and q are in the algebra.

I lived and breathed algebraic logic during those years, during faculty meetings, during my after-lunch lie-down, during concerts, and, of course, during my working time, sitting at my desk and doodling helplessly on a sheet of yellow paper. On one such doodling occasion I was impressed, again, by the unity of mathematical thought. I needed to straighten out a seemingly elementary Boolean identity, but I couldn't—I was stuck— I didn't see the trick. Just to doodle in a different way, with no directed motive, or so I thought, I pulled out a book from among the ones on my desk—my little introductory book on Hilbert space—and I flipped through some of its pages at random, or so I thought. Bingo!, the pay-off: there on p. 58 of the Hilbert space book was the Boolean trick I needed for algebraic logic, an elementary argument that I would have sworn I had never seen before. (The argument proves that a projection-valued measure is multiplicative.)

The theorem that gave me the most trouble was the climax of Algebraic Logic II. That theorem was the culmination of the preliminaries that the theory needed, the justification of the axioms; it asserts that the models that motivated the definition of polyadic algebras are indeed sufficient to represent the polyadic algebras that are pertinent to logic. I remember the evening when I got over the last hurdle. It was 9 o'clock on a nasty, dark, chilly October evening in Chicago; I had been sitting at my desk for two solid hours, concentrating, juggling what seemed like dozens of concepts and techniques, fighting, writing, getting up to walk across the room and then sitting down again, feeling frustrated but unable to stop, feeling an irresistible pressure to go on. Paper and pencil stopped being useful—I needed a change—I needed to do something—I pulled on my trenchcoat, picked up my stick, and mumbling "I'll be back", I went for a walk, out toward the lake on 55th street, back on 56th, and out again on 57th. Then I saw it. It was over. I saw what I had to do, the battle was won, the argument was clear, the theorem was true, and I could prove it. Celebration was called for. It was almost 10 p.m., and the grocery store I just passed was getting ready to close. A nasty drizzle was coming down and the puddles reflected lights and looked beautiful. Among the outdoor displays at the grocery store, about to be taken inside, were some scraggly flowers. I looked in my pockets and found 98 cents. I asked the clerk whether I could buy some of those scraggly leftovers for that, and he grinned cooperatively—sure, they cost a buck, you can have them for two cents less. I spent my 98 cents and took the flowers home to my wife. Damp and happy I proposed that we have a beer to celebrate.

The representation theorem was the climax of my work on algebraic logic, but it was not the end. The representation theorem made it possible to recognize the Gödel completeness theorem as the semisimplicity theorem for polyadic algebras, but the algebraic treatment of the Gödel incompleteness theorem (that's the famous one about "you can't ever prove everything") remained to be done. Almost everything I was doing in the middle 1950's had that for its main purpose. Published papers (including Algebraic Logic III and IV), masses of manuscripts never prepared for publication (Algebraic Logic V, 375 pages of it still sitting in my filing cabinet), Ph.D. students (Galler, Le Blanc, Daigneault), lecture courses and other spin-offs (such as an AMS Summer Institute in logic, and such as a little book on Boolean algebras)—it was all aimed at trying to capture the algebra of Gödel's quintessentially formal inspiration.

I know it's there, it's waiting to be found, and Algebraic Logic V cleared away some of the underbrush. The formalism (the free-group-like approach) is dispensable; the question is about the recursive number theory. To my taste that's one of the most brilliant ideas in mathematics, and one of the ugliest and least elegant. I went all through it in Algebraic Logic V, via the concept of "Peano algebra", a special kind of polyadic algebra that can mirror the number-theoretic technique characteristic of Gödel's proof. The mirror is perfect; every step of the proof can be expressed in every Peano algebra. The conclusion is that the ideal of refutable propositions or, equivalently, the filter of provable propositions in a free Peano algebra is not maximal. It is natural to form the quotient modulo that ideal and to use the language of universal algebra to describe the result. Since an algebra with no non-trivial ideals is customarily called "simple", and since the Gödel theorem is, in effect, the statement that in the quotient algebra just mentioned non-trivial ideals do exist, the algebraic formulation of Gödel's crowning achievement is that number theory is not simple.

Polyadic algebra can do more than number theory: it can "do", which means copy, imitate, mirror in *all* detail, every applied predicate calculus, and, in particular, set theory. There are mathematicians who believe that all mathematics is a logical consequence of set theory, or, in any event, that that is true of all extant mathematics; I am one of them. Set theory, in turn, is a logical consequence of the Zermelo–Fraenkel axioms. Since it is, in principle, easy to formulate those axioms in the language of polyadic algebra, it follows that the kind of algebras that would be specified by such a formulation, let me call them ZF algebras for now, would be a perfect mirror of all extant mathematics (and possibly of all conceivable mathematics as well).

What's going on here? Using the ordinary language and procedure of mathematics, nothing more involved than groups with operators, I define a special kind of structure—polyadic algebras—and, as instances of that kind of structure, I encounter the ZF algebras. Everything that I did to get there can be mirrored, completely mirrored, with nothing left out, within ZF algebras; in particular I could, if I wanted to, discuss groups with operators within such algebras, and I could, if I wanted to, discuss ZF algebras within such algebras. The latter discussion would stand to the former in exactly the same relation as the former stood to the establishment of the theory of ZF algebras in the first place. The ZF algebras discussed within ZF algebras would be a perfect mirroring of a perfect mirroring of all mathematics. Do you see the Quaker Oats phenomenon—the familiar cereal box with a picture of a Quaker holding the familiar cereal box that has on it a picture of a Quaker, etc.? It's there, and it might give you a feeling of discomfort—are we in the presence of that unpleasant disease called "infinite regress"? I don't know. It doesn't worry me, it doesn't frighten me, but I admit that I don't see what implications the picture has. Is there anything wrong with this kind of infinite regress? Or, alternatively, is there anything good about it—does it provide a view that I didn't even know I wanted to see? I don't know, but I do think that the phenomenon is a lot of fun to contemplate.

Logic students and logicians

Of my three Ph.D. students in logic, I have already mentioned the first one, Galler (1955), whom I inherited from Marshall Stone. His thesis studied the relation between polyadic algebras and cylindric algebras and showed that those systems accomplished, each in its own way, what they set out to accomplish—namely, to represent the "pure" first-order predicate calculus and the predicate calculus with equality, respectively. The last one was Aubert Daigneault (1960), who was my student in a strange sense only, but I am glad he was my student in some sense.

I met Daigneault during one of my Institute years, 1957–1958, when he was a graduate student in Princeton working with Church. What he wanted to do, however, was far from Church's usual interests. In that respect he was an unusual Ph.D. candidate—he found his own thesis topic (having to do with algebraic concepts and constructs such as auto-morphisms and tensor products). Church was willing to sponsor the thesis, but, not feeling at ease with the subject, he asked me to be its un-official reader and de facto judge. I said yes, of course, and, when the time came, I read Daigneault's work much more carefully than I usually read theses whose growth I had been watching for years. It was a good thesis, more in the spirit of pure algebra than my own papers, and I learned a lot from it. It was not all algebra—it made contact with the work of the philosophical logicians as well as the formal ones. One of its accomplish-ments, for instance, was to formulate a result of Beth's on the theory of definition as the analogue of a known fact in group theory: in a free pro-duct of polyadic algebras with amalgamation the intersection of the factors is precisely the amalgamated part. I was impressed. I wrote a careful and detailed critique for Church's use, and Daigneault's, and, with less pain than usual, I could add one more name to my list of disciples.

Sometimes a student chooses a Ph.D. supervisor for the "wrong" reason and then finds himself (perhaps to his own pleasant surprise) doing something quite different from what he expected. That's what happened to Leon Le Blanc, my only real all-my-own logic student. I gave a lecture at the University of Montreal once, and this bright young man was introduced to me. He was the best student in their real variable course that year, he enjoyed measure theory and had even made a small publishable observation about it—I was a known measure-theorist, so could he come to Chicago to work with me? Sure, I said, why not.

A few months later, when he finished his undergraduate work, Leon came to Chicago, but of course he wasn't ready to start a thesis—there was a lot of mathematics to learn first, courses to take, requirements to satisfy. By the time he was ready for a thesis, I wanted to have nothing to do with measure theory—I was deeply embedded in polyadic algebras. Result: Leon wrote a thesis on polyadic algebras. The change was not as abrupt as the telling of it, and the change was voluntary on his part. He first knew about me as the author of the book he studied in Montreal; when we met and got to like each other he came to know me in a different capacity, and my interest in and enthusiasm for algebraic logic rubbed off on him. He wasn't making a sacrifice: he was swimming with the stream. Besides (as I often tell advisees) what you write your thesis about doesn't commit you to a lifetime contract. You spend a couple of years on your thesis, and you had better not try to do anything else during those years—but once they are finished, you can (and you should!) do something else.

Leon's thesis was about non-homogeneous polyadic algebras. The idea is that the set of "variables" comes divided into different "sorts" intended to vary over different parts of the structure under study. Thus, for example, in Peano arithmetic some of the variables might be used for integers and others for sets of integers, and in pure second-order logic there might be "individual" variables and "functional" variables. Leon extended a substantial part of polyadic theory to non-homogeneous algebras, and then, instead of going on to do something else, continued to work on algebraic logic for several years. He stayed with the subject till his untimely death (from a brain tumor) and made lasting contributions to it.

I kept trying to spread the gospel not only among my own Ph.D. students, but everywhere I could, every chance I got. I spent two quarters in Berkeley in 1953, and gave a series of lectures on polyadic algebras. For some reason (curiosity?, courtesy?) Gerhard Hochschild was a faithful attendant. During the same terms he was lecturing on Lie groups, and, of course, I was one of his most faithful attendants. A little later, in the spring of 1955, I gave a course on algebraic logic in Chicago. It was a surprisingly large class, considering the subject—it had 17 students. Moe Hirsch was one of them (a topologist of renown), as was Jack Towber (an algebraic number theorist), and Mike Morley (the only pro in the lot—a genuine logician).

There weren't many conferences, jamborees, colloquia, and workshops in those days, and the few that existed were treasured. The AMS Summer Institutes were especially effective and prestigious, and I decided that it would be nice to have one in logic, especially if it were at least partly algebraic. It was a brash decision. I had no stature as a logician, I had no clout, I wasn't a member of the in-group; all I had was the brass (willingness to stick my neck out) and the drive (willingness to do the spade work). I phoned Tarski, I phoned Kleene, I phoned Rosser; I made up lists of names and lists of subjects; I wrote circular letters, I tried to design plausible budgets, and I filled out application forms. It worked, because Tarski's and Kleene's and Rosser's fame and high standards made it work—but I started it and I was proud and glad when it worked. It was a great summer in Ithaca in 1957. About 75 or 80 people participated in the Summer Institute, including Michael Rabin (who spoke about finite automata), Haskell Curry (who spoke about combinatory logic—what else?), Georg Kreisel (Gödel's interpretation of Heyting's arithmetic), and Martin Davis and Hilary Putnam (who reported on joint work on Hilbert's 10th problem). I talked about polyadic algebras (what else?), and the short summary of my talk in the Proceedings volume of that Summer Institute is, in fact, the only publication in which I gave some technical clues to my hopes and dreams about Peano algebras.

The proceedings volume ended with the "Recessional":

> If you think that your paper is vacuous,
> Use the first-order functional calculus.
> It then becomes logic,
> And, as if by magic,
> The obvious is hailed as miraculous.

The passport saga

The middle 1950's were the best part of my life and the worst. As a teacher and as a contributor to the research literature I was well established, and my work on algebraic logic was giving me a lot of satisfaction and a feeling of accomplishment. I crossed the magic boundary of 40 in 1956. I was getting more and more involved in the sort of mathematical community service that was soon to become the main part of my work, almost my raison d'être. At the same time the sins and follies of my youth were catching up with me and giving me a very hard time indeed. My life was not in danger, and neither were my health and my physical comfort, but I was psychologically more upset and insecure then than at any other time before or since.

My most conspicuous and legally documentable sins were to have signed some petitions about Israel Halperin, to have questioned the right of the F.B.I. to run security checks on mathematical fellowship applicants, and to have refused to be a spy in Uruguay. There were other and older and vaguer sins that belonged to my early 20's: I joined too many of the wrong organizations and knew too many of the wrong people.

Is Halperin is an old man now, even older than I, living in quiet retirement from his job at the University of Toronto. He was one of my predecessors as von Neumann's assistant, and he has been a fervent admirer of von Neumann all his life. He has also been a fiery, radical, activist leftist, and, for that reason, he has been in serious trouble with the authorities of both Canada (of which he is a native) and the U.S. Along with many other mathematicians, I signed a couple of petitions concerning Halperin. The time was in the 1940's, and I have forgotten exactly what the petitions asked, but, roughly, they had to do with getting him what we called a square deal. They might have been addressed to the Canadian authorities and asked for a fair trial, or they might have been addressed to the American authorities and asked for fair consideration of his request to enter the U.S. for one year so that he could accept the fellowship he had been awarded. I hardly knew him; we met only once before that time (around 1939, when he was leaving Princeton and I was just arriving). The second time we met was at the 1953 summer meeting of the AMS in Kingston, Ontario. We talked for ten minutes, about von Neumann and about the heat wave.

As for the F.B.I. security checks, they had to do with fellowships awarded by the Atomic Energy Commission for the study of pure science. The fellows were to have absolutely no access to classified information, and the description of the fellowships stated explicitly that it was not intended that at the end of their training they become "atom scientists". Despite this, the F.B.I. was instructed to conduct a loyalty check on each prospective fellow. The procedure seemed inappropriate to me and to some of my colleagues at Chicago; we thought it established dangerous precedents, tending to upset the traditional freedom of scientific inquiry. To express how I felt, I typed up a short statement (less than 200 words) and the next time an F.B.I. agent came to me asking about an applicant I gave him a copy. The climax of the statement said that in protest against the investigation of candidates for fellowships in mathematics, I refuse to cooperate in such loyalty checks "except for cases in which it is clear that access to classified information is involved".

All this was in 1951 or before, and, except for chatter at lunch and at tea, it took a total of less than an hour of my life. By 1952 I had forgotten all about it, and I was busy preparing for the 1954 International Congress in Amsterdam. The year 1954 was to be the 25th since I left Europe, without ever returning. I was eager to go, to participate in the Congress, to spend a few months in England, and, quite possibly, to visit Hungary.

H. R. Pitt was an ergodic theorist at the University of Nottingham, and he and I started to correspond about the possibility of my visiting there for a few months after the Congress. In the late autumn of 1952 I went into high gear: I applied for whatever might become available in Nottingham, and, at least for a short time, in Cambridge; I applied for a Fulbright grant, and for a travel grant (from the NSF, via the Society); I reserved a stateroom on the S. S. Ryndam, I applied for a passport, and I enlisted Pitt's aid in finding me a suitably large and suitably warm place to live in from September 1954 till about March 1955.

Everything succeeded—well, almost everything. Pitt found a house; he described it by telling about the number of fireplaces on the ground floor and adding that the upstairs was heated by the Gulf Stream alone. More than a year in advance I had a firm offer from Nottingham (£500, labelled as contribution to expenses), an official and definite but unpaid invitation to visit Cambridge, a travel grant from the NSF ($500), and a letter from the Fulbright committee telling me that the only hurdle I still had to pass was the formal one—my name was on the list they had forwarded to the Department of State for final approval. I had paid the membership fee for the Congress, and bought the banquet tickets for both my wife and me. The S.S. Ryndam was scheduled to leave on August 21, and I was all organized and ready to go on May 1.

The next week my laboriously constructed world collapsed. The letter from the Department of State arrived on May 6; it said that my application for a passport was denied. Reason: "it has been alleged that you were a Communist." The letter went on to tell me my rights: I could request a hearing and I could be represented by counsel.

What do I do now? What could I do? Of course I requested a hearing, and two weeks later, late on a Friday afternoon, I received notice that the hearing was to take place the following Tuesday, in Washington. On the advice of my senior colleagues I had made contact with a lawyer in Washington (recommended by the counsel of the University of Chicago), and I had told him as much of the case as I could over the phone.

When I arrived at the lawyer's office Tuesday morning he had already been in touch with the passport office. They told him that the main thing they had against me was my failure to cooperate—with the F.B.I. and, in Uruguay, with the U.S. Embassy. I told the lawyer the facts of almost my entire life—everything I told here and much, much more, in much greater detail. I told him also that I was willing to answer any questions, old or new, under oath, except for one kind—I said that I would refuse to answer questions asking me to reveal the names of the people whom I knew, or thought, to have been communists.

The lawyer asked, and was granted, a twenty-four hour postponement of the hearing. The time gained was spent in consultations. He talked, several times, by phone, to the officials of the passport office. I talked, by phone, to my colleagues in Chicago, and, in person, to the acting scientific adviser to the Secretary of State.

On the basis of our consultations the lawyer's conclusions were unequivocal. Here is what he said. (1) Unless I do name names, in the present circumstances and with the present attitude of the passport office, it is certain that I will not get a passport. (2) The F.B.I. files concerning me are, legally speaking, hearsay. My sworn statements at a hearing would become a permanent government record. Their existence might force the government to take action that in turn would result in a lot of publicity, of a type familiar by then, and in the expenditure of a lot of my time and money. (3) If the temper of the country and the legal requirements for obtaining a passport were to change in a few years, then, with no hearing behind me, I would be in a position much better than the present one.

From his conclusions the lawyer distilled his advice: call off the hearing, withdraw the passport application, go home, and forget about going to Europe for a while. The lawyer's arguments (which I phrased in my words, not his) seemed convincing to me. I accepted his advice. I withdrew the passport application, I cancelled the hotel reservation for Amsterdam and the car deal I had hoped to complete there, I returned $500 to the NSF, and I concentrated even more than before on algebraic logic.

I can no longer recapture my emotions of thirty years ago, and, being emotions, they were not reasoned. I felt hurt and rejected, I felt upset and angry, I felt indignant and guilty. I did not think I was guilty of any crimes. I knew that I had hurt no one and violated no loyalties. I knew without doubt, from my several interviews with F.B.I. agents, that the names I was asked to "reveal" were already known; my refusing to name them again was pride—a desperate insistence on saving self-respect—or just plain cussed stubbornness.

The only action I could take, the only thing I could do that even faintly resembled a cure, a revenge ("I'll show them!"), a solution, was to leave, to go away, to tear up roots again. I looked into the possibility, but on re-reading the correspondence of those days I now find that I liked my second roots just fine and was not really anxious to sink a third set. The correspondence was with Canada. I wrote a dozen or so letters, some to friends and acquaintances, and some to unknown chairmen and deans: "is there any chance of a job for me at your university?" The answers ranged from polite interest to friendly enthusiasm, they resulted in a few nibbles but no offers, and, once the dust settled, I was back where I started: personally pained, professionally ambitious, and mathematically still driven.

The story has an end but not a climax. The years went by, the McCarthy hysteria died down, and the time came to make arrangements to go to the 1958 Congress in Edinburgh. In 1957, acting on legal advice, I asked for a hearing at the Passport Office—to close the books, to tie up loose ends—and everything went smoothly. The hearing was almost a farce. There were only three of us in the room, my lawyer and I and one passport official, a fattish man who was eager to get the formalities over with so that he could go down the hall to the office party. He coughed apologetically, and explained that he hadn't really had a chance to look at the file yet. Well—let's get down to business—he glanced at my affidavit, asked one technical question, and gave me an ordinary passport application form—just fill this out, please, and we'll go on from there. Goodbye now.

I asked my lawyer whether I hadn't wasted my time in coming to Washington. No, definitely not, he said. The passport people liked that I made myself available to them, and they liked to feel that they could have asked all the questions they wanted.

Six weeks later, with no further fuss, I had my passport. I even got a travel grant again! The travail was over; I had rejoined the human race.

Service

Research was exciting and the passport affair was discouraging, but the normal business of academia lumbered along neither on high peaks nor in deep valleys but in the firm ruts of tradition and convention.

Mathematics has to be not only thought about but administered, papers and books have to be not only written but edited, and students have to be not only taught but supervised. I was willing, even eager, to be a part of it all; I never said no to a chore, and I was sufficiently reliable at cleaning scholarly stables that bit by bit I was handed more and more shovels.

Why did I accept them? Why does anyone? To mutter about conscience is stuffy and is probably not an accurate description of the true motive, but there is something to it. There is at least this much to it: we tend to disapprove of the super-geniuses, real or self-styled, who never serve on a committee, refuse to referee, and are of very little help to the generations of students coming along behind them. That's not right, we feel, and once we hear ourselves express that view we are less likely to be churlish about the next chore that comes along. If, God help you!, you turn out to be good at chores, then you're stuck. We all like to do what we do well. We all like the reward of feeling that we've accomplished something—a reward that is guaranteed to come with jobs that require perspiration only. Research cannot be done without sweat, but it needs inspiration too, and to a smaller extent so does teaching. The "service" that a dean, or an editor, gives to the university or to the profession needs nothing but wisdom, experience, and patience—qualities that we are a lot surer of possessing than insight, originality, and creative talent. And, besides, all those "thankless" hours are usually pretty well thanked: you are famous in the bush-leagues and you feel important. There is an unkind one-word possible summary of all this: the word is cop-out. We do it to be able to say—to have a justification for saying—that we have no time for research, for scholarship, for creative work. Why did I accept more and more stables to be cleaned? Because I was nearing 40, reaching it, and then passing it, and, consciously or otherwise, I was far from sure of being able to continue research work. That's not what I said to myself at the time, but that's my a posteriori self-psychoanalysis now, twenty-five years later.

"Service" used to be a good word, but it has become a shibboleth of university administrators. It used to be taken for granted that if you use the library sometimes or if you publish a paper every now and then, then you are a part of the library-using or paper-publishing community. It followed, as a matter of course, that you would help the library (by, for instance, keeping an eye on the new books coming out and advising the library about which ones to acquire), and that you would help the publishers of journals (by, for instance, being a considerate and conscientious referee). You should no more expect to be rewarded for such activities than you are rewarded for bathing every now and then or for turning your grades in at the end of the term. Such activities are not good in themselves, they are merely indispensable; not doing them can have catastrophic effects.

The slogan "research, teaching, and service" is balderdash; it's the failed scholar's excuse for shuffling papers. A university exists to create and to transmit knowledge—not to design and administer placement exams, not to arrange times and places of classes, not to keep student records, and not even to raise and distribute money. All such things are to some extent necessary, and we must all do some of them. They are actions of social service. Failure to perform them puts more on the rest of us, and is therefore antisocial—but they are not positive contributions to be rewarded. A poor teacher may have a place in a university (if he is a great creator), and a second-rate research man has his place (if he is a great teacher). But a second-rate research man who is a second-rate teacher has no more right to be on the faculty than a janitor, a secretary, a budget systems analyst, a dean of Latino affairs, or a vice president for administration. The work of such people is necessary—well, some of it is anyway—but it should not for that reason be regarded as a promotion, looked up to, revered, and overpaid.

Since time is finite, worse luck, we must make choices: which kinds of service do we have the talent and the training to provide? My choice has always been on the departmental level and the professional one; the in-between level of the college and the university I chose to by-pass. I have never served on something like the college promotion committee, but I think I have been a member of every conceivable kind of departmental one: library, colloquium, graduate policy, curriculum, and fellowship, to mention but a few. As far as the profession goes, I couldn't possibly remember all the committees of the AMS and the MAA that I have served on, and nobody would want to know even if I could. Here is a short partial list: Committee to Prepare a Pamphlet to Aid in Reading Russian Mathematical Articles, Committee to Select Hour Speakers for Western Sectional Meetings, Committee on the Doctor of Arts Degree, Committee on Prizes, Committee on Publication Problems, and Nominating Committee.

"Service" is, fortunately, not synonymous with "committee work". The service work that I found most rewarding was supervisor and editor. By "supervisor" I mean director of Ph.D. students—and perhaps that's not service but teaching; by "editor" I mean not only editor but also referee and reviewer—everything other than mathematical writing that has to do with mathematical writing.

As for Ph.D. students, I have already mentioned the ones I had at Chicago, with two exceptions. They are, in reverse chronological order, Carl Linderholm and Moe Schreiber. I list them in that order because even though Carl was the last of my Chicago students to finish (1963), his work belongs to my ergodic period, which came long before the operator period. Carl generalized theorems about measure-preserving transformations to theorems about transformations that may change the positive values of the measure, but do not change its zero values in either direction, direct or inverse. He was a Chicago student, but he accompanied me to Michigan when I moved, and did his last year's work there. That sort of thing is negotiated every now and then: when a professor changes location, his new university has to provide a home for his students, at least temporarily, to prevent them floundering alone and abandoned at the old. It worked fine. We were together in Ann Arbor, while Carl was finishing his thesis; soon after it was finished, and Carl left Ann Arbor, the University of Chicago invited me back for an afternoon so that I could act as the chairman of his Ph.D. final exam committee. Not long after that Carl and his family emigrated; he has been living and working at Reading all these many years. He became famous for a brilliant, witty, extended mathematical in-joke, a book called *Mathematics Made Difficult*. The book treats high school trigonometry, for instance, from the point of view of category theory, for instance—I recommend it highly.

Moe Schreiber became a good friend during his Chicago thesis days (Ph.D. 1955), and is one still. He gave me one of the rare deep surprises that thesis supervisors get. I suggested to him that he study the unitary power dilations of strict contractions and look at them as spectral invariants: what information about a contraction can we infer from knowledge of its unitary power dilation? I didn't believe Moe the first time when he, hesitantly, told me the answer. None, he said. All those unitary power dilations are the same. They are, that is, all unitarily equivalent to the unilateral shift of infinite multiplicity. Here is an analogy, a limping but partially suggestive one. If all I know about a human being is that for part of the year he has only 5 or 6 daylight hours and for another part of the year he has 15 or 16, then I don't know much, but I know something—he lives in a latitude far from the equator, near to (but not at) one of the poles. If, however, what I know is that his years have 365 days in them (all right, all right—except for leap years), then I know nothing at all—he could live in Anchorage or in Caracas for all I can tell. Schreiber's theorem (as it quickly became called) is that the spectral information conveyed by the unitary power dilation is like the 365-day information: it says something about everybody but it doesn't say anything about anybody.

Editing

As for editorial work, it crept up on me in small steps. It took me hardly any time at all at the beginning. An efficient hour or two every now and then was enough for knocking out a referee's report for the Transactions, a book review for the Bulletin, or a formal letter inviting someone to contribute a volume to the Ergebnisse series. It took hardly any time at the beginning; it takes most of my working time now.

The small steps were more and more requests to referee papers and review books. In the 1950's fourteen of my reviews appeared in the Bulletin, and only three of them had to do with functional analysis (Zaanen, Achieser–Glasmann, and Dunford–Schwartz). The rest ranged over nearly all of pure mathematics, including set theory and logic (Fraenkel and Tarski), integration (Bourbaki and Picone–Viola), algebra (a radically new freshman text by Levi, and lattices by Hermes), topology (or rather topological dynamics by Gottschalk–Hedlund), and some curious bypaths (such as Tietze's *Gelöste und ungelöste Probleme...* and Pólya's *Mathematics and Plausible Reasoning*). At the same time I was a frequent contributor to MR (I had to be to get the reduced subscription rate) and an occasional reviewer for statistics and logic journals. As a result I received some fan mail and whatever the opposite should be called ("hate mail" is too strong). When I chided Fraenkel for his prolixity and old-fashioned notation, he wrote to complain and defend himself. We exchanged a couple of letters as a result and we parted friends; "I do hope to meet you in Amsterdam" was the end of Fraenkel's last letter to me. Abby Robinson thought he got short shrift in my Journal of Symbolic Logic review, and told me so, but later he wrote a friendly letter. "Since I believe that in Rochester I expressed some dissatisfaction because you had not considered the later parts of my book *Théorie metamathématique des idéaux* in your review in the J.S.L. it is only fair that I should thank you now for the full report on *Complete Theories* in the Math. Revs."

The real editorial work started in 1958 when I replaced Kakutani as one of the editors of the Proceedings. The other three editors were Ralph Boas, Irving Kaplansky, and Hans Samelson. We split the work among us by fields; I was in charge of real and functional analysis (with a strong assist from Kaplansky in the latter), and "other fields", which meant mainly probability, logic, and some of the crank papers in number theory.

I was used to typing—papers, books, letters, syllabi—and except for the occasional odd job (dittoing exams or typing class notes that were to be mimeographed and then distributed in large numbers) I hardly ever needed a secretary; when I became editor, those days of freedom were over. From then on (for the last twenty five years or so) I have needed help to get the work done—and I both blessed and cursed the day when it all started. A good secretary can be a big help—but to work with one takes time as well as saves time. You have to organize things in advance and explain how they should be done; you have to adjust your priorities and your timing to not only your own whims, preferences, illnesses, vacations, and other conflicts, but to someone else's too, and you keep feeling pressure whether the work is heavy or light. Will Beverly be able to finish this before Friday? Did I give Jaynel too much to do—should I request work-study help for her? Do I have enough for Karen to do?—it's demoralizing to do nothing, and it sets a dangerous and possibly incurable precedent to lend her part-time to a colleague. It started with a one-tenth time secretary (there are ten half-days in the working week and I had enough work to fill one of them), and has now reached the point where I usually need eleven tenths—except for those unpredictable lulls when nothing comes in, no manuscripts are submitted, no letters demand an answer, and we turn in desperation to reorganizing the files.

During the years I was a Proceedings editor (1958–1963—by tradition everybody served two three-year terms), I seem to remember processing something like 120 papers each year and accepting about 50 of them. The work was fairly evenly spread among the four editors. One of us was managing editor (the title rotated), but, except for a small amount of official correspondence with the Providence office of the AMS, he had no more to do than the rest. We operated independently and autonomously; we each decided what to accept and reject and there was no appeal from our decision.

One year when I happened to be managing editor one of the routine parts of my duties was to collect the annual statistics. Average number of papers received per month, average number accepted, estimated number of pages—that was the sort of thing needed—and I dutifully wrote around and asked each of my editorial colleagues to send me the information. Eldon Dyer was one of the editors that year, and he conscientiously replied. He sent me a table with twelve rows, numbered from 1 to 12 to indicate the months, and with the appropriate set of columns for papers, acceptances, pages, etc. At the bottom of each column he wrote its sum and then its average—including, very helpfully, the first column. I learned that the sum of the months $(1 + \cdots + 12)$ was 78 that year, and their average was 6.5.

The Proceedings in those years printed about 1000 pages each year, so that, on the average, each of the four editors was responsible for 250 pages. I looked recently at some of the volumes of the Proceedings in the 1980's. The number of pages per year has more than doubled—there were 2160 pages in 1981—and the number of editors has risen to 18. That works out to an average of 120 pages per year per editor, and it just goes to prove that the world is going to the dogs in a handbasket and the younger generation is getting soft.

How to be a big shot

Question: how do you get to be a big shot, a member of the in-group in the AMS or MAA? Answer: learn the rules, and do *something*.

By "learn the rules" I do not mean "be a conscious toady, and obsequiously imitate the reactionary establishment"—I just mean learn the rules. Learn when and where the meetings are, what their announced purpose is, and what really goes on at them, and learn the table of organization—who is in charge and how do decisions get made? It's easier to change or bend or break the rules if you know what they are. In my case I guess (and it is just a guess) that my first nomination to member at large of the Council (in 1950) came about because I was being difficult—I was trying to change, bend, or break the rules by which the Cambridge Congress was being organized.

By "do something" I mean do something, anything, that is genuinely, honestly, sincerely professional. Write papers and books and go to the meetings to talk mathematics with people of similar interests; teach classes, prepare syllabi, and write books and go to the meetings to talk with people of similar interests; pose and solve elementary problems, organize problem seminars for students and faculty, and go to the meetings to talk with people of similar interests. Get and stay in touch with the profession.

When I was beginning to do these things, during the first ten years after my Ph.D., I wasn't even aware that I was doing them. I went to the two big meetings each year (paying my own way) because I wanted to talk with people of similar interests, and between meetings I wrote long mathematical letters to anybody who was willing to receive and reply to them. I was fascinated by the profession, and I was eager to learn how it functioned. I was 32 years old when I was appointed to my first AMS committee (the one concerned with the Russian reading pamphlet), and that's considered old nowadays—wordly-wise deans-to-be start bucking for such connections and getting them in their middle twenties. The reason I was appointed was that I spent many many hours during the preceding year learning mathematical Russian, and, at meetings, I kept discussing the problems I ran into with people of similar interests. Once the word got out ("Halmos was asking me about something just like that last week"), the official step was almost automatic ("How about Halmos—why not put him on?").

Young people seldom realize (and I certainly didn't) how aware the older generation is of them and how much the older generation wants them to come in to help push and tug and turn the machinery. It's not an excluding conspiracy ("we're in, let's keep them out"), but a shanghaiing one ("grab them and put them to work"). When I was young I was surprised at being grabbed; now that I am at the other end, grabbing is my hardest task. As editor of a series of books, my job is not to reject submissions, but to invite and encourage them—the more come in, the greater my success (and the greater my income—let me not be coy about that). As editor of a journal, my hardest job is to search for the editors of the future, who can be recognized in the present by their ability and willingness to be reviewers and referees.

I have known several people who were out and kept asking how to get in. They wanted very badly to be in, but they were doing it very badly.

They didn't bother to learn the rules and the only "something" that they did about getting in was to want it.

When I became a member of the Council, I went to its meetings; I took them seriously and I participated in them. It was fun in a way, but far from pure fun—the meetings could be long (adjournment past midnight was not unheard of) and horribly boring. A few of the early meetings I went to were small—12 to 15 people—and they have been increasing ever since; nowadays they can run from 50 to 60 or more. Some of the people in the room are support staff (the Executive Director of the AMS, the Executive Editor of MR, secretaries, and various financial and publication experts), but most of them are there just to talk. The stated purpose of the Council is to make scientific policy (the Board of Trustees holds the purse strings), but in the last twenty years decisions about meetings, colloquium volumes, and journals took a poor second place to the social-political fulminations of ardent special interest groups. The AMS has been a publishing company and a convention bureau during most of its existence, and, to a growing extent, an employment agency; the modern trend is pushing it in the direction of a trade union, a financial lobby (squeeze all we can out of the NSF), a civil rights group concerned with problems of discrimination against blacks and women, and a political action committee fighting apartheid in South Africa and anti-semitism in the U.S.S.R.

It seems reasonable, and experience seems to show, that the Society cannot effectively be all these things. The same person can, to be sure, be in favor of high-quality publication, be anxious to get summer money, and be against anti-semitism. One person, however, is not likely to have the scientific judgment to make a good editor, the political knowledgeability to make a productive lobbyist, and the energy, zeal, and charisma to make an effective activist.

When the AMS was young and eager and idealistic about mathematics, there was but little problem in finding its officers. The concern of the Society was with scientific subjects only; the only reason to belong to it and work for it was dedication to mathematics. The nominating committee tried to draw the emerging people into Society activities, the ones who showed the greatest promise of mathematical judgment and the greatest willingness to work. A mild effort was made to avoid obvious unbalances, such as too many algebraists or too many Californians, but no attention was paid to sex, race, age, religion, rank, income, or political zeal. With the change of emphasis came changes of procedure. Nowadays candidates offer themselves to the electorate, and campaign on various odd platforms, such as not having tenure or being unemployed.

Are the new procedures more democratic? Superficially the answer seems to be yes, but the proof has some gaps in it. The point is that of the 20,000 or so members of the Society usually less than 15 per cent do the voting, and that means that a determined letter writer who can swing as many as 100 votes can tilt the outcome of the contested parts of an election. Are the new concerns of the Society more appropriate, more desirable? The answer depends on whom you ask.

But enough of this philosophizing; there is something about Council procedure that I'd like to tell about. I was on the Council in the early 1950's, and, a few years later, in 1957, I was nominated and elected to it again. In 1958 I became a Proceedings editor, a job that carries Council membership with it, so that for a while I was on the Council in two capacities. That sort of dual membership is not unusual but it can lead to complications. Everett Pitcher became Secretary of the AMS in 1967, and the following year he discovered the most amusing (and most mathematical) complication. He sent a memo to the Council; it read, in part, as follows.

Perhaps only a collection of mathematicians would create a set of bylaws with built-in mathematical traps. However in section 4 of Article IV, I think we have done just that. I quote.

"Section 4. All members of the Council shall be voting members. The method for settling matters before the Council at any meeting shall be by majority vote of the members present. If the result of a vote is challenged, it shall be the duty of the presiding officer to determine the true vote by a roll call. In a roll call vote, each Council member shall vote only once (although he may be a member of the Council in several capacities), and he shall state before the vote in which capacity he votes. The group consisting of the four Associate Secretaries shall have one vote, and it shall be divided equally among those who vote as Associate Secretaries. Each of the eight Publications Committees shall have one vote, and it shall be divided equally among those who vote as members of the respective Publications Committees. All other members of the Council shall have one vote each. Fractional votes shall be counted."

Here are two curious features of this rule, illustrated by Councils of hypothetical composition subject to a rule such as Section 4 above. The Council in each case is choosing between candidates α and β.

Example 1. In order to gain his will, a member does not necessarily declare himself as a member of the committee which yields him the largest vote.

Council consists of

Comm. A	Comm. B	Comm. C
x	x	c_1
a_2	b_2	c_2
a_3	b_3	c_3
	b_4	c_4
		c_5

Member x wishes to see α elected. He knows a_2, a_3, c_1, c_2 will vote for α and b_2, b_3, b_4, c_3, c_4, c_5 will vote for β. Then x accomplishes his will by declaring he is voting for α as a member of Committee B, thus casting $1/4$ vote. If he were to vote as a member of Committee A, he would cast $1/3$ vote but β would win.

Example 2. The order in which voters declare their voting capacity is significant.

Council consists of

Comm. A	Comm. B	Comm. C	Member
x	x	c_1	d
y	y	c_2	
a	b	c_3	
		c_4	
		c_5	

Here a, b, c_1, c_2 vote for α and c_3, c_4, c_5, d vote for β. Suppose x wishes α to win and y wishes β to win. If x and y declare for the same committee, then α wins while if x and y declare for different committees, β wins. That is, whichever of x and y is the second to declare his voting capacity can do it so as to win.

Example 2 shows that the rules for voting may not be stable. For on a roll call vote one may change his vote before the total vote is recorded; voters x and y could alternately change their vote to the point of absurdity. I doubt that Robert considers this case.

How to be an editor

As the years went by, my editorial involvements kept increasing. After being an author in the University Series in Higher Mathematics published by Van Nostrand under the editorship of Marshall Stone, I became one of the editors of the University Series in Undergraduate Mathematics published by Van Nostrand under the joint editorship of J. L. Kelley and myself. Still later Fred Gehring and I collaborated in editing a Van Nostrand paperback series, and I became connected with Springer-Verlag as one of the editors of its prestigious Ergebnisse series.

How does an editor of a book series earn his keep? Does he read manuscripts with a red pencil in his hand looking for misplaced commas and split infinitives? No, sir!, not in a million years; he earns his keep as far as the publisher is concerned by being an advertisement and an advisor.

As for advertisement, I was becoming known as the author of some reasonably successful books (on measures, and vectors, and operators) plus a moderate-sized bagful of research papers, and that was enough for Van Nostrand, and later Springer. What's the idea, I sometimes wonder? If you're an author looking for a publisher, do you think that the success and prestige of the editor's name will somehow rub off on your book and will help it to sell better? Publishers, at any rate, believe that some such mechanism is at work. As for being a publisher's advisor, what a series editor mainly does is not so much directly advise, the way a referee advises a journal editor (although there is some of that), but he spends his time looking for, asking for, receiving, and accepting or rejecting the manuscripts of books.

The Van Nostrand series of paperbacks was something I had been trying to get started for years; I wanted an American version of the marvelous little books that de Gruyter had been publishing in Germany under the general title Sammlung Göschen. The little Knopp books on complex functions were in that series, and the Hasse volumes on "higher" algebra, and many others. They were slightly smaller in area than the smallest of the paperbacks on drugstore racks (much smaller than the large ones) and not more than a half inch thick. They fit easily into most pockets, they were well printed and well bound, and the expository quality of many of them was very high. I found them a pleasure to hold in my hand, attractive to look at, and, for a combination of all these reasons, easy to learn from. They were, incidentally, ideal to pack for a trip, and handy to have with you if you needed to do some of your reading while straphanging on a bus.

No publisher would listen to my Sammlung Göschen propaganda, but eventually Van Nostrand listened to my counterpropaganda about the prevalence and unsatisfactory character of mimeographed notes. They distributed an advertising leaflet I wrote under the title "A note about notes...", and in the early 1960's the Van Nostrand Mathematical Studies, the paperbacks, finally got started. Here is how the first two paragraphs of my advertising copy read.

During the past ten or fifteen years a considerable part of mathematical communication took place via "notes". This not very descriptive word has come to mean something quite specific: it usually means a collection of mimeographed sheets, somewhere between 75 and 300 of them, badly reproduced, loosely stapled, orally advertised, cheaply sold, and quickly exhausted. The result is unsatisfactory from everyone's point of view. Mathematics departments grumble at finding themselves in the publishing business, departmental secretaries complain of the overload of work, authors find that they have done half the work on a book with only a small fraction of the recognition and none of the financial reward, and readers struggle with material that is easy to damage and hard to read. An unwanted consequence of the informal distribution is the creation of in-groups, consisting of people who have access to all the current notes on a subject; lone-wolf scholars in non-central institutions, and their students, frequently have no chance to learn of the contents (or even of the existence) of certain notes.

Despite its manifold disadvantages there are reasons why the system continues. One reason is that notes offer a convenient medium of publication, somewhere between the research journal and the printed book. In notes we may find classical material presented from a modern point of view, important surveys of recent, rapidly developing fields of mathematics, and proofs of 'well-known' but, in fact, inaccessible folklore. A second reason is that authors, who may be unwilling to take on the work and the responsibility of a full-fledged book, leading to the oppressive finality of the printed word, are much more willing to dash something off, something obviously informal and ephemeral, as a trial balloon. Still another advantage of the system is its speed: a set of notes can be reproduced quickly and read while the contents are still timely.

The VNMS was an attempt to keep the advantages of the notes system but eliminate its disadvantages, and the attempt was quite successful. Among the most popular items were Lyndon's *Notes on Logic*, Mackey's complex variable lectures, and Chern's complex manifolds. The series kept going till Van Nostrand started being sold, absorbed in conglomerates, and being resold, and then, like all the rest of Van Nostrand's good mathematics, it peacefully died.

When Gehring and I started a more elementary series for Springer, we sent letters to a few people we thought would make good authors (quite a few—about a hundred in one mailing that I remember). Here is how the letter with Gehring began. "Dear Whoever: Have you been writing any good books recently, or do you know anyone who has? We are the editors of a new series, the Undergraduate Texts in Mathematics published by Springer-Verlag, and we would like to publish the books of high quality that, in recent years, most publishers have been avoiding."

The Ergebnisse der Mathematik und ihrer Grenzgebiete is one of the yellow series that Springer has been publishing for a long time. The Grundlehren der Mathematischen Wissenschaften in Einzeldarstellungen is the big yellow series and the Ergebnisse the little one. The Grundlehren can be long, monumental, scholarly, comprehensive, encyclopedic, and frightening reference works; the Ergebnisse are intended to be short, polished, sharply focussed, specialized, and up-to-date research tools. My mental picture of a Grundlehren author is a graybeard near the end of his career summarizing what he learned in the last forty years, but I see an Ergebnisse author as a young one who, to be sure, is no longer "out", but having recently come "in" is now a missionary, eager to make converts and establish a school. When my name started appearing at the bottom of the title pages of Ergebnisse volumes, first as one of the dozen or so advisors, and later as a member of the policy-making editorial board of three, the area that I was assigned to be in charge of was Reelle Funktionen, liberally interpreted.

Commercial publishers in this country always looked for, and nurtured, and fostered new manuscripts, but nevertheless most of what they published came to them unsolicited from hopeful authors. The Ergebnisse series used to acquire its material by invitation only—well, almost only. To be invited to contribute to the series was regarded as an honor, almost like a prize, and the invitations were generally accepted. Such things had begun to change before I came aboard, but much of the old tradition was still alive. I could and did reject unsolicited submissions on the ground that we don't operate that way, and I wrote many letters of invitation.

M. M. Day, 1971

One of my successful Ergebnisse ideas was Mahlon Day's book. Sometime before the middle 1950's I approached him and invited him to contribute a volume. His answer was an immediate, friendly, and enthusiastically loud no. No, he was too busy to write that book, and besides he had been planning to write another type of book, a much bigger book, partly a modernization of Banach (by then twenty years out of date) and partly an encyclopedia like Dunford–Schwartz (then not yet in existence). I argued with him, and, for once, I found the right argument. "By all means", I said, " by all means, Mahlon, write that big book—but perhaps the best way to do so is first to write it in compressed form, as an outline,

as a survey, or, in other words, as an Ergebnisse report—and then, when
you have something like that in your hand, something solid, go on from
there and make the big book grow out of the little one." Mahlon wasn't
really convinced, but he was then, as he has been all his life, a hard
working ambitious mathematician. He went to work, hard work, and in
no time at all—approximately three years later—he was finished, and
I was turning the pages of the manuscript of *Normed Linear Spaces*,
Ergebnisse number 21. The rest is history—three editions and more than
a quarter of a century later *Normed Linear Spaces* is still one of the
most highly respected classic references of the subject.

I hope I have illustrated at least in part what it means to say that
a book series editor is an advertisement; let me say a few words about
how such an editor acts as an advisor. Would "judge" or "evaluator" be
a more descriptive word? What happens most often, with say the
Graduate Texts in Mathematics (the most successful series I have ever
worked on), is that an author submits a manuscript, and it is up to us
(the small editorial board, currently consisting of Fred Gehring, Cal
Moore, and myself) to decide whether to publish it or not.

A fat brown jiffy bag appears on my desk, full of some 500 pages of
paper with typing on them. I sigh, and I reach for my red pen. Oh, yes, I
use it all right—not for corrections and improvements, but for marking
the virtues and faults so as to remember them, collect them, balance them,
and base my verdict on them. The first thing I look for is the title page
(you'd be surprised how often it isn't there): is the subject suitable for
GTM and is the author likely to do it justice? An unsuitable subject
might be one that's too elementary (it's possible but not likely that a subject is too advanced), or too special, or already too well represented in
the literature. As for the author: I know many mathematicians, personally
or through their work, and if the author is someone I know, I can pretty
well predict whether the book will be good or not. Example: if this is the
author's second book, and if the first one was a success (or a flop), I am
ready with a first approximation to my prediction.

The next thing I look for is the table of contents. How is the material
divided, how are the parts organized, which subjects other than that of
the title does the book have contact with? Immediately after that I look
for the preface, and if it isn't there I mutter dark imprecations; the author
has just failed a test with serious effect on his grade point average.

Don't you do something very like what I am describing when you pick
up a new mathematics book in the university bookstore? What is it
about?—is it something you want to learn?—who wrote it?—is it someone you know anything about, good or bad?—what's in it?—and, last
but certainly not least, is it for you?, is it at the right level?, and is the
promised treatment one that you're likely to approve of and profit from?
Audience, level, and treatment—a description of such matters is what
prefaces are supposed to be about—and if knowing about those aspects

of a book is important to a prospective reader, it is much more important to the editor, the evaluator, the judge.

The time has now come to start reading the manuscript itself, but not yet its content—I'm still trying to evaluate the form. Is the manuscript clean—well enough prepared so that a typesetter who is not a mathematician can cope with it? Are the spelling, punctuation, idiomatic usage, diction, and syntax correct? Is the presentation informal enough to be digestible and at the same time dignified enough not to ruffle too many feathers?

If the answers are yes, then I can begin to pay attention to what exactly is in those 500 pages. For instance: are there any problems, preferably good problems, interesting, and challenging, and doable problems? No? Bad. Also: is the organization Landau-like (definition, lemma, proof, theorem, proof) or is it conversational, explanatory, motivated, smooth? The latter, we hope. Is the material all theory, or is it interlaced with examples and applications? The latter, we hope. Is the exposition clear or fuzzy? The former, we hope, we all very much hope. (If the subject is one I know, then I think I can judge whether the exposition would be clear to a tyro; if I do not know the subject, then I am the tyro, and my judgment of the quality of the exposition can count as at least one vote on its success or failure.)

Have you noticed that something is missing from all the many hurdles that I've put the manuscript through? Have you noticed that I haven't yet said a single word about the quality of the mathematics in the book? I left that to the end because although it is perhaps the most important hurdle, it is the one that any particular editor is the least likely to be able to judge, but nevertheless the one that is the easiest to come to a decision about. Is the mathematics correct, in the sense that the statements are true and the proofs are valid? Is the treatment old and trite or new and bright? How can I tell, if I happen to be a geometer but the subject is epsilontic analysis? I cannot, and that's where I need help. The editorial board as a whole has more mathematical expertise than any one editor, but even so it doesn't have enough to cover all mathematics. No problem—take the manuscript down the hall, or mail it across the country, and ask an expert you can trust: is it true and is it new (or, in any case, new enough to justify making it available to the possible buyers and readers of GTM)? The "reviewer" (remember?—that's what a commercial referee is called) gets a modest fee for his report, and the editorial board is now ready to come to a conclusion.

Journal decisions, made by large and democratic boards depending on busy and dilatory referees, sometimes take many months or several years to arrive at. Book decisions are much faster. I frequently do my part (sometimes with and sometimes without evaluating the correctness of the mathematics—that depends on whether or not I know the subject) in an hour—yes, one hour, plus perhaps ten minutes to dictate my report into a machine. A month, or maybe two, is enough for the editors to correspond with one another and, if they can get along without an outside expert, to come to a conclusion. With outside experts the timing can double, or worse.

There is more to publishing books than the part that series editors are involved in, and they are involved in more than I have just described. There are procedural questions (isn't it about time we arranged a face-to-face meeting?, which of two proposed expert reviewers should we consult?, if this manuscript is not suitable for GTM, can we use it elsewhere, or do we just reject it outright?), there are financial questions (this book is good but risky—shall we gamble?, this book is a sure loser, but the loss won't be great and are our relations with the author worth it?), and there are editorial questions (the book is very good, but the author is not at home with English—where can we find a good line-by-line editor?, the subject and the treatment are good, but the author is a slob—can we get a copy editor to change wrong words and commas into right ones?).

I'll stop this report on what book series editors do by recording my general impression that, by and large, the book business operates better than the article business. Not only is book publication much speedier than article publication, but it is generally a lot less expensive (yes, that's what I said, less expensive), and of higher quality.

As for expense: journal publication is often subsidized (by, for instance, the universities that publish journals or the foundations that pay page charges), and, if it weren't for that, many journals are so inefficiently run that they would go out of business quickly. Books are more often commercial ventures, intended to make a profit, but despite the loud howls that mathematicians sometimes emit, the prices are not out of line. Pricing policies vary from one publisher to the next, and some publishers seem to be in the highway robbery business, but most of them are reasonable. Suggestion: the next time before you howl about the book, judge its cost not by the number of dollars you have to take out of your pocket but by comparing some objective measures. (A good rough approximation is cents per page.)

As for quality, that's just a feeling I have. I think that, with extremely rare exceptions, even the book of lowest quality is likely to be correct most of the time, and to be well enough organized and expounded that studying it would add to your mathematical wealth. Articles are more often wrong and very often so badly done that reading them is more work than it is worth. They must exist—don't misunderstand me—I am not advocating that we stop publishing current research papers and go back to the days when "publish" was synonymous with "publish a book". No, papers are absolutely necessary, but books are better.

Have you been writing any good books recently? Send them to me— I'll publish them.

Recent progress in ergodic theory

As a larger and larger fraction of me was becoming editorized, the world of mathematics went on for other people and, somewhat surprisingly, even for me.

So, for instance, despite my preoccupation with algebraic logic, I kept being interested in and teaching ergodic theory. I took special pains with the version of the course I taught in the summer of 1955, and that turned out to be lucky. Soon afterward the Mathematical Society of Japan started issuing a series of little books that looked as if my ergodic notes might be just right for. There was a brief exchange of correspondence, the matter was arranged—all I had to do was deliver the manuscript. That meant revise (easy) and re-type (not so easy). Realizing that I'd be dealing with compositors whose facility with English was not maximal, I tried to go to extraordinary lengths to avoid errors—mine and theirs. Here is an example: what does a Japanese typesetter know about the rules of English syllabication? What will he do when a word has to be broken at the end of a line? I knew from experience that the answers were "mighty little" and "probably something dreadful". In the cover letter that I sent along with the manuscript, I asked that word separation should be kept to a minimum, and later, when proofsheets arrived, I spent an extra minute on each page, looking down the right-hand margin. Incidentally, as an extra safety measure, I had typed the entire book without ever breaking a word at the end of a line.

Ergodic theory was in my bloodstream; I was addicted; it was hard to let go. In 1948 I delivered an invited address on ergodic theory at the November meeting of the Society in Chicago, and in 1960, when I was invited again, this time for the summer meeting in East Lansing, I chose the title "Recent progress in ergodic theory". There had indeed been some spectacular recent progress, namely the Kolmogorov–Sinai solution of a long outstanding conjugacy problem and the Ornstein solution of the equally old and at least equally difficult problem of invariant measure.

The first of the Kolmogorov and Sinai papers appeared late in 1958, but I heard about it a few months earlier, in March, through a letter from Kolmogorov. He wrote from Paris, where he was visiting, and where he had just seen my ergodic theory book. He wrote to tell me that he had introduced into ergodic theory a new invariant called entropy (inspired by the known concept of that name in thermodynamics), and that the new concept could be used to solve one of the problems that I had been propagating. In vague, non-technical language the problem could be formulated as follows: is there a measure-theoretic way to distinguish between the passage of time in two experiments, namely the infinitely repeated independent tosses of a loaded coin and the same thing for an honest coin? In technical terms the question might look like this: if T_1

is the shift transformation on the Cartesian product of a bilaterally infinite sequence of copies of a probability space consisting of two points with different masses, and if T_2 is the same thing except that the two points have equal masses, are T_1 and T_2 conjugate? (That is: does there exist a measure-preserving transformation S such that $S^{-1}T_1 S = T_2$?)

There were good heuristic reasons for predicting either answer; the experts were too stumped even to agree on their guesses. The two stochastic processes T_1 and T_2 ("stochastic" is a learnedly elegant way of saying "random") induce unitary operators that are equivalent (which means that they are conjugate elements of the group of unitary operators); that was evidence for guessing yes. Examples were known of transformations that were unitarily equivalent but not measure-theoretically conjugate (von Neumann and I had constructed some); they seemed to gainsay the positive evidence. What if T_3 is the shift built with a probability space consisting of three points with equal masses—are T_2 and T_3 conjugate? In this case the transformations to be compared are homeomorphisms as well as measure-preserving transformations, and although the underlying topological spaces are homeomorphic, the mappings T_2 and T_3 are not conjugate homeomorphisms; that was evidence for guessing no. As I remember, von Neumann's guess was no, and the entropy answer proved him right.

Entropy is, roughly speaking, a measure of information; the entropy of an experiment is the average amount of information we can gain from one performance of it. The amount of information that can be gained from two independently performed experiments is the sum of the two separate amounts. In other words, entropy changes multiplicative input (independence) into additive output; it's not surprising that it behaves logarithmically. Sure enough, the entropies of T_2 and T_3, the 2-shift and the 3-shift, turned out to be log 2 and log 3 respectively; since entropy is a conjugacy invariant, it followed that T_2 and T_3 could not be conjugate.

I was pleased and excited; I asked Kolmogorov to send me reprints or preprints or whatever he could as soon as he could. He did; he sent me copies of his own first Doklady note, and of subsequent ones in which Sinai improved and extended the Kolmogorov definition and Rohlin applied it to compact abelian groups. I worked on the material very hard, and I came to understand it pretty well. I gave lectures on it at Chicago, and I wrote a set of notes, "Entropy in ergodic theory", that were circulated in mimeographed form for several years. (They weren't really "published" till more than twenty years later; the printed version appeared in the second volume of my *Selecta* in 1982.)

The conjugacy problem was on my mind and it worried me for years before the Kolmogorov–Sinai solution, but I never really worked on it; the invariant measure problem was one I worked very hard on. Eberhard Hopf gave a pretty condition that was necessary and sufficient for the existence of a finite measure invariant under a given transformation and pleasantly related to a given non-invariant measure. For all anybody knew, however, an invariant infinite measure always existed—the invariant measure problem was to find out whether it did or not. I succeeded in extending Hopf's condition to the infinite case, and the extension was just as pretty as Hopf's, but it wasn't effectively usable—for all anybody could tell, it was always satisfied.

The answer came in 1960; Don Ornstein gave a fiendishly ingenious example of a transformation that wasn't potentially measure-preserving in either the finite or the infinite sense. The ingenuity went into the construction of the underlying measure space. The points could be thought of as "infinitely large integers" with expansions similar to decimals but based on not one radix but infinitely many of them, growing very rapidly. The space was, in effect, an infinite adding machine. The transformation itself was easy to describe (to get the image of x, form $x + 1$). Given the input (a point x) and the instruction (add 1), the combinatorial structure of the machine went to work in such an intricate way that no measure had a chance of remaining invariant under it.

Recent progress in ergodic theory was recent indeed and progress indeed.

Writing for a living

Do you like the Scientific American? I keep thinking that I should, but I don't seem to be able to. It's shiny and colorful and attractive, but most of the articles are too hard for me. They go into details about matters biological, and they assume that I know and care more about chemistry and seismology than I can absorb. The unhappy collaboration of authors and editors produces an improbable result, namely articles that are difficult and boring at the same time. The effect is produced by experts who have no notion of how to put the right words in the right order to explain their beloved subject to a layman and by firmly instructed editorial hacks who rigidly replace everything truly individual and challenging by the gray, smooth, inoffensive, and absolutely uniform house style. The result is that articles on astronomy, paleontology, and economics all read alike—they all tell me more than I want to know but not what I really want to know.

I have subscribed from time to time, and then, disillusioned, allowed the subscription to lapse, and occasionally, when a truly exciting piece appeared, I hunted in the newstands to find the one issue. I enjoyed Martin Gardner's columns and I was impressed and instructed by Philip Morrison's erudite book reviews, but most of the rest left me cold—and slightly guilty for feeling that way.

I wrote for the Scientific American twice. The first time was a behind-the-scenes description of Bourbaki (May 1957), and while I was bothered by the inflexible house rules, the subject was interesting enough, and I did a good enough job that the result was satisfactory. The second time I was invited: a whole issue was to be devoted to innovation in this and that, and would I write on innovation in mathematics? I tried, but the result was a catastrophe (September 1958). My original version (I wish I had kept a copy) was ruthlessly edited by a junior editor; the published version is watery, badly organized, badly expounded. It is as if I tried to popularize the statement and the solution of Hilbert's fifth problem (the one about locally Euclidean groups being Lie groups) by explaining the elements of trigonometry: pointless, ineffective, and dishonest. That's not in the article, it's only an analogy, but what is in is just as bad, or worse. Thus, for example, the article "explains" the contributions of James Clerk Maxwell by references to "the theory of functions of a complex variable (a variable containing the imaginary number i, the square root of minus one)". I didn't write that—so help me!—but horrified and exhausted I allowed it to stand, and I should be shot.

Much of the article is not my diction, not my arrangement, not my style—it upset me and made me feel ashamed when it appeared. I wrote to Dennis Flanagan, the dictatorial master-editor in charge of policy, to say "never again", and, of course, that's how it was—I never wrote for him again and he never asked me to. But he is still in power, and the Scientific American is still shiny and colorful and attractive, and how can I be sure that my critical attitude is right? The magazine seems to be a success—it keeps coming out and people keep buying it—and I cannot quarrel with success. But I don't enjoy reading it.

Being furious with Flanagan was not a full-time occupation. At about the same time as I was writing my Bourbaki piece, I had begun thinking about a second edition of FDVS. Van Nostrand was interested in putting out a commercial version, although back then, in the dark ages of the 1950's, it still wasn't clear whether linear algebra was the right stuff to foist off on defenseless undergraduates. I let myself in for several months of slave labor when I undertook the second edition, without knowing, when I started, whether I'd find the effort either intellectually or financially profitable. The final score is in now and the answer is: intellectually no; financially yes, a little, but was it worth the time?

I rewrote FDVS by rewriting it, that is re-copying it, every single word of it, in longhand. That was the only way I could keep my attention focused on the details. So far as I remember, every word of the original Princeton edition is in the second edition; the difference between them is the new material. I put in a little field theory in the beginning, and I had to learn multilinear algebra (alternating forms and the like) to treat determinants honestly, but most of my effort went to the exercises (well over 300 of them). I invented them, I stole them, I begged my friends for them—I considered them important. The printed version appeared in 1958 (a quarter of a century ago as this is being written), and I am pleased at how it stayed alive. It keeps selling, at about the rate of 800 copies each year—not a run-away bestseller, but definitely alive.

People sometimes ask (and sometimes do not dare to ask) how much money there is in being a mathematical author and editor; this is as good a time as any to offer at least a partial answer. My 800 copies of FDVS bring in royalties of about $2000. Author's royalties are usually between 5% and 15%, often rising from a low rate to a higher one after the first few thousand ("break-even") copies are sold; 10% is the most common figure. Editorial royalties vary from less than 1% to around 4%, again rising with sales; the most common figure is 2%. If a book has more than one author (many books do, but most don't), or if a book is published in a series that has an editorial board of more than one person (and most editorial boards are like that), the royalties are split. Typical example: it takes you two years to write a book that appears in a series with an editorial board of three people; it sells a total of 4000 copies, at $20 each, spread over five years, and then, sales having dropped off to essentially zero, it brings tears to your eyes by going out of print. Your total income will have been $1600 a year for each of the five years that the book stayed alive, and each member of the editorial board will have got $100 for your work during each of those years.

Speaking of second editions, I have a piece of advice to offer to any author who is thinking of writing one: don't. If the book is at the calculus level, and if you are avaricious, then I might just have given you some bad advice. Usually, however, to write a second edition is just like writing a book—it takes just as long and it takes just as much out of you—but it is nowhere near so rewarding. By "rewarding" I mean here to refer to the intellectual stimulation and emotional feeling of accomplishment that creating something can give you. The writing of second editions is just not creative enough. I know; I've done it. Twice.

The Institute again

In the later 1950's, while I was still a happy Chicagoan and a patriotic one, there were forces at work that were to result ultimately in my leaving Chicago. I loved a lot of people there, and vice versa, but I didn't love everybody, and vice versa. I continued to spend my quarters out of residence at far away places (England, Scotland, Seattle, Italy), and I received nibbles (and sometimes offers) of jobs (often chairmanships) from several parts of the country (Urbana, Illinois—Gainesville, Florida—Pullman, Washington—Iowa City, Iowa—Rochester, New York). The period I am talking about is approximately the four-year interval between 1957 and 1961.

They were good years, all of them, and they began well: in August 1957 I started a year's leave of absence, which I spent mostly at the Institute. The housing project next to the Institute grounds was brand new then—in fact it wasn't quite finished either physically or administratively. The electric "radiant heat" system worked sometimes—but during the bad blizzard it gave up, some of the unused apartments got so cold that the water froze in the flush-toilets, and Fred delivered huge armfuls of firewood to each occupied apartment. If you wrapped up in a blanket and huddled in the living room, near the fireplace, you might have trouble trying to get some work done, but you could stay warm enough. Fred was wonderful: he was a black giant, friendly and efficient, in charge of all maintenance problems. The Institute was feeling its way in the landlord business, and not all the visitors pronounced the initial attempts successful. Some apartments had two families assigned to them for a few confusing and worrisome hours, and the reservations of some visitors must have fallen under a desk in a secretary's office, and temporary emergency quarters had to be provided for a few uncomfortable days. I was lucky; 79 Einstein Drive had no major problems and I loved it.

Oppenheimer was the director of the Institute that year, and it was part of his job to invite every temporary member to his office for a ten-minute welcome interview. I don't think he enjoyed that part of his job. When I was called in for my ten minutes, we shook hands, and he began the conversation with "What's new in algebraic logic?". To me that sounded like a phony formality; he didn't know what algebraic logic was, and he didn't want to know what was new in it. To ease the atmosphere, I smiled and said "You were briefed very well". He didn't smile, the atmosphere remained sticky, and we finished the formalities formally. Much later, a couple of decades later, I told this story to Joe Doob, and he said "Well, that was rather a nasty crack, wasn't it?" I was amazed. I saw what Joe meant, but I had never thought of that view of the matter before. I didn't think I was being nasty, but perhaps I was. Was I?

It was in 1957, in Princeton, that I became healthy. I had passed 40 and I was beginning to feel it. Except for the occasional fierce game of tennis in my teens (I was not good at it), I never exercised; in my 30's I drove instead of walking and lay down instead of driving whenever I could. I was a moderate social drinker, which meant that I had a drink or two or three every day and was tipsy on the average once a week. I was a heavy smoker, which meant between 20 and 40 cigarettes a day, and more on party nights. I had been trying to keep my weight under control for the past few years, but I still had a sweet tooth and hated healthy vegetables; my idea of a perfect meal included large quantities of boiled lobster dipped in melted butter, a couple of slices of banana cream pie, and heavily sugared coffee with sugared whipped cream on top. A regimen like that, combined with the pressures of professional life and an apparently congenital case of chronic hypochondria, had predictable effects. I had symptoms. I had colds, I had headaches, my throat felt scratchy, my stomach was nervous, and my heart beat too fast. I was sure I had everything: tuberculosis, brain tumor, lung cancer, ulcers, and the last stages of total cardiac degeneration.

All this was not sudden and new, but it was getting worse. I was used to consulting the medical profession about every twinge, and used to being told that there was nothing much wrong—"just take these pills for a couple of weeks, and let me know if you're not better". What was different in Princeton was that the cumulative psychological effect was greater than before, and that I was lucky in finding a sensible doctor. Deane Montgomery recommended him to me (I can't even remember his name!). He knew I was only a visitor in Princeton, he didn't care a hoot about me, and he didn't mind being honest and tough. He X-rayed me, he took my blood, he peered into my eyes, ears, and nose, he pounded me and pinched me, and he made me assume uncomfortable and undignified positions. When it was all over, he said, in effect: "There is not a damn thing wrong with you, Halmos; why don't you just go out and take a long walk or something?"

I was ready for that advice. I was already a weight watcher (not an intelligent one, but at least a conscious one), and I had already been trying for a couple of years to give up smoking—now I started exercising too, and I became healthy. I walked, at a moderate pace, five minutes the first day (once around the circular driveway in front of the Institute), and added a minute or so every other day. I have kept up walking as exercise, systematically, compulsively, ever since. Now I walk a minimum of one hour every day, covering a minimum of four miles. On weekends I can snatch more time from other things and am likely to walk 8 or 10 and sometimes 15 or 20 miles a day. I try to eat a sensible diet, I haven't smoked in over 25 years, and I feel pretty good. Thank you, Dr. Whatever-your-name-was.

The Institute was larger in 1957 than ten years before—new buildings and new professors had begun to sprout and the population seemed to be about twice as large. There were twelve professors in the School of Mathematics instead of the founding six, plus Veblen who was emeritus. There were also three permanent members: Jimmie Alexander, who demoted himself to that rank long ago, plus Julian Bigelow and Herman Goldstine. The latter two were there because of von Neumann; they had worked with him before, and their job was to help him design and build his computer. By now von Neumann is long dead, and Herman worked for IBM for many years (without, however, giving up his permanent membership); the last I heard, Herman was back in Princeton, and both he and Julian still have offices at the Institute.

The common room of Fuld Hall, the Institute for Advanced Study, 1947. *On the floor, from left to right*: S. Minakshisundaram, I. E. Segal, E. H. Spanier, F. I. Mautner, S. Kaplan, *Behind*: R. A. Leibler, P. R. Halmos, W. F. Eberlein, H. Rubin.

The members of the Institute were geographically and professionally as mixed and as stimulating a group as during my earlier stays there. There were people from Eastern Europe (Sibe Mardešić and C. D. Papakyriakopoulos), from Western Europe (de Rham, Leray, Lorenzen, and Serre), from Israel (Aryeh Dvoretzky), from the left (Chan Davis), from Georgia (Tom Brahana—the topologist son of my erstwhile algebraic mentor, Roy Brahana), and from physics (Jeremy Bernstein). There were about 85 temporary members in the School of Mathematics that year (that number includes the physicists); the joint was jumping.

Jeremy Bernstein wasn't famous for his writing yet, but just chatting with him at lunch and at tea I could see that he was a man not only of charm but also of culture. Later he wrote many book reviews for the New Yorker, mainly of scientifically oriented books, and then profiles, and articles, and books. He blew a mean trumpet. Paul Lorenzen was at least as much a philosopher as a mathematician: he and I could talk some about logic. He looked and in some ways was as German as a Hollywood casting director's notion of what a Prussian army officer was like under Kaiser Wilhelm: tall, ramrod straight, handsome, formal, stiff. A monocle would have looked perfectly natural on him. He invited us over to his apartment, two doors down from ours on Einstein Drive, for a drink and a chat one evening; 8 o'clock, he said. When 8 o'clock came we were at home, getting ready to stroll over, but intending to wait for another 10 or 15 minutes so as not to arrive impolitely early. At 8:03 our phone rang. "This is Lorenzen. Did you forget? Aren't you coming?"

The Institute was good to me. My office was larger than it had been the time before, but I shared it with Dvoretzky. That caused me no difficulties, but it might have caused him some: I was beginning to work as Proceedings editor, the Institute gave me secretarial help, and when I was dictating, poor Dvoretzky couldn't concentrate on his work too well. We worked out the timing amicably—I tried to dictate around him and he tried to think around me.

I had many lunches with Veblen that year. Veblen was 77. He still had his office, and came to it almost every day, but he didn't have much to do. He had become almost blind, and he missed his administrative responsibilities and powers. He still looked beautiful—tall, white haired, erect, dressed in his four-button coat. His memory was perfect and he loved to reminisce. He was especially fond of reminiscing about his old days in Chicago, where he got his Ph.D. in 1903. I was from Chicago and I was interested—and, besides, I liked the old gent. He was by no means a boring monologist—he was as eager to hear about Chicago now as he was eager to tell about Chicago then—but he was lonesome and needed to fill the time. Frequently after lunch we went to his luxurious office, lounged in his soft leather chairs, and swapped Chicago stories.

Boolean algebras and sets

At the end of that Institute year I went to England and, mainly, to Scotland, to attend the 1958 Congress in Edinburgh. I had several friends in Britain, people whom I had met at meetings, or who had visited Chicago—they and my natural anglophilia made me feel very much at home immediately. I became a fervent scotophile. I loved Edinburgh in particular and the southern part of Scotland in general. I have been to Britain often since then, and I still have a special love for Scotland and the Scots.

The whole trip was pleasant to experience but not likely to be of interest as reading matter. I am a personal kind of traveler: I go for people (gossip and mathematics), not for scenery (mountains and landscapes) or culture (cathedrals and museums). We rented a car the moment we landed and took five minutes to get used to driving on the wrong side of the road; after that it was one affectionate reunion after another, up and down from Southampton to Scotland and back. Bill Cockcroft in Southampton, David Kendall in Oxford (he didn't move to Cambridge to become Mr. English Statistics till later), Peter Swinnerton-Dyer in Cambridge, H. R. Pitt in Nottingham—so it went.

Edinburgh itself was the same thing raised to a large power: many old friends and many glamorous new acquaintances. As for old friends: nine members of the Chicago mathematics department were there, more than were usually in residence in Chicago at any one time—Albert threatened to call a faculty meeting. As for new ones, of greatest interest were the ones from behind the Iron Curtain. Alexandrov was there (in fact they both were—P.S. and A.D.), and Sobolev, and Kurosh, and Shafarevich, and several from Hungary (Kalmár, Turán) and Poland (Marczewski, Sierpinski)—meeting people like that is the main thing international congresses are good for. Paul Alexandrov spoke perfect German, and A.D. used English; with Sobolev it was French, and with Kurosh fractured German. Polaroid cameras were not universally known yet, and when someone took a picture of Kurosh with one and made him a gift of it, he was surprised, impressed, and touchingly grateful—"spasibo, danke, danke, spasibo" he kept saying. The Hungarians insisted on speaking Hungarian with me, and the Poles spoke every language.

Nagy was not there; the 1956 troubles were not long enough past and he did not stand in good enough favor with the powers in Hungary to be allowed out. He sent his manuscript to Smithies, the hardest working functionary in charge, and asked that I present it. I did, of course; I talked about polyadic algebras for me and about operator dilations for him.

It was a well attended Congress, a good one; for me it was the first time in Europe for 28 years and the first time in Britain ever. Just the same, after a year away I was good and ready to go back home and dig in: the Institute is great and a Congress can be exciting and stimulating, but home was where my heart was, and I could think better about mathematics and about everything else sitting at my own desk, and with my familiar books and reprints within reach where my muscle memory could reach for them without even looking.

As far as mathematics goes, although I was still giving reports about ergodic theory and although I was leaning more and more toward operator theory, some of my heart was still in algebraic logic. A considerable part of my time and energy was spent on logic and its spin-offs. An early spin-off was a seminar I conducted on mathematical machines—that was even before I went to the Institute. Logicians do many different kinds of things: they do set theory, for instance, and formal syntax, and model theory (the interpretation, the semantics of the formal syntax), and, since Gödel, they have been doing more and more of computability. How do you define computability? Answer: via recursive number theory, or the so-called λ-calculus, or Turing machines—they all come pretty much to the same thing. I wanted to learn what that mysterious thing was and I persuaded a few students that they too wanted to learn more of it. They included Mike Morley and Anil Nerode (both of whom were already well on the way to being pros at that stuff), and the seminar chugged along quite well for a while, and even produced some notes, with some research ideas, that I keep saying I'll return to some day—if I live to be 99 and have returned to all the other things I keep saying that about.

Two other logic spin-offs were Boolean algebra and set theory. Boolean algebra was a course I gave in 1958, when I came back to Chicago from the Congress; a few years later my notes came out as one of the little Van Nostrand books. I have always loved that subject: it seems clean to me. It seems clean and neat and beautiful, like all mathematics, but, at the same time, it is an idealized, simplified, reduced, purified picture of mathematics. It has, to be sure, its dirty parts—as does everything, including mathematics—it *can* get complicated and ugly, and full of "large" cardinal numbers, and, of course, even the easy and beautiful parts are not trivial—the subject has deep theorems and challenging unsolved problems.

The main thing that attracts me to Boolean theory is that it is in a sense a discretized photograph of all mathematics through a telescope turned the wrong way. Almost everything in mathematics has its zero-one version in Boolean theory. (I wanted to say "everything in mathematics", but cowardice made me put in that "almost".) Where analysis treats functions, Boolean algebra treats sets (characteristic functions); the polynomial equations, matrices, and ideals of algebra have their small-scale Boolean versions; and the Boolean manifestation of general topology is that curious compromise between the finite and the continuous called the totally disconnected. I mention one example: the Boolean version of the Gelfand representation theory for commutative C^*-algebras is the Stone representation theorem. Indeed, Stone is an easy corollary of Gelfand, and the skeleton of the main steps in the Gelfand proof is visible in the Stone proof. More is true: the Gelfand proof can be arranged so that the Stone proof is its first step, the one that effects the main construction.

I am convinced (by faith) that if I knew everything about Boolean algebras I would be very close to knowing everything about analysis—or, to be a little more precise about such a vague article of religion, that I would be as close as a person who knows everything about measurable sets is to knowing everything about measurable functions. Close, yes, but not yet there. Was the method of exhaustion used by Archimedes to find the area under a parabola close to the theory of Lebesgue integrals such as $\int_0^1 x^2 dx$? Yes, it was close, but it took over 2000 years to close the gap.

In the spring of 1959 I was invited to a glamorous visiting professorship at the University of Washington, in Seattle. I taught algebraic logic (of course), and I got to know many members of the far-Western circuit who are not frequently visible elsewhere. Ed Hewitt was away that term, and for one quarter I was Ed Hewitt: I lived in his house, ate his breakfast cereal, drove his Porsche, worked in his office, and dictated to his secretary.

Washington had a curious tradition then: visiting faculty were treated just like permanent faculty in many more ways than is usual. When, for instance, job applications were under consideration, the letters of the applicants were read out at the departmental meetings to which I was invited, and my vote was solicited. On one occasion the departmental library had an unexpected surplus of a few hundred dollars, and the problem of what to do with the surplus was discussed in open meeting. My motion to buy a hundred copies of *Finite-Dimensional Vector Spaces* did not pass.

In the course of the years I got used to doing my "private" work (thinking and writing) at home in the morning (I was envied by all the patres-familias with pre-school children), and my "official" work (teaching, preparing, advising, corresponding) at the university in the afternoon. During the mornings of my Seattle months (they were called spring, but the weather was warm and wet—it was the soggiest summer of my life) I perched on the edge of Ed Hewitt's bed and wrote *Naive Set Theory*. I had recently learned a little about axiomatic set theory, and I was all gung-ho about spreading the word. I wanted to reach the student who didn't know and the reader who didn't like the formal manipulations of logic that frequently intervene. The result was a book on axiomatic set theory from the non-logical, non-formal, naive point of view.

I've been teased about my use of the word "naive", and translators (into Dutch, French, Hungarian, Italian, and Portuguese) have mistranslated the title in as many different ways as the languages they used. The Dutch called it "intuïtieve", the Italians "elementare", and the Hungarians used the word "elemi", which can mean "elementary" but can also mean something like simple, rudimentary, and just a wee bit stupid. The Portuguese used "ingênua", and the French avoided the issue by using no adjective at all—they just called it an introduction. The only courageous translators were the German ones: they called the book *Naive Mengenlehre*. My reason for the word was innocent. I didn't set out to find a provocative title (which I almost always do when I give a lecture); I was merely laboring under the misapprehension that "naive" was a bona fide technical word in set theory, meaning nothing more or less than the antonym of "axiomatic". I can't remember where I got the idea, but I think it was from one of the founding fathers—either Cantor himself, or Russell, or a later writer of some historical paragraphs contrasting the points of view of Cantor and Russell on the one hand and Zermelo and Fraenkel on the other.

Writing NST was fun and easy. It seemed clear that I had to start where I started, and that I had to take the next step in a uniquely determined direction, and the third step suggested itself after the first two. I felt that I had no choice, I had no organizing to do—the subject organized itself. The structure of the whole book was created while I was perched on the edge of Ed Hewitt's bed the first day, and from then on the book wrote itself. It took six months all told; three months in Seattle and three months polishing afterward. That's a record, for me; none of my other books was done that fast. I wonder: what would have happened if Hewitt's house had had a proper study, so that I didn't have to use a bedroom to work in?

The spin-offs stopped spinning after a while. I left logic as I had left ergodic theory, and moved on, permanently, to other subjects and other places. The move to operator theory was easy and natural. It was not a planned revolution, but a gentle step-by-step evolution—one little paper led to another. The move from Chicago was something else again; that was a wrench.

Farewell

I loved Chicago when I was in high school—I loved riding the streetcars all around town, I loved the lake, I loved the long straight streets, their beautifully systematic Cartesian coordinate system, and the rich, exciting, and full shops downtown. When I came back, years later, I

P. R. Halmos, 1949

relished the "victory" of the return, and I kept loving Chicago. I especially admired and loved the University of Chicago, and I still do, but gradually, as I lived there for fifteen years, I fell out of love with the city. It was noisy, dirty, and dangerous, and getting more so as the years went by.

Those things would probably have worried me no matter what, but they were not enough to make me leave; the main reason for my leaving was Saunders Mac Lane. He was the chairman of the department for six years, 1952–1958; and he didn't think much of me. We got along—there was no enmity, there was no feud—and he used to come to my office frequently to talk to me about the problems of his job. He liked to operate that way—think by asking questions—and I enjoyed our sessions and was flattered by them. Later, when I had decided to leave and was saying goodbye, Saunders, no longer chairman, told me that he too enjoyed them, and he thanked me; he said that my sympathetic ear and my views and my suggestions were a help to him. Perhaps he was just being polite, but I don't think so. When I said that he didn't think much of me, I meant mathematically.

S. Mac Lane and S. Eilenberg, 1967

Mac Lane's ambition as chairman was to build "the best goddam mathematics department in the world" (his words), and every second-rate mathematician he got rid of and could replace by a first-rate one was a step in the right direction. He thought I was second-rate. He understood Kaplansky's algebra, and he was impressed; he thought he understood my logic, and he wasn't impressed. He didn't understand my ergodic stuff and my operator stuff, and there was no reason why he should want to. There was no direct way he could get rid of me, but he tried hard to use indirect ways: keep my rank down as long as possible, keep my salary raises minimal, and recommend me for jobs elsewhere (often chairmanships) every time he had a chance to do so. I caught on, of course, and I was uncomfortable. I couldn't be and I wasn't "fired" from Chicago, and I didn't leave till three years after Mac Lane stopped being chairman, but I have always felt that in effect he fired me.

The department was changing at Chicago during the Mac Lane years. The pre-Stone-age people were out of it; they were too old, or over the hill for other reasons. Stone himself was being resented more and more: "he gets paid much more than he is worth, he travels too much, he is never here, he is of no further use to the department". I always got along with him very well and I am still grateful for all he did for me, but the power structure, both in the central administration and in the department, were beginning to turn against him. The power structure in the department consisted of the other people (other than Stone, that is) in their mid-forties and above, and it was becoming plain which ones of the rest of us would be the power structure five to ten years later. The 1959–1960 salaries (expressed in K's, where K stands for kilo-buck, or $1000.00) indicate something: Stone 20; Schilling 9.5 and Graves 12; Albert, Chern, Mac Lane, and Zygmund—all 16; and Kaplansky and I—both 13. Stone's *first* derivative was nearly zero, and so was my *second* derivative.

The chairmanship offers that kept coming in didn't attract me. I am good at paper work but not at people work, and much of being a chairman is the latter. With that I am uncomfortable even when I succeed; I am just not cut out to be a chairman. The non-chairman offers that I got had other things wrong with them—till, that is, I got friendly with Dan Hughes.

Dan had one of the temporary instructorships at Chicago for a couple of years, and when he left he went to Michigan. Michigan was the best mid-western state university, I think, and still is, and in those years it was relatively rich. The mathematics department was expanding and Dan suggested that they hire me. Somewhat to his surprise and mine "they" listened to him. George Hay was the Michigan chairman, and around the middle of August in 1960 he wrote and asked whether I wanted to be considered for a professorship. "Why not?", I said.

Michigan was good, quite good, but it just wasn't in the same league with Chicago. The mathematics department was much larger than Chicago's (bad), much of it was deadwood (bad), and the quality of the student body, both undergraduate and graduate, was way below that of Chicago's (bad). But the University of Michigan was respectable and respected—the best in its league, as I said—and that was good; I had visited there often before and had friends and acquaintances (good); and it was in a pleasant, clean, small town where good walks were possible and the crime rate was a small fraction of Chicago's (very good).

The negotiations didn't take long; in September, a month after the first nibble, I had a firm offer in my hand. They wanted me badly. My 1960 salary at Chicago was slated to be 14K; Hay offered me 20K for 1961, plus a generously paid three-month summer research appointment to begin with, plus a half-time teaching load for the first year (three hours a week), plus moving expenses, plus a teaching assistantship for my student Carl Linderholm, plus a cushy large office, plus a half-time secretary. Michio Suzuki, the group theorist, was being invited at the same time. I knew him slightly and liked him a lot; looking forward to being his colleague was a definite attraction. When that fell through, I was asked to suggest other possibilities. (My list consisted of Hy Bass, Ioan James, Victor Klee, Leopoldo Nachbin, and Ed Spanier. None of them came, but it was a good list.)

In other words, I was being offered a chance to change from a small frog in one of the best ponds to a very large frog indeed in an acceptably good pond. Adrian Albert was the chairman at Chicago that year, and he wanted me to stay. The first thing he did when I told him about the Michigan offer was to demand to see the dean that very day. The dean became visible in the late afternoon, and Albert phoned me at home as soon as he left the dean's office; my salary was to be raised from the budgeted 14K to 17K effective immediately. (Effective, that is, a year before my Michigan job was to begin.) Mac Lane was not happy, Adrian told me later; paying me that much "upset the salary balance of the department", and, of course, anything more would be totally unthinkable.

The day before Thanksgiving, on November 23, 1960, I wrote to Hay. "Dear George: O.K.—it's official. In other words: thank you very much for all your efforts; I accept the invitation."

When I told Albert my decision, I told him also that I wasn't leaving because of *his* chairmanship but because of Mac Lane's. He said he knew that; as Mac Lane's successor he had been punished more than once for administrative decisions that he hadn't made and never would have made. Chicago has hired many new people in the many years since I left, some much better than I and some not as good. My leaving has not harmed Chicago an iota, but, just the same, I am not humble enough to be convinced that my departure raised the quality of the Chicago mathematics department.

Why did I leave? Was it a cop-out? Was I running away from the challenge of the major leagues? I don't think so, but, obviously, I am not likely to be able to give an objective answer to the question. I saw many more reasons for leaving than for staying, the choice was mine, and when I made up my mind, I made the change with zest and enthusiasm—but ever since then, in odd moments, I've been wondering whether I did the right thing.

PART III
SENIOR

CHAPTER 12

How to teach

Shifting gears

Even the most junior departmental secretaries scared me when I first arrived in Chicago—they represented officialdom. If I asked one to do a five-minute duplicating job, and she said she'd have it for me the day after tomorrow, I said yes, sure, all right. I didn't really know how long the job should take, and I was brought up never to argue with the government. By the time I left Chicago I had learned that secretaries were human and that they were not always right ex officio.

Just before and during my last Chicago summer I did more administrative work of the kind that needed secretarial help than I had ever done before; it had to do with a little conference. It was a pretty effective conference; a good group of operator theorists were brought together to meet and to teach each other the latest goodies; it was not a waste of money. It came into being, however, for that purpose only—well, not exactly to *waste* money, but to spend it. I had an NSF grant that still had another year to run, and it was either impossible or, in any event, unusual and messy to transfer it to Michigan. Albert wouldn't even dream of returning the money to Washington—*that* would be wasting it, he said. Spend it, he said, quick; arrange a conference, or something.

Even a little conference takes a lot of arranging, and an impromptu one needs more. Secretaries wrote letters, made phone calls, looked up bus, train, and plane schedules and fares, made reservations in hotels and dormitories, and I found myself in charge, having to make decisions, being the boss. I have never really lost my eastern European notion that officialdom has to be treated gingerly, with the cautious respect halfway between that due to a non-commissioned officer and to a lunatic. That summer I learned, however, that a civil servant (such as a conference arranger or a secretary to one) is not necessarily the same as a saber-rattling subaltern in the Austro-Hungarian monarchy.

"Conference" may not be the right word, because the affair lasted more than just a few days; it lasted, in fact, from the middle of June to the middle of August. The crowd was small, about 20, but enthusiastic. We met for formal seminars twice a week, and for informal get-togethers over lunches, coffees, pizzas, and beers at least twice a day. That's when I first met Joe Stampfli, and that spectacularly promising "grandson" of mine, Charlie Berger (he was Moe Schreiber's student). Svetozar Kurepa was there, near the beginning of his career, and so was Nagy. Gianco Rota hadn't yet become king of combinatorics, and Chan Davis, Dick Kadison, Cal Moore, and John Wermer were not yet the famous names that they have become since then. We talked about single operators and algebras full of them, we talked about shifts and states and dilations and commutators; it was a satisfactory and successful summer.

When it was over, and I moved to Ann Arbor, my usually resigned but occasionally rebellious respect for authority had diminished but not disappeared. I still had (and still have) the attitude that while it may be all right to break the rules sometimes, it is imperative first to know what they are. George Hay, my new boss, expected me to be a difficult prima donna, and he was astonished at the number of times I came to see him during my first semester not to demand privileges but to learn procedures. Can NSF money for travel be used to pay a colloquium speaker? Can evening make-up classes be arranged for the ones the AMS meeting made me miss? Is it customary for paper-graders to take some of the office-hour load?

When I started my Chicago job I felt insecure but young and eager; when I left I was middle aged and established. I don't mean that I felt tired and ready to slow down. Almost the opposite: I was ready to convert some tentative gestures of the earlier years to ways of leading the academic life that were still new to me. I wasn't mathematically dead yet, but I became more and more involved in professional service (committee and editorial work), and, as far as teaching goes, in Ph.D. supervision and the Moore method.

As for mathematics, in the two decades after I left Chicago (1961–1980) I published about two dozen papers, and the best ones among them, the top six or eight, are definitely comparable in quality with the ones in the 40's and 50's. The first one of these "second wind" papers is on shifts. The basic work on that subject was done by Beurling for the so-called unilateral shift of multiplicity 1; the extension to shifts of finite multiplicity is variously attributed to Helson, Potapova, and Lax. My contribution was to formulate the statement for shifts of infinite multiplicity, and to offer a considerably simplified proof for the general case, all multiplicities.

As for teaching, I kept doing it of course, both elementary and advanced courses, but I was getting blasé; very few aspects of that part of my work could surprise me any more. I continued to enjoy classroom teaching and to think that I did a pretty good job, but the novelty had worn off. That might be an important reason why I was attracted by the challenge of the Moore method.

The Moore method

I can't remember when I became a Moore convert. I heard about R. L. Moore while I was still in graduate school; Felix Welch knew him at Texas and told stories about him, many legends were in general circulation, and I kept looking at Moore's blue Colloquium book hoping to make sense out of it. I never succeeded. I have always been fascinated by set-theoretic topology, and that's what the book seemed to be about, but the language was ugly and the theorems seemed much too special. Here is a sample (from p. 33 of the first edition).

"If M is a closed point set and G is an upper semicontinuous collection of mutually exclusive closed point sets such that every point set of the collection G contains a point of M and every point of M belongs to some point set of the collection G and K is a subset of M and W is the collection of all point sets of G that contain one or more points of K and no continuum of W contains a point of $M - K$ then W is a region with respect to G if and only if K is a domain with respect to M."

Moore felt the excitement of mathematical discovery and he understood the relation between that and the precision of mathematical expression. He could communicate his feeling and his understanding to his students, but he seemed not to know or care about the beauty, the architecture, and the elegance of mathematics and of mathematical writing. Most of his students inherited his failings as well as his virtues (diluted, of course); only the greatest, such as Wilder and Bing, could overcome the handicap of being a Moore student and become genuine mathematicians.

Moore insisted on using his own terminology, which was not that of the rest of the topological community, and he was inflexible about some of his mathematical attitudes. He wouldn't say "consider two points p and q..." for fear that in a degenerate case p could coincide with q, and then the number of points wouldn't really be two. He would never agree that the intersection of two sets is always a set, because, for him, there was no such thing as the empty set. These are forgiveable quirks of language. Worse than that, however, he was usually intolerant about every part of mathematics other than his own: algebra and analysis were different subjects, competitors, enemies. That's bad: that made him a lesser mathematician than he had the talent to be, and it made his less talented students much less well educated and useful than they could have been.

R. L. Moore, 1966

He was a Texan, almost the fictional prototype. He spoke Texan, he was politically rigid, he had strong prejudices, he stood up when a lady entered the room, and (the story goes) he wouldn't accept students who were black, or female, or foreign, or Jewish. (At least a part of that is false: he had women Ph.D. students, notably Mary Ellen Estill Rudin. So far as I know he did not have a black student.)

When I first heard about him, I was interested in the legend but emotionally neutral toward it. I have always had an ambivalent love-hate feeling to the South (Texas included). Insofar as it represented old-fashioned gentility and courtesy, and soft speech and warm weather and mint juleps at the race track, I admired it and wanted to be a part of it— but when it represented prejudice and bigotry and hatred and violence I was repelled. The gentle southern belle, with her melodious accent and youthful grace I found attractive, of course; but the superannuated ex-belle, the middle-aged or elderly harridan, selfish, demanding, grasping, and domineering—from her I turned and ran as fast as I could. Moore, the educated well-spoken Texan mathematician extraordinary, was a hero of mine; Moore, the mathematically outmoded and ethnically prejudiced reactionary power, was a villain.

An effective hero, a productive villain, everyone must admit. He turned out a record-breaking number of Ph.D.'s in mathematics; they loved him and imitated him as far as they could. He did it by what has come to be called the Moore method. It is also called the Texas or

Socratic or discovery or do-it-yourself method. I wish I could have seen it in action, instead of just hearing and reading about it, and, later, experimenting with modified versions of it. I got to know Moore slightly, and I did see the method in action, once, for one hour. When I passed through Austin one time, I screwed up my courage and knocked at Moore's office door. He received me politely and was willing to talk about his method. I was fond of the old man—he might have been bad and wrong in some ways, but he was human and strong, he was pleasant and interesting to talk with, and altogether I liked and admired him. (I enjoy remembering that when I took his photograph he insisted on combing his hair first—there was always a comb in his pocket—and he warned me that he would not smile—he explained that trying to smile made him look silly.) He seemed willing to put up with me. He invited me to stop by again, and I did.

I saw him that way, in Austin, perhaps four or five times. On one of those occasions he invited me to attend the first meeting of a calculus class he was teaching, and that was the only time I saw Moore in action. (I saw the movie about him, Challenge in the Classroom—in fact I saw it four times—but that's seeing him in action at a distance, not the real thing. Still, it's very good—I recommend it warmly.) He was good all right. He could draw the students out, he could keep things moving, he didn't bore anyone, his personality dominated the room.

What then is the Moore method? It begins with a preliminary interview: each prospective student must first go to see Moore and answer questions about his mathematical background. If he has already taken a course similar to the one he wants to be admitted to, if he knows too much, then the answer is an automatic no—he's out. A basic rule of the game is that all the players must start at the same place, and, in particular, must be untainted by the terminology, the notation, the methods, the results, and the ideas of others. A second rule is that the student, once he is admitted, must actively think, not passively read. Moore, so the story goes, would put students on their honor *not* to read about the subject while they were studying it with him, and, to lessen the temptation in their way, he would remove from the library all the pertinent books and journals. Now the class is ready to begin.

At the first meeting of the class Moore would define the basic terms and either challenge the class to discover the relations among them, or, depending on the subject, the level, and the students, explicitly state a theorem, or two, or three. Class dismissed. Next meeting: "Mr. Smith, please prove Theorem 1. Oh, you can't? Very well, Mr. Jones, you? No? Mr. Robinson? No? Well, let's skip Theorem 1 and come back to it later. How about Theorem 2, Mr. Smith?" Someone almost always could do something. If not, class dismissed. It didn't take the class long to discover that Moore really meant it, and presently the students would be proving theorems and watching the proofs of others with the eyes of eagles. One of the rules was that you mustn't let anything wrong get past you—if the one who is presenting a proof makes a mistake, it's your duty (and pleasant privilege?) to call attention to it, to supply a correction if you can, or, at the very least, to demand one.

The procedure quickly led to an ordering of the students by quality. Once that was established, Moore would call on the weakest student first. That had two effects: it stopped the course from turning into an uninterrupted series of lectures by the best student, and it made for a fierce competitive attitude in the class—nobody wanted to stay at the bottom. Moore encouraged competition. Do not read, do not collaborate—think, work by yourself, beat the other guy. Often a student who hadn't yet found the proof of Theorem 11 would leave the room while someone else was presenting the proof of it—each student wanted to be able to give Moore his private solution, found without any help. Once, the story goes, a student was passing an empty classroom, and, through the open door, happened to catch sight of a figure drawn on a blackboard. The figure gave him the idea for a proof that had eluded him till then. Instead of being happy, the student became upset and angry, and disqualified himself from presenting the proof. That would have been cheating—he had outside help!

Since, as I have already said, I have not really seen Moore in action in a serious mathematical class, I cannot guarantee that the description of the method I just offered is accurate in detail—but it is, I have been told, correct in spirit. I tried the method, experimented with it, kept running into tactical problems, worked out modifications that seemed to suit the students and the courses I was faced with—and I became a convert.

Some say that the only possible effect of the Moore method is to produce research mathematicians—but I don't agree. The Moore method is, I am convinced, the right way to teach anything and everything—it produces students who can understand and use what they have learned. It does, to be sure, instill the research attitude in the student—the attitude of questioning everything and wanting to learn answers actively—but that's a good thing in every human endeavor, not only in mathematical research. There is an old Chinese proverb that I learned from Moore himself:

I hear, I forget; I see, I remember; I do, I understand.

One time at the University of Michigan I taught a beginning calculus class by something like the Moore method. It was a small class (about 15 or 16 students) and it was called an honors class. The designation did not mean that the students were geniuses, or even that they were mathematics majors—all it meant was that they were brighter than average, and, as it turned out, cooperative and pleasant to work with. The size of the class was right—at least for me; if it had been twice as large I don't think I could have handled it à la Moore. I have had Moore classes of 3 or 4 students and some of 18 or 20, and the latter, as far as I'm concerned, is the absolute maximum that the method works for. In one recent class of more than 40 I was able to get quite a bit of participation from the students and thus to avoid the straight sermon system—but the participation came from about half the class only. The other half sat stolidly no matter what I did—I didn't even learn the names of most of them before the term ended (or they mine?). Incidentally, too few students is also a disadvantage—2 is too few, and usually so is 3 or 4. It takes a critical mass to generate the kind of cooperative competition that makes the method work—the ideal class size is between 5 and 15.

I don't have a story of spectacular success to report about my honors calculus class. It went well enough, I am sure I didn't do them any harm, and at the end of the term they were at least as ready for the next course as I could have made them with any other method. To be sure, that was the first (and, I'm sorry to say, the only) time I tried Moore on a class at that level. I have faith that with more experience I could have made the method work for ordinary classes.

Another time I used the Moore method on an honors class in linear algebra with about 15 students. The first day of class I handed each student a set of 19 dittoed pages stapled together, and I told them that they now held the course in their hands. Those 19 pages contained the statements of fifty theorems, and nothing else. There were no definitions, there was no motivation, there were no explanations—nothing but fifty theorems, stated correctly but brutally, with no expository niceties. That, I told them, is the course. If you can understand, state, prove, exemplify, and apply those fifty theorems, you know the course, you know everything about linear algebra that this course is intended to teach you.

I will not, I told them, lecture to you, and I will not prove the theorems for you. I'll tell you, bit by bit as we go along, what the words mean, and I might from time to time indicate what this subject has to do with other parts of mathematics, but most of the classroom work will have to be done by you. I am challenging you to discover the proofs for yourselves, I am putting you on your honor not to look them up in a book or get outside help in any other way, and then I'll ask you to present in class the proofs you've discovered. The rest of you, the ones who are not doing the presenting, are supposed to stay on your toes mercilessly—make sure that the speaker gives a correct and complete proof, and demand from him whatever else is appropriate for understanding (such as examples and counterexamples).

They stared at me, bewildered and upset—perhaps even hostile. They had never heard of such a thing. They came here to learn something

and now they didn't believe they would. They suspected that I was trying to get away with something, that I was trying to get out of the
work I was paid to do. I told them about R. L. Moore, and they liked
that, that was interesting. Then I gave them the basic definitions they
needed to understand the statements of the first two or three theorems,
and said "class dismissed".

It worked. At the second meeting of class I said, "O.K., Jones, let's
see you prove Theorem 1", and I had to push and drag them along
before they got off the ground. After a couple of weeks they were flying.
They liked it, they learned from it, and they entered into the spirit of
research—competition, discouragement, glory, and all.

If you are a teacher and a possible convert to the Moore method,
don't make the mistake that my linear algebra students made: don't
think that you'll do less work that way. It takes me a couple of months
of hard work to prepare for a Moore course, to prepare the fifty
theorems, or whatever takes their place. I have to chop the material into
bite-sized pieces, I have to arrange it so that it becomes accessible, and
I must visualize the course as a whole—what can I hope that they will
have learned when it's over? As the course goes along, I must keep preparing for each meeting: to stay on top of what goes on in class, I myself
must be able to prove everything. In class I must stay on my toes every
second. I must not only be the moderator of what can easily turn into
an unruly debate, but I must understand what is being presented, and
when something fishy goes on I must interrupt with a firm but gentle
"Would you explain that please?—I don't understand."

I am convinced that the Moore method is the best way to teach that
there is—but if you try it, don't be surprised if it takes a lot out of you.

Moore and covering material

The two biggest obstacles to the success of the Moore method (or,
for that matter, of teaching of any kind) are students who don't want to
be there and students who want to be somewhere else. The two are not
the same thing; let me explain.

By students who don't want to be there, I refer to required courses.
If a student comes to me and asks my help to learn something that I
already know, I am overjoyed. I am sure I can teach it to him, whatever
it is, and I expect the process to be pleasant for both of us. If, however,
he comes to me and says "I don't really want to know this stuff, but it's
required that I get a C in it before I can go out and make a lot of
money", then I'm unhappy.

Some universities used to require a mathematics course of all pre-med
students (sometimes more than just one), for no reason except to serve

as a sieve. The number of people preparing for admission to medical school was greater than the number that could be admitted—one way to lower the number of competitors was to flunk some in a mathematics course. Another reason for requirements is typified by chemists. They have to know some physics, and to learn the physics they have to learn some mathematics—it is a required course for reasons they don't fully understand and are inclined to resent. Similarly, students in the business school are required to learn some calculus, most of them for no reason at all and a few others for psychological reasons only: when, later in life, they meet an integral sign in an article on economics, they shouldn't be frightened out of their wits. A class of students who take a course because of such requirements is a sad and discouraging class. The highly motivated ones will work all right, and since most of them are not likely to be stupid, they will learn something, but the first prerequisite for the learning process to be both pleasant and effective— namely curiosity—is lacking, and that ruins it all. It ruins the teaching, it ruins the learning, it ruins the fun.

I dream of the ideal university, full of students who are full of intellectual curiosity. The subset of those among them who take a mathematics course do so because they want to know mathematics. They may be future doctors or chemists or executives in a shirt factory, but, for whatever reason, they *want* to find out what this mathematics stuff is about, and they come to me free willing and ask me to teach it to them. Oh, joy! If that really happened, I'd jump at the chance, and, at every chance I got, I'd use the Moore method. As matters stand, in *this* world, we teachers can only do the best we can with what we've got. Some of us do it in a good-natured fashion and some not, but we all teach a lot of mathematics as a trade. Under certain circumstances you remove parentheses, and here is how you do that; at other times you replace x by $\tan \theta$; and at still other times, very rarely, you choose δ to be $\varepsilon/3\sqrt{m}$—follow the rules, and you won't go wrong. It's a living, and for those of us teachers who like mathematics, it leaves time to study the subject and think about it between classes, in the evening, and on weekends—it's not a bad life.

In any event, the Moore method is not certain to work with students who don't want to be there. Its second main obstacle (students who want to be somewhere else) is smaller than the first, but frightens some teachers anyway: how can you use the Moore method in Math. 307 and still be sure that your students will be prepared for Math. 407 next year? Isn't it true that you don't "cover" as much material that way?

Yes, it's true; no, it's not true. If the course description insists that you teach your 307 students each of a list of 31 theorems—teach in the sense of state, prove, and ask about on examinations—so that they'll be prepared to have the next 31 theorems taught to them in 407, then hurry up and start lecturing—it's better than getting fired. If, however, 307 is advanced calculus, say, and 407 is "baby real variables" (the story is the same for any other pair of courses in sequence), then ask yourself (and have your department ask itself) just what is the purpose of the game. Surely the course description is not an end in itself. The reason for the 31 theorems in 307 is to put the student in position to understand the theorems in 407, and the reason for them, in turn, is to understand some other theorems (for the mathematician) or to understand and apply them (for the chemist and the economist). To be able to understand and to be able to use—that's what we really try to teach, and that's what the Moore method does better than any other.

From my experience and from conversations with other users of the Moore method, orthodox and reformed, I conclude that the business of "covering" material is a red herring. Other methods cover more in less time, some say; Moore is a luxury we can't afford. I'm sure that's wrong. I feel strongly about it, strongly enough to interrupt the discussion of the Moore method, and say something about "covering" things.

There is no use trying to teach Calculus 2 to those who haven't learned Calculus 1—that seems to be true enough. Let me ask something, however, Moore or no Moore—is it true that the teacher of Calculus 1 always covers the six chapters that the course description demands and that, therefore, the students who passed Calculus 1 are ready to begin Chapter VII? Listen to almost any teacher of Calculus 2 at almost any university griping about his predecessor, the one who taught Calculus 1 last year: "Why, he didn't even mention partial fractions, and I don't know what he thinks he taught them about trigonometric substitution, but they sure don't know it. I had to spend three weeks doing first year stuff—no wonder I got behind." Description or no description, Moore or no Moore, the material that course N covers is never—hardly ever—what course $N + 1$ needs.

It's hard to teach a non-trivial course in probability to people who cannot use the exponential function. I taught one once, and the prerequisite was plainly stated (Calculus 2, or whatever it was called), and the prerequisite was supposed to have treated e^x. Five of my 25 students knew that it was its own derivative, and two of those had heard the rumor that it was essentially the only function with that property. Since, however, none of them had any facility with infinite series (and were uncomfortable even when I "reminded" them of the formula for the sum of a geometric series), the first two or three weeks of the probability course were in effect the last two or three weeks of Calculus 2. (Infinite series were in Chapter XI of the calculus text, and "nobody ever gets that far".)

That has nothing to do with the Moore method—it just seems to be a fact of life. What can be done about it? What seems to work is an intelligent syllabus for each course. An intelligent syllabus is one that tells you what tools you *must* have to be able to enter the course and cope with it, and what you should take with you when you leave it, so as to be able to cope with the next one. Both the start and the finish should be described explicitly. It is not good enough to refer to one book, the principal text, and say "the course consists of the first six chapters". Much better: "The indispensable prerequisites are the following ten theorems or techniques or formulas. (1) Integration by parts, as in Text *A*, pp. 81–93, or Text *B*, pp. 7–11, or Text *C*, pp. 207–234...." Similarly: "The *minimum* that the course is designed to cover is the following set of 31 theorems. (1) Term-by-term integration, as in ..., etc."

The syllabus should be explicit but not rigid. It should, to begin with, not try to fill every minute of the term; it should leave some time, say one third of the total, to the teacher's discretion. That is realistic, and it replaces the atmosphere of a cram session by something like the spirit of higher education. Every teacher has, and should be encouraged to have, private prejudices about what are the prettiest, or most useful, or most important parts of the course, and should be given the chance to insert those individualistic prejudices when they are appropriate. Teachers aren't, and shouldn't be, interchangeable parts of a machine, and students should have a chance to learn about different points of view. Another way that a syllabus could be rigid, and obviously shouldn't be, is to be thought of as something fixed for all time. Mathematics changes, fashions change, applications change, books change— syllabi should change as often as necessary. Every year is probably too often, and every ten years is probably not often enough—curriculum committees can easily stay on top of such matters.

Would syllabi have helped my probability course? Probably, in two ways. If a student sees the exponential function, and infinite series, in the minimum list that his calculus course should cover, he might act as part of his teacher's conscience. "When are we going to do infinite series? Is that really important? Will I need it when I take Statistics 3 next year?" He might, if the calculus teacher doesn't teach series, look it up and learn it by himself—improbable, in our modern university atmosphere, but not totally impossible. That's one of the two ways: the student is given a chance to act responsibly about the calculus syllabus. Emphasis: syllabi are not to be just secret score cards that the teacher knows about—they must be in the hands of the students. The other way a syllabus can help, and might have helped the probability course, is as a warning to the prospective student: if you haven't heard of exponentials and series, you're likely to be in trouble. Learn them first, or don't take probability.

A syllabus is a sheet of paper—it doesn't, by itself, work magic. One reason it could work, however, is that it spreads the responsibility, it helps the student and the teacher share it. It's no tragedy if not everything is covered, so long as everyone knows just what was supposed to be covered. Gaps can be filled, if we know what they are.

Many of my colleagues are pessimistic (or are they just realistic, or are they cynical?) about the likelihood that students will look up something in a book and learn by themselves. Students, many of us seem to think, are stupid and badly trained, and they must be spoon-fed. Sometimes that may be true, but so long as we believe it to be always true, we'll never change any of it. If business school students are too stupid to learn calculus except by spoon-feeding, and, even then, if what we teach them is just a bunch of mechanical rules, is it really wise to try to teach it to them? Some books are badly written and hard to read, and some students are stupid or badly trained—but if we always act "realistically" and refuse to try to improve matters, we'll almost certainly go from bad to worse.

Syllabi, intelligently designed ones, would be a great help for advanced courses, graduate courses, as well as for elementary ones. Graduate students do come to us "free willing"—much more than most undergraduates—and they are somewhat less frightened of studying by themselves. Syllabi would help keep the courses straight (you better know what the closure of a set is before you sign up for the complex variable course), and they would be a great help with the part of the examination system that spans several courses. Most Ph.D. programs have a qualifying examination as one of their major parts, and students frequently complain that they had algebra from Jones last year, but Robinson, who is teaching the course this year, emphasizes completely different topics, and he is the one who made up the exam. If both Jones and Robinson were bound (at least 2/3 bound) by a good syllabus, some of the grounds for the complaint might go away.

Syllabi help the teacher too, even the dilatory one, the one who finds himself two weeks from the end of term with six weeks of the syllabus still to go. You *can* present the last 12 theorems in two weeks instead of six, and you might even do it better than if you had six weeks to plough through all the computations. Here is how. For each theorem, describe the context it belongs to, the history that produced it, and the logical and psychological motivation that makes it interesting; state it, perhaps roughly, intuitively at first, and then precisely; and, finally use it—show what other results it makes contact with, and what follows from it. Take any of your favorite theorems and imagine yourself doing that—the fundamental theorem of projective geometry, or of Galois theory, or even of calculus, or, if you prefer, some results of a less sweeping kind, such as, say, the simplicity of the alternating groups, or the construction of the Cantor function. Your students will have to do without seeing you put your carefully prepared complicated proofs on the board. Will they really miss them very much? Proofs can be looked up. Contexts, histories, motivations, and applications are harder to find—that's what teachers are really for.

If you do what I just described, then, in part, you are already using a modified Moore method. It doesn't matter why you were behind the syllabus—because you told too many jokes in class, because you got mixed up in your earlier proofs, or because you cajoled the earlier proofs out of your students à la Moore. There is my point; the defense rests. What I just said is that with a little elasticity you can even adapt the Moore method to "covering" a previously fixed amount of material. You'll be covering a lot of it a lot better than your lecturing colleague (namely, the part that your students were led to discover and prove by themselves), and you'll be covering a part of it, all but the details of the proofs at the end, well enough so that your Moore-trained students are not likely to miss your doing for them what you've trained them to do for themselves.

How to be a pro

Anna, one of the mathematics department secretaries at Chicago, complained to me once when we met at a party and she had already consumed several gin and tonics. She was a good typist, and she didn't have much trouble picking up the art of technical, mathematical, typing —but she hated every minute of it. You have to look at every symbol separately, you have to keep changing "typits" or Selectric balls, you have to shift a half space up or down for exponents and subscripts, and you have no idea what you're doing. "I'm not going to spend my life flipping that damned carriage up and down", she said.

Another time Bruno, a well-known mathematician, asked me: "How did you manage to make your ten lectures at the CBMS conference all the same length? Isn't it true that some things take longer? When I did it, some of my talks were 45 minutes, and some 75."

More recently still I asked Calvin, a colleague: "Can you give a graduate student a ride to this month's Wabash functional analysis seminar?" "No", he said, "you better get someone else. I've been away, giving colloquium talks twice this month, and that's enough travelling."

I didn't make up any of this (except the names). Do you see what these three stories have in common? To me it's obvious, it jumps to the eye, and it horrifies me. What Anna, Bruno, and Calvin share and express is widespread and bad: it's the "me" attitude. It is the attitude that says "I do only what's important to me, and I am more important to me than the profession."

I think an automobile transmission mechanic should try to be the best automobile transmission mechanic he has the talent to be, and butlers, college presidents, shoe salesmen, and hod-carriers should aim for perfection in their professions. Try to rise, improve conditions if you can, and change professions if you must, but as long as you are a hod-carrier, keep carrying those hods. If you set out to be a mathematician, you must learn the profession, every part of it, and then work at it, pro-fess it, live it as best you can. If you keep asking "what's there in it for me?", you're in the wrong business. If you're looking for comfort, money, fame, and glory, you probably won't get them, but if you keep trying to be a mathematician, you might.

An artist must know his medium and make the best possible use of it. If you set out to paint a miniature, don't complain that you don't have enough room. The size is an intrinsic part of the medium and you adapt to it the way you adapt to breathing on a mountain top—you have no choice. That's why Bruno's question sounded to me like asking why Michelangelo's statues aren't in the key of C or why Thackeray didn't write in cobalt blue. A vital part of the definition of each creative art is the medium in which it is expressed. An hour's lecture is, by its most common definition, a 50-minute performance, not a chapter in *your* book or a complete exposition of *your* recent work on harmonic analysis. When you agree to give a one-hour lecture, you do your best to give a one-hour lecture, making allowances for the level of the audi-ence, estimating the probability of interruptions by questions, and organizing the material to fit the medium.

I didn't preach to Calvin, but if I had done so, my sermon would have said that you support an activity such as the Wabash seminar (I'll tell more about that seminar later) without even thinking about it—it's an integral part of professional life. You go to such seminars the way you get dressed in the morning, nod amiably to an acquaintance you pass on the street, or brush your teeth before you go to bed at night. Sometimes you feel like doing it and sometimes not, but you do it always—it's a part of life and you have no choice. You don't expect to get rewarded if you do and punished if you don't, you don't think about it—you just do it.

A professional must know every part of his profession and (we all hope) like it, and in the profession of mathematics, as in most others, there are many parts to know. To be a mathematician you have to know how to be a janitor, a secretary, a businessman, a conventioneer, an educational consultant, a visiting lecturer, and, last but not least, in fact above all, a scholar.

As a mathematician you will use blackboards, and you should know which ones are good and what is the best way and the best time to clean them. There are many kinds of chalk and erasers; would you just as soon make do with the worst? At some lecture halls you have no choice—you must use the overhead projector, and if you don't come prepared with transparencies, your audience will be in trouble. Word processors and typewriters, floppy disks and lift-off ribbons—ignorance is never preferable to bliss. Should you ditto your preprints, or use mimeograph, or multilith, or xerox? Who should make decisions about these trivialities—you, or someone who doesn't care about your stuff at all?

From time to time you'll be asked for advice. A manufacturer will consult you about the best shape for a beer bottle, a dean will consult you about the standing of his mathematics department, a publisher will consult you about the probable sales of a proposed textbook on fuzzy cohomology. Possibly they will be genteel and not even mention paying for your service; at other times they will refer delicately to an unspecified honorarium that will be forthcoming. Would they treat a surgeon the same way, or an attorney, or an architect? Would you? Could you?

I am sometimes tempted to tell people that I am a *real* doctor, not the medical kind; my education lasted a lot longer than their lawyer's and cost at least as much; my time and expertise are worth at least as much as their architect's. In fact, I do not use tough language, but I've long ago decided not to accept "honoraria" but to charge fees, carefully spelled out and agreed on in advance. I set my rates some years back, when I was being asked to review more textbooks than I had time for. I'd tell the inquiring publisher that my fee is $1.00 per typewritten page or $50.00 per hour, whichever comes to less. Sometimes the answer was: "Oh, sorry, we didn't really want to spend that much", and at other times it was, matter-of-factly, "O.K., send us your bill along with your report". The result was that I had less of that sort of work to do and got paid more respectably for what I still did. My doctor, lawyer, and architect friends tell me that prices have changed since I established mine. The time has come, they say, to double the charges. The answers, they predict, will remain the same: half "no" and half "sure, of course".

A professional mathematician talks about mathematics at lunch and at tea. The subject doesn't have to be hot new theorems and proofs (which can make for ulcers or ecstasy, depending)—it can be a new teaching twist, a complaint about a fiendish piece of student skullduggery, or a rumor about an error in the proof of the four color theorem. At many good universities there is a long-standing tradition of brown-bag (or faculty club) mathematics lunches, and they are an important constituent of high quality. They keep the members of the department in touch with one another; they make it easy for each to use the brains and the memory of all. At a few universities I had a hand in establishing the lunch tradition, which then took root and flourished long after I left.

The pros go to colloquia, seminars, meetings, conferences, and international congresses—and they use the right word for each. The pros invite one another to give colloquium talks at their universities, and the visitor knows—should know—that his duty is not discharged by one lecture delivered between 4:10 and 5:00 p.m. on Thursday. The lunch, which gives some of the locals a chance to meet and have a relaxed conversation with the guest, the possible specialists' seminar for the in-group at 2:00, the pre-lecture coffee hour at 3:00, and the post-lecture dinner and evening party are essential parts of the visitor's job description. It makes for an exhausting day, but that's how it goes, that's what it means to be a colloquium lecturer.

Sometimes you are not just a colloquium lecturer, but the "Class of 1909 Distinguished Visitor", invited to spend a whole week on the campus, give two or three mathematics talks and one "general" lecture, and, in between, mingle, consult, and interact. Some do it by arriving in time for a Monday afternoon talk, squeezing in a Tuesday colloquium at a sister university 110 miles down the road, reappearing for a Wednesday talk and a half of Thursday (spent mostly with the specialist crony who arranged the visit), and catching the plane home at 6:05 p.m. Bad show—misfeasance and nonfeasance—that's not what the idea is at all. When Bombieri came to Bloomington, I understood much of his first lecture, almost none of the second, and gave up on the third (answered my mail instead)—but I got a lot out of his presence. I heard him hold forth on meromorphic functions at lunch on Monday, explain the Mordell conjecture over a cup of coffee Wednesday afternoon, and at dinner Friday evening guess at the probable standing of Euler and Gauss a hundred years from now. We also talked about clocks and children and sport and wine. I learned some mathematics and I got a little insight into what makes one of the outstanding mathematicians of our time tick. He earned his keep as Distinguished Visitor, not because he is a great mathematician (that helps!) but because he took the job seriously. It didn't take his talent to do that; that's something us lesser people can do too.

Mathematical talent is probably congenital, but aside from that the most important attribute of a genuine professional mathematician is scholarship. The scholar is always studying, always ready and eager to learn. The scholar knows the connections of his specialty with the subject as a whole; he knows not only the technical details of his specialty, but its history and its present standing; he knows about the others who are working on it and how far they have reached. He knows the literature, and he trusts nobody: he himself examines the original paper. He acquires firsthand knowledge not only of its intellectual content, but also of the date of the work, the spelling of the author's name, and the punctuation in the title; he insists on getting every detail of every reference absolutely straight. The scholar tries to be as broad as possible. No one can know all of mathematics, but the scholar can succeed in knowing the outline of it all: what are its parts and what are their places in the whole?

These are the things, some of the things, that go to make up a pro.

Musings on teaching

A big part of my profession is to profess it—that is, to be a professor, a teacher—and I've done a considerable amount of professing, both in the classroom and out. Are there any tricks of the trade, and have I picked up any?

No, not really. Some things are obvious: to teach a subject you must know it, you must love it, you should know your students and take them seriously, and you must whenever possible anticipate and avert their difficulties and misunderstandings. As a teacher you are *in loco parentis*, and you must be ready with realistic advice on more subjects than your complex variable course prepared you for. "Can I take linear algebra before advanced calculus?" (Why not?) "Do I need all this math. stuff for a major in computer science?" (If you need to ask such a question, you shouldn't be in computer science.) "Will it hurt my chances if I drop out for a year and go to Europe?" (You bet. Dropouts never finish—well, hardly ever.)

Much administrative nonsense has been said and written about the mechanical quotidian aspects of teaching. A typical example is the argument between semesters and quarters: adherents of each are convinced that the other is an invention of the devil. The quarter system makes good use of all twelve months instead of wasting three of them, says one side; the quarter system, with three registrations and three exam periods in nine months, instead of two, is clumsy and inefficient, says the other. A semester is long enough to cover a decent amount of material, say some, but quarters make it possible to offer greater variety, say the others. As far as I'm concerned it's stuff and nonsense, all of it. I have worked with both systems for many years, and I am convinced that educationally, intellectually, it makes not a whit of difference which you use. Just tell me what the subject is, when the term starts and ends, and how often and for how long classes meet in between, and I can plan a course to fit those conditions. Sure, I won't "finish" analytic geometry in ten weeks—do you ever finish anything in fifteen? [This particular conflict, incidentally, looks as if it is doomed to become moot before long— quarters are on the way out and the streamlined semester, between Labor Day and Christmas, is winning nearly everywhere.]

If the Master Schedule Designer in the Administration Building is told that the university is changing from, say, semesters to quarters, he does his job by counting minutes. Calculus 1 used to meet for three periods of 50 minutes during each of 15 weeks, less Thanksgiving— that's 2200 ($=((3 \times 14) + 2) \times 50$) minutes. To give the same course in 11 weeks, we have to make it four hours a week—it's that easy. A similar problem arises when we want to offer a concentrated course "covering" both Calculus 1 and Calculus 2 during the 12-week summer session. The arithmetic says that we must allow 366.66 ($=4400/12$)

minutes per week. A good approximation is to arrange a 75-minute class for five days each week from the end of May through the middle of August, excepting only the fourth of July.

I am not poking fun at an imaginary goblin—such calculations are often performed and decisions are based on them. If the campus grows too large to make it possible for thousands of students to change classes in ten minutes, all we do is change the definition of an hour—and I didn't invent that either. At some places the first class period of the day is from 8:00 a.m. to 8:50, the second from 9:05 to 9:55, the third from 10:10 to 11, and so on. Result: the classroom hour remains 50 minutes, the interval becomes 15, the total number of minutes in each course (presumably established by divine fiat a century ago) stays the same as it has always been, and confusion reigns.

Stuff and nonsense, all of it. What it takes to learn a difficult challenging subject (calculus, say) is measured in maturity, not minutes. It takes time for new ideas to become a part of you, to seep into your subconscious so that you can remember and use them effortlessly. For most people it takes several months to learn calculus—a year, more or less—and, within very broad limits, the number of minutes is totally immaterial.

After we have divided the year into the right number of weeks, and the weeks into hours consisting of the right number of minutes, the main problem still remains: how do we teach mathematics? How can we teach abstract concepts and the relations among them, how can we teach intuition, recognition, understanding? How can we teach these things so that when we are done our ex-student can not only pass an examination by naming the concepts and listing the relations, but he can also get pleasure from his insight, share it with others, apply it to the "real world", and, if he is talented and lucky, be vouchsafed the discovery of a new one?

The answer is that we cannot. The only way I know of for an individual to share in humanity's slowly acquired understanding is to retrace the steps. Some old ideas were in error, of course, and some might have become irrelevant to the world of today, and therefore no longer fashionable, but on balance every student must repeat all the steps— ontogeny must recapitulate phylogeny every time.

What then can we do to earn a living? Can the mathematician of today be of any use to the budding mathematician of tomorrow? Yes. We can point a student in the right direction, put challenging problems before him, and thus make it possible for him to "remember" the solutions. Once the solutions start being produced, we can comment on them, we can connect them with others, and we can encourage their generalizations. Almost the worst we can do is to give polished lectures crammed full of the latest news from fat and expensive scholarly journals and books—that is, I am convinced, a waste of time.

You recognize, I am sure, that I am once more advocating something like the Moore method. Challenge is the best teaching tool there is, for arithmetic as well as for functional analysis, for high-school algebra as well as for graduate-school topology. When I was teaching freshman algebra I challenged my students by asking them to solve the quadratic equation $cx^2 + bx + a = 0$. Some of them became upset and angry ("unfair—dirty trick"), but all of them (except the best two or three) learned something from the twist. Another time I asked them to graph $y = x^5 - 5x^3 + 5x$, with similar results. (As x goes through the values -2, -1, 0, $+1$, and $+2$, so does y; half the graphs on that quiz were straight lines.)

Here is a bit of innocent fun that is not much of a challenge, but most calculus students seem to enjoy it. Partly as integration drill and partly to make a point about the use of "dummy variables", I'd call on several students, one after another, and demand that they tell me what is $\int \frac{dx}{x}$, $\int \frac{du}{u}$, $\int \frac{dz}{z}$, $\int \frac{da}{a}$, and then, as the clincher, I'd ask about $\int \frac{d(\text{cabin})}{\text{cabin}}$. Some of them would grin amiably and shout out "log cabin", and they were surprised when I told them that I didn't agree. The right answer (as I learned when I was learning calculus) is "houseboat"—"log cabin plus sea".

At the same time, by the way, I'd take advantage of the occasion and tell my students that the exponential that 2 is the logarithm of is not 10^2 but e^2; that's how mathematicians use the language. The use of ln is a textbook vulgarization. Did you ever hear a mathematician speak of the Riemann surface of ln z? And speaking of vulgarizations, did you ever hear a mathematician pronounce "-3" as "negative three"?

These things remind me of some examination questions, on a different level, that deserve the legendary status they have attained. At an oral Ph.D. qualifying exam one time, the story goes, it was Azumaya's turn to do the questioning. "Galois?", he said, and that's all it took; the candidate was off and running. A similar story concerns a written exam that Zorn gave to a class in advanced calculus. He wrote it on the board; it looked, in toto, like this:

$$e^x.$$

Zorn denies the story, but I still like it; doesn't it give you an idea? According to Zorn's version it was a graduate course, and the exam looked like this:

$$\sin z.$$

One more, this time about Tamarkin. On a Ph.D. oral he asked the candidate about the convergence properties of certain hypergeometric series. "I don't remember", said the student, "but I can always look it up if I need it." Tamarkin was not pleased. "That doesn't seem to be true", he said, "because you sure need it now."

Then there is the Titchmarsh story, which is not about exams, but has the same kind of texture. Titchmarsh announced one autumn that he would offer a connected series of lectures, a double course as it were, lasting not one academic year but two. The first half of the series ended in April, and the second half began on schedule in October. The audience was assembled, Titchmarsh strode in, picked up a piece of chalk, and said "Hence, ...".

But let's get back to teaching by challenging. An intrinsic aspect of the method at all levels, elementary or advanced, is to concentrate attention on the definite, the concrete, the specific. Once a student understands, really and truly understands, why 3×5 is the same as 5×3, then he quickly gets the automatic and obvious but nevertheless exciting unshakable generalized conviction that "it goes the same way" for all other numbers. We all have an innate ability to generalize; the teacher's function is to call attention to a concrete special case that hides (and, we hope, ultimately reveals) the germ of the conceptual difficulty.

To teach "real variables" by beginning with the Peano axioms, proceeding through the rational numbers as equivalence classes of ordered pairs of integers, and then passing to the real numbers via Dedekind cuts—the classical method—is a crime. Begin with interesting facts instead of obvious axioms and artificial definitions. Begin with examples of strange numbers (the Liouville transcendental numbers are accessible the first day of the class), strange closed sets (the Cantor set is hard to beat), and strange functions (the characteristic function of the irrationals has its uses). Look at examples and counterexamples, look at continuous and discontinuous functions, look at uniform convergence and its opposite. Make such things friendly and familiar, the same way as we all become friends with 3, and 5, and 15 in elementary school—and *then* raise the question of what could have been a good starting place to get here from. It takes forever to go from minus to plus infinity; it's much better to work toward both ends from the middle.

The hardest part of teaching by challenging is to keep your mouth shut, to hold back. Don't *say*; ask! Don't replace the wrong A by the right B, but ask "where did A come from?" Keep asking "Is that right? Are you sure?" Don't say "no"; ask "why?".

How to supervise

Hilbert once said, or so the story goes, that supervising a Ph.D. thesis is just like writing a paper under exceedingly trying circumstances. He was right, of course—more than half right anyway. Directing a Ph.D.

student is a kind of teaching, perhaps the most challenging and interesting kind. It is part of the job of being a mathematician, and it is like many parts of many jobs in that the done thing is to grumble about it. It is, however, a bit of an honor to be approached by a student and asked to be his thesis supervisor—and it's bad form, very bad, for the professor to do the approaching. It happens, of course, and it often works (what graduate student would dare say no?), but if you do, you can expect to be frowned upon and gossiped about. (I'll be right there, frowning and gossiping.)

Lecturing is one of the ways of teaching, perhaps the easiest way and certainly not the best. I am sure we all know some simply awful lecturers—and one of the worst I ever knew is one of the best teachers I ever knew. He and I were colleagues for several years, and he, the worst lecturer, usually attracted the most talented students and turned out the best Ph.D.'s. He inspired them, annoyed them, and goaded them on to higher achievement; he fought with them and fought for them, and, like the irritating grain of sand in the oyster, he produced pearls.

A curious relation, the one between a Ph.D. candidate and his advisor. You might think that a student likes finite groups, say, so he seeks a supervisor who is an expert on finite groups. That happens, sometimes, but usually life is not so simple. To begin with, a student chooses the graduate school he goes to partly by where his parental home is, partly by the advice of his undergraduate teachers, partly by where he was able to get a fellowship or assistantship, and in a great part by chance. Once he is in a graduate school he *can* change and go somewhere else to find the ideal supervisor—but he is not likely to do so. He'd have to face a new bureaucracy, admission requirements, out-of-state obstacles, and a different examination system. The force of inertia is great: better stay put. He must, therefore, choose his supervisor from the mathematically active people where he happens to be—if no one cares about his finite groups, or if the one who does is too busy to take him on, he must go into differential topology instead.

Young and middle-young professors, eager to be thought important, are eager to acquire Ph.D. students. It is points in their favor if they succeed—it becomes part of their dossier. If you do *not* have any Ph.D. candidates working with you, that too is noticed when your seniors discuss raises, tenure, and promotions.

The hardest job of the thesis supervisor is to get the student started —to suggest a problem for the student to work on. Some supervisors hand each student a theorem and say "prove it"; others go to the opposite extreme. Example: a Columbia student goes to Sammy Eilenberg's office and says "Sir, I'd like to write a Ph.D. thesis". Says Sammy: "Why don't you?". I should add, in all fairness, that, according to another version of the story, that used to be the Columbia attitude before Sammy changed it. Another possibly apocryphal quotation is attributed to Marshall Stone. "If I had any good problems", he's alleged to have said, "I'd work on them myself."

It's not easy to find, to discover, to invent(?) good new problems. I have known people, several of them, who, once they were lucky and stumbled across a problem that promised to be fruitful, acted like misers—they hoarded it, and kept it secret, they were deathly afraid that someone would steal it from them. I don't think that's either morally or strategically right. In my experience it's far better to "cast thy bread upon the waters: for thou shalt find it after many days" [Ecclesiastes 11/1]. Time after time new ideas occurred to me even as I was explaining a vague problem possibility to a student, and mentioning such possibilities to colleagues at coffee time is much more likely to result in gifts than in theft. Somebody's casual reaction to your question can suggest a brand new direction for you to look in—and lo! your vague idea leads, a few months later perhaps, to a theorem, with a grateful footnote giving credit to your coffee mate's insight.

Norman Steenrod used a charming phrase when he described how he conducted preliminary discussions with would-be thesis students. It's not a question of giving away a problem, he said; it is just pointing to an interesting "area of ignorance".

Once a student is started on a problem, there is a delicate line that the advisor has to tread, halfway between the busybody jinni in the bottle and the discouraging power of darkness. I have known both extremes.

It's not kind to a student to write his thesis for him—it's almost as bad as to compete with him. Trjitzinsky went to one extreme, as I mentioned before, and Nate Coburn (a Michigan colleague, the father of my student Lew Coburn) had to hold himself in not to go that far. Nate had multiple sclerosis and taught from a wheelchair for many years. Mathematically he was an applied analyst. He told me once that he never suggested a problem to a student till he was sure that he himself could solve it—and, of course, the only way he could be sure of that is to have solved it already.

I dare not name the competers I have known. One of them would be generous enough to suggest a research problem to a student, most likely one that he himself had worked on, and possibly one that he had laid aside for a while, but as soon as a student began attacking it his own interest was rekindled, and he stayed up nights to be sure to get to the finish line before the student did. He didn't produce many Ph.D.'s. Another one did his discouraging only a shade more subtly. One time he gave a student one of his papers to read, a joint work written with a junior collaborator, and told the student to try to generalize it. After a few weeks the student succeeded. The new result was not earthshaking, but it was a step, a step in the right direction, and the student felt good about it. He felt good, that is, till the next afternoon, when he told his advisor what he had done. The advisor's reaction came seven seconds later. "Of course", he said, striking his forehead with the heel of his palm, "how could we have been so stupid as to leave that out!" A little while later the student changed advisors.

The folktale along these lines is about the professor who shifts gears rapidly and repeatedly when a student tells him a result. "That's trivial!" is the immediate reaction; "no, wait a minute—oh, no, it's false —I have a counterexample"; ... "hang on—oh, sure—yes, I knew that —Joe and I proved it three years ago on the plane as we were going to the San Antonio meeting."

Advisors are not the only ones who can make difficulties; students contribute their share. I have had two students who wanted to change problems twice, twice each. One was the one who "proved" that the Toeplitz spectral inclusion theorem and the subnormal one are not related; "give me something else to do", he demanded. The other decided that he just didn't like his problem. "I haven't been studying operators for two years just to spend my time looking for 3×3 matrices", he complained. (His problem had to do with the relation between the spectrum of a matrix and the spectra of its dilations; it was natural to begin by trying to get a feel for the problem in low-dimensional cases.)

In my experience theses seldom (never?) end with a solution of the original problem—they have the tendency, after a few months, to veer sharply to the left and up, and zigzag off in a totally unexpected direction. Yuri Rainich, a Michigan colleague, told a story like that about a student whom he had set to work on some complicated but dull calculations in differential geometry. The student kept plugging away all right, but he wasn't enjoying it; whenever he thought he could afford a few hours away from that stuff he went back to a "trivial" puzzle of his own, something about Boolean algebras. When, after a year, Yuri became more and more concerned about his student's slow geometric progress, the story came out. Yuri was interested, he was encouraging, and he assured the student that the Boolean question was not all that trivial. Happy ending: one Ph.D. thesis produced on a new characterization of Boolean algebras, and one fresh Ph.D.'s outlook on mathematics kept from staying permanently sour.

More Ph.D. students

Don Sarason was my first Ph.D. student at Michigan. He started out to be a physicist before he saw the light, and he turned out to be a joy to work with. My technique for conferences with Ph.D. students attained its final form with Don, and it has been working well ever since. "Come and see me in my office", I'd say, "two o'clock Monday afternoon, every Monday. Let us both save that hour from two to three for each other, even if it turns out that we have nothing to say. In the worst case we'll just sit and stare at a blank sheet of paper for an hour, glumly and silently, and try to think."

The worst case never happens. I'm not talking about the initial conferences, the ones during which we agree to try to work together and pick the "area of ignorance" to look at. I'm talking about what happens at a typical conference after that. What actually happens varies somewhat from student to student, but, with all students, most of the hour is spent asking questions, and most of them are asked by me. It can happen, but it is rare, that the student arrives Monday afternoon and begins with "Is it true that ...?" or "How do you prove that ...?" If he is stuck, then chances are he cannot even think of the right question, and if he isn't stuck, he is eager to report progress.

My questions are unprepared brainstorming. "What would happen if we insisted that the difference be not only small in norm but also compact? Does this work for projections only, or does it go the same way for every partial isometry? Could the two conditions be realized simultaneously?" The questions are not always intelligent, but they are, I hope, pertinent subquestions of the main one the student is trying to answer.

D. E. Sarason, 1962

Don Sarason is a quiet man; he never uses eight words when seven will do. He is one of the smoothest and clearest lecturers I know and he has an extraordinary sense of time. He knows almost to the minute how long it will take him to explain something. In an hour lecture he needs to fine-tune his performance by glancing at the clock on the tower across the street just once, about three quarters of the way through the material; if there are 12 minutes left to go, he'll finish in $11\frac{1}{2}$. His clarity and his timing must have been congenital; he had them when he was a student and he hasn't improved since—there was no room for improvement. When I met him he had a full beard and long black hair, worn pony-tail style held by a rubber band in the back; he wore blue jeans and sandals. Twice after that he became square—face shaved, hair cut—but he reverted to the conservative hippie style both times and settled down to it permanently. He is not only talented, but conscientious, diligent, and trustworthy. If he says he'll finish the bibliography in three weeks, he will do so, he will work hard at it, and he'll have the references completely right—publishers, pages, dates, and all.

When I brainstormed questions at him, he nodded intelligently and silently, signifying that he understood. If I didn't remember to raise the questions again the following week, Don wouldn't mention them either. If I did ask "what about those power hyponormal operators?", he'd say, "Oh, yes, that's trivial", and show me his ingenious non-trivial answer. I still haven't decided whether Don was my best student or Errett Bishop; fortunately I don't have to.

Don is intelligent and quick, and although he doesn't waste words, he never fails to say what must be said. He has strong opinions (about politics, for example), but he doesn't insist on foisting them off on you.

Once he had occasion to write down a part of the story of the first time I invited him home to dinner, but he didn't tell it all. What he left out was the sequel. The day after the dinner he stopped by the house with a present of a gorgeous box of candy for my wife. She was out at the time, so I said I'd give it to her when she got home. A couple of days later my wife asked me whether I had told Don how much she liked the candy. I hadn't, but I made a note to do so, and the next time I saw Don I told him, "My wife says I'm a slob—I should have thanked you for the candy." "That's all right", he said; "my mother told me to do it."

If I can understand a thesis it's probably not very good. I couldn't understand Sarason's difficult techniques of automorphic functions on annuli—it was fine. In some cases the calculated "weaning" process that I put students through is make-believe, but in Sarason's case it came easily and naturally. Weaning means that near the end, when I realize that a student has produced a thesis or is within a few inches of it, I set out to give him self-respect as a mathematician. I ask him questions, genuine questions, questions that I really want to know the answer to. I defer to his judgment. I say "How do you think this problem should be attacked? What do your techniques yield? What other important open questions remain?" It takes a while, but it works—gradually students begin to realize that they are no longer students, they are mathematicians, and the realization is an important part of their training.

They must come to understand that what they have done is their work, and that if they are going on to do something else—as they should—they can, as they must, stand alone.

I have said before, more than once, that after your thesis you can and you *should* do something else. Why should you? Answer: because that's when you really begin to fly, that's when you solo, that's when you become a mathematician in your own right. If a student writes a thesis on the calculus of variations when he is 25, and keeps publishing papers on the calculus of variations till he is 65, he may be a sound mathematician, but he is almost certainly not a first-rate one. Originality, adventurousness, continual learning and expanding—all these things are necessary conditions for mathematicianship. Joe Doob treated them almost as if they were sufficient too. I have seen several letters of recommendation he wrote (one of them about me), and his highest praise is not "good" or "strong" or "deep", but "the second paper is on a subject different from the thesis".

Don was my first Michigan student and Peter Rosenthal was my last. It was fun to watch Peter grow from a fumbling beginner to a confident pro. As he was working on his thesis, proving bits and pieces, he would keep wondering: when do you stop—when do the results add up to a thesis? I told him that that's part of my job; I have to try to judge when the results add up to an integrated whole, a measurable contribution that deserves to be accepted and published. He accepted that—but he beat me to the draw. Later—about six months later as I now remember —he came to me one day and said "I think that's it—I think with that problem solved, I've now got a thesis". He was right.

I treasure a letter I got from Peter. Once we were finished with the student-teacher business, I told him to lay off the "Professor Halmos" stuff—but he found it hard to promote our relation to a first-name basis. In a letter there is no way out—you do, or you don't—so he began it this way: "Dear P-P-P-P-Paul: There! I've said it!"

It's only tangentially relevant, but this story is very much like another that I found touching. It was at a party where Sammy Eilenberg was the guest of honor, and Dave Harrison, very near the beginning of his career, was one of the guests. Sammy's name is Sammy and he is never referred to any other way—to talk about him as Samuel would be like talking about President Carter as James. At the party, after several drinks, Dave came up to the group I was with, beaming, and he announced happily: "I did it, I did it: I called Sammy Sammy".

Between Sarason and Rosenthal came Steve Parrott, a loner. I knew Steve, but only out of my peripheral vision; he was "just a graduate student". The first personal contact we had was about a minor catastrophe. Several of us kept our private coffee mugs on hooks in the common room, and Parrott one day broke my mug. I was annoyed, of course, but at the same time I didn't want to make this shy-seeming, quiet, and obviously embarrassed young man feel any worse. I growled at him, therefore, with an obvious hyperbole, hoping that that would more likely put him at ease than a polite formula: "Well, Steve, there goes your Ph.D. career!" The tone of my voice must have been acceptably close to what I intended to achieve: he smiled and not too much later, in the same academic year, he asked if he could work with me. I said yes.

Lew Coburn, Eric Nordgren, and Bob Kelley started after Sarason, but they all reached the thesis stage before Don finished. As a result the five of us (Don and I, and the three younger candidates) could arrange a private seminar that I remember with pleasure—it was a pleasure then, and it seemed to be of value to all of us. We'd get together at a blackboard, and the thesis writers would explain to one another what they had been trying to do and what they were stuck on. (These sessions were in addition to the private sessions I had with each of them separately.) They were working on different problems, of course, but all the problems had to do with operator theory, and they could all understand what the others were talking about. (Years later when I told this to a non-mathematical colleague—Anna Hatcher, a linguist—she was horrified. Who is writing whose thesis?, she demanded to know. She couldn't seem to imagine that a free exchange of half-baked ideas could take place without leading to plagiarism. Neither could R. L. Moore, when it comes to that.)

CHAPTER 13

To Sydney, to Moscow, and back

Sydney 1964

As a mathematician I had been doing quite a bit of travelling—a conference here, an invitation to give a colloquium lecture there, and a meeting of the AMS somewhere else—but most of it was of the one or two night variety, involving flights lasting two hours or four. About the middle 1960's a few bigger opportunities came along and I could see a little more of the world. It began with a bang in 1964 with a trip to Sydney in January.

Professional trips take arranging, sometimes six months in advance and sometimes a couple of years. The Sydney trip began with a letter from T. G. Room inviting me to be the principal lecturer at the 1964 Summer Research Institute of the Australian Mathematical Society. He wrote in June 1963; the SRI was scheduled to begin on January 7, 1964, and to last for five weeks. My immediate answer was a sincerely regretful no. I was, after all, a working man; while January 7 was between semesters at Michigan, the five-week period would have cut almost four out of the 15 weeks of the second semester—a big gash. I was scheduled to teach honors calculus and a graduate real functions course. For two or three lectures—a week or so—you can always arrange for a friendly substitute, but four weeks is beyond the limits of most friendships, and, besides, that much substitution gives a course an undesirable patchwork effect.

One of the reasons I was invited, in addition to whatever merits as a lecturer Room might have ascribed to me, was that I had met Room and his family when we overlapped in Princeton in 1957. He was a reserved, correct, formal Englishman. He lived in Australia twenty years

T. G. Room, 1964

(more?), and was proud of how much he had learned about his adopted country, but as far as the Aussies were concerned he always remained a pommy. (I don't know why that word means "English", but the important thing is that its connotations are pejorative: a foreigner, an imperialist carpetbagger, not one of us.) His wife and his daughters were charming, friendly extroverts. I was especially in love with Geraldine when she was seven; I wish I could still remember all the words of "Waltzing Matilda" that she taught me.

Room wouldn't take no for an answer, and he half won. The Fulbright people were willing to subsidize two halves instead of one whole imported lecturer as originally planned (despite the double airfare), and I eagerly agreed to go for $2\frac{1}{2}$ weeks instead of five. My other half was Ralph Gomory whom I had met, but not then; he arrived in Sydney a day or two after I left. When the program was finally arranged there were three "lecture courses"; 12 lectures about operator theory by me, 10 lectures about combinatorial problems by Gomory, and 4 lectures about algebraic number fields by Kurt Mahler whose presence didn't cost any extra airfare—he was based in Australia at the time. Jack McLaughlin and Arlen Brown agreed to begin my courses in Ann Arbor and keep them running till I got back. There was red tape, of course: Fulbright application, medical exam, financial arrangements, plane and hotel reservations, but that's normal. Things like that always go a little bit wrong in the middle and always come out essentially right in the end.

I left Ann Arbor about 10 o'clock on Wednesday morning, New Year's day 1964, and arrived at the Fiji airport, in Nandi, at 4 a.m. on Friday—there was no Thursday that week. The most spectacular intermediate stop was in Honolulu at what the locals called midnight. I had been there before, on a visit to my brother and the mathematics department of the University of Hawaii, but the contrast between the frigid, dark, nasty winter that I came from and what Honolulu offered was striking and wonderful. It was warm, balmy, spring weather, about 70°, girls were walking around in muu-muus, barefoot—the palm trees swayed gently and the air smelled of fresh flowers. This is for me, I thought; this is where I want to live.

After a weekend in Fiji I arrived in Sydney at what they called 7:00 o'clock Monday morning, hoping that there is something more to life than travelling by air. Room and Geraldine met me, and after breakfast and a few hours of midday sleep, I rejoined the human race. The SRI officially started next morning, but the first lecture (mine) was at 4:00 p.m.

There were about 50 or 60 people in attendance at the SRI, most of them from Australia of course. Bernard and Hannah Neumann were there, as was Tim Wall. Janko was there, the simple group man, but he attended only the first one of my lectures; he said that he didn't think that learning about Hilbert space would help him determine all finite simple groups. George Szekeres was one of the senior people present, and there were many young people whom I enjoyed meeting and talking shop with: David Asche, Don Barnes, Michael Butler and his wife Sheila Brenner, Sam Conlon, and Geoffrey Eagleson.

Lectures, lunches, seminars—in some sense all conferences are isomorphic and there's not much to say about one that can't be said about them all. The mathematics I learn at one meshes into what I already know, increases my experience, and makes me, I hope, a better mathematician, but I don't remember the occasion, the day, the place, the source. What I sometimes remember is what went wrong. Two things went a little wrong with my lectures at the Sydney SRI in 1964: the level and the notes.

I asked Room about the right level in our preliminary correspondence. "I assume", my letter said, "that the level of the talks should be such as to reach intelligent and educated mathematicians who just happen to be working in another field. In other words, I assume that the audience will not consist of a group of experts on Hilbert space, and that it will not consist of mathematical beginners either. Are these assumptions reasonable?" Room replied: "You have gauged the level of the audience to a nicety." Thank you, Professor Room, but you're too kind. In fact, I missed my aim. Every culture has its mathematical folklore, and the one shared by the Australian mathematicians of those days was different from the one in America. The Australians knew more algebra than most Americans and less topology and functional analysis. The experts in, say, functional analysis knew the same amount on both continents; the trouble was that the analytically innocent Australian bystanders, and their students, were less likely to have heard of the principle of uniform boundedness than their American counterparts, and I, of course, was used to the American folklore level.

It was clear halfway into my first lecture that I should shift to a lower gear, and I did. Room had an inspiration: would I be willing to add a few "tutorial" sessions to my job, so that those who wanted to could ask background questions more elementary than they would dare to raise at a lecture? Would I ever?! Great! It worked. About 15 people came to the two tutorials that were scheduled (out of my total audience of 40), and after that the audience seemed to be much more with it.

The notes story has to do with a heavy package that I sent to myself from Ann Arbor to Sydney, containing reprints and notes. The notes were an embryonic version of what later turned into my Hilbert space problem book, and my Sydney lectures were to be based on them. The size and the weight of the package made it almost impossible to take it with me on the trip, and Air Express promised that they would deliver it in ample time. As a safety measure I did pack in my suitcase about 20 pages of a rough outline; if worse came to worst, I said to myself, I could use the outline to buttress my memory and make do without the detailed notes.

Worse came to worst all right. My first lecture was on January 7— no notes. Eight days later—half way through the course—still no notes. In a way it was good to have to do without them. It made my performance more extemporaneous, and, besides, it served as an excuse. I told my listeners that my notes had not caught up with me, and I was able to blame every misprint, every confusion, every time I got stuck on being forced to talk without a crutch. Eventually Room came to the rescue again. He called the post office—no, it should have been Air Express. He called Air Express International—no, it should have been the air express division of the U.S. Railway Express Agency. He called Qantas—no, it should have been Pan American. Late afternoon on January 16 he finally located the package—oh, yes, sir, it arrived in Sydney on January 11—no, it left the States on time, on January 11—oh, very well, if you insist we'll send it out to you! I waved the package at my audience next morning; it was greeted by a round of applause.

Some of the seminar talks I heard were dull and some were not. One was given by a middle-aged Ukranian refugee who never took a step without his mother; the toothless little old lady in a large hat sat in the front row while he lectured. The lecture was awful. He had filled the blackboard with symbols before he began and he read them to us, confusing *g* and *j* as he did. One chap in the audience told me afterward that he had turned his hearing aid off. He told me also that a friend of his tried to get him to drink less; "alcohol", the friend said, "makes the ear worse". It took the hearing-aid chap a few seconds to organize his reply. "On the whole", he finally said, "I think I prefer what I drink to what I hear".

I enjoyed Sydney as a mathematician, and I enjoyed it at least as much as a tourist. I had a pleasant little suite in a "private hotel"—that seems to mean a hotel that has no public bar. The hotel was built on a plan familiar in England: to get to my room you go upstairs and left, then two steps down and right, and then (seemingly) back up and do it again. The room was air-conditioned—and it had to be: the temperature outdoors reached 112° one day. The yellow pages in the telephone directory were pink. Television reception was extraordinarily good; more dots pers square inch then we have. I watched a marvelous performance of "Man for all seasons" one evening: one hour and twenty minutes uninterrupted by a single commercial.

If I had any tendency to get homesick for Hungary, Sydney would have made it worse. My hotel was near King's Cross, a sort of Australian Greenwich Village, and I heard more Hungarian spoken there than I have heard anywhere since I left Budapest. When I spelled my name at the laundry, the proprietor spoke to me in Hungarian; the shop that had sides of beef hung out front was labelled MÉSZÁROS, not BUTCHER. The speakers of Hungarian were of all ages: the older 1938 refugees from Hitler and the younger 1956 refugees from Khrushchev. I bought the "SIDNEY-I MAGYAR ÉLET" ("Sydney's Hungarian Life), a poor, sad and angry little newspaper. It was angry at Khrushchev (of course), it was angry at the U.S. because we talked to Khrushchev, and it was only a little bit angry at Sidney; it was nostalgic about and angry at Hungary.

Budapest 1964

I thought about visiting Hungary both before the war (1937) and after (1957); it might have been possible, but I was scared. I wasn't so much scared of being drafted (I was 21 in 1937 and Hungary isn't impressed by American naturalization papers) or scared of the secret police (after all, as an American in 1957, I was on the "wrong" side of the Iron Curtain)—I was just scared. I felt psychologically insecure about visiting the scenes of my childhood, about re-establishing contact with long estranged cousins and aunts, and about the problem of re-adapting to a culture and a language that I never really had time to acquire before. My uneasiness (fear is too strong a word) manifested itself in expressions of dislike stronger than I really felt: I kept saying that I didn't like Hungary, I didn't like Hungarians, and I didn't like the language. When I was invited to address the Bólyai János Mathematical Society in Budapest in 1964, I felt ambivalent: sure, it would be exciting and interesting, but no, it would be uncomfortable and unpleasant. Shaky decision: I accepted and I went.

The trip took eight days, portal to portal—the "closed" week of spring recess, containing the Sundays at both its ends—that way I didn't have to cut any classes. I started out inimically, and returned the same way, I thought—but I was wrong. Something made a deeper impression on me than all the annoyances that seemed at first to justify my attitude. I swore up and down that I'd never go back—but I did, a year later, and several times since, and it wasn't so bad. But I'll begin by saying how bad it was.

Almost 32 years had elapsed since I last saw Hungarians en masse, and I was horrified when I saw them at the Amsterdam airport, boarding the MALÉV plane to Budapest. What I saw was primitive aggressiveness, the very opposite of fabled English upper-class courtesy. They screeched at each other loudly across the departure lounge, and little old ladies pushed and shoved to cut ahead in the queue. The disorder caused a pile up at the rear door through which we were entering the plane—over here, over here—no, better by the window!—where's Molly?—let's be together—there's an empty one! The poor stewardesses kept trying to calm everybody—please be seated, there's lots of room, there's lots of choice, please be seated—but no one listened. The idea is that there is never enough of anything, the bastards in the world are out to grind you down, and you have to screech, claw, and shove to get anything. I know—I have some of it just from my first 13 years in that hungry culture, and I recognize it when I see it.

My first contact with Budapest after we landed was the cavernous, cold, poorly lit customs shed in charge of what appeared to be three high-ranking army officers, two male and one female: polished high boots, khaki uniforms, Sam Browne belts, flat caps, epaulets, and stars. They were the luggage inspectors on whose whim the immediate future of all arriving passengers depended. A mob was milling around with no system, no guidance, no alphabet. The suitcase of the man next to me didn't arrive; he seemed relieved to find me to tell it to in English. A bewildered Dutchman tried to explain that he was delivering some delicate X-ray instruments to a doctor, but the problem was beyond the linguistic capacities of the inspectors. After an hour I screwed up my courage and asked when it's my turn. "Oh, you're not finished yet?—all right, let's see"—and I was waved through one minute later. My hosts, represented by László Fuchs and Paul Révész, had been waiting all this time. I had met both of them at conferences in the U.S., and I was very glad to see them again. They were young fellows then, obviously talented and ambitious mathematicians. Fuchs immigrated to the U.S. later, wrote THE book on abelian groups, and settled at Tulane; Révész became one of the leading figures in Hungarian mathematics, an expert in hard analytic probability, and a tireless world traveller.

They ensconced me at the Gellért Hotel—I remembered the famous name and its elegant reputation from my childhood. We had supper. Odd system: the head waiter takes your order, another one brings your food, and a third one, in-between in rank, takes your money. When you are ready to leave you call him, and tell him what you had; he scribbles numbers on a slip of paper, and tells you how much to pay. The price includes a percentage for service, but you are expected to give a tip in addition to that.

In my bathroom I found a small piece of soap—the tiny motel size— and, as it turned out, that was supposed to last me a week. The wash bowl drain was clogged—it took forever for the water to ooze away. The lampshade was torn. The bed had a puff instead of blankets, of an area slightly smaller than me; any part that stuck out from under it during the night became quickly chilled.

There was a dance in the ballroom downstairs; I recognized most of the tunes. One was "All of me, why not take all of me", another was "Yes, sir, that's my baby", and a third (honest!) "Rock of ages". Neither that nor the streetcar traffic right outside my window could keep me awake the first night.

Slowly, slowly memories started to come back. The little smoky foggy wisps that do not deserve to be called memory became something solider as I walked around the once familiar parts of town and talked with relatives. Instead of wisps they became disconnected puddles, but still formless, like mud. If I had nourished them and cultivated them, they might have become the real thing—but mostly they just crumbled into dust. Here is where an aunt bought me some hot chestnuts, if I turn left there I'll come to a park—and didn't I meet a Chinese visitor at my uncle's house, one who could speak Esperanto? I looked at the large apartment building I once lived in. The number of names listed by the front entrance was twice what it used to be—every apartment had been split. The elevator was out of order, the stairway was damp, large chips of plaster were missing from the wall, an old bathtub had been abandoned in the courtyard. Outside, the main boulevard was dark and deserted. (Fuchs said "it's different from what you remember, isn't it— everything is so much lighter now?"—and I didn't have the heart to answer him.) Everything was run down at the heels, shabby, dingy, poor. The general impression I got was unpleasant—the whole city (the whole country?) seemed backward, dirty, militaristic, totalitarian, quintessentially Eastern European.

My cousin Frank was the relative I knew best, and he told me how things were. He was the chief administrative engineer (in German, the impressive sounding Herr Generaldirektor) of a largish concern (annual budget of over 100 million dollars); he owned a car (a small Opel) and had the use of a company car (Russian make, looked like a middle-priced five-year-old Chevy, driven by a chauffeur, a young farmer boy in a sweater). His standard of living was below that of an assistant professor at Wisconsin twenty years younger than he. The only reason he could afford the Opel was that he had worked in Cairo for six years and got paid in pounds sterling. The upkeep was a strain; he certainly couldn't afford a garage for it. He and his wife and two college-age children lived in a four-room apartment, four small rooms: a combination living room and study, a master bedroom, and one bedroom each for the boy and the girl. He admitted that he was exaggerating when he said that he couldn't afford to half-sole his shoes, but he insisted that he really couldn't afford the new winter overcoat that he needed. His wife was a school teacher, working mornings one week and afternoons the next for 8 forints an hour (less than 50 cents exchange value). A cleaning woman came to help her five or six times a month for a morning; she got 10 forints an hour.

School children must learn Russian, but the accepted attitude is to dislike that rule and grumble about it. Kalmár, the logician, told me about the time he was invited to lecture in Warsaw, and the question of languages arose. Many Poles understand both Russian and German, but Kalmár, of the older generation, never learned Russian, and German, which he could speak, had become extremely unpopular in post-war Poland. He solved the problem by beginning his lecture, in German, something like this: "I deeply regret that I cannot speak Polish, but I hope you'll find it acceptable if I deliver this lecture in the language of our East German comrades."

Most educated Hungarians can read and speak English, and in pre-war times, when I lived in and later visited Hungary, English newspapers and books were easily available at all newsstands and bookstores. In 1964 I was told that you could still find them, and I hunted hard, but without success. The New York Times I bought in Amsterdam lasted me a few days, but when I finished it, I was out of luck. I happened to have it in my briefcase the day I first visited the Mathematical Research Institute. I was shown around, and, in particular, I was taken to a large, pleasant, airy room, containing a table with several newspapers on it, from many spots around the world; the only English one was the London Daily Worker. Ah, I said in all innocence, I can make a contribution to this room—and I put my Times on the table. My host, the one who was in charge of me that morning, became embarrassed. I was slow—I didn't know what was wrong—it took me several minutes to catch on. First he made a joke, but then he made it clear: he didn't want to be held responsible for leaving the Times there. I gathered that people would think that to be a joke in bad taste

—like leaving the Daily Worker at the Elks Club or at a Baptist church supper.

Having to speak Hungarian most of the time bothered me. I found the language harsh, and I found it imprecise; the many nice distinctions that I had learned in English (such as therefore, hence, because, so, and consequently) all came out the same when I tried to make them in Hungarian. Things improved as the days went by—I would suddenly remember a word I hadn't used in 35 years, but the chances were that I would use it wrong. And then, of course, there were all the words that I didn't know because they didn't exist when I was a child. An example came up when I went to reconfirm my plane reservations. The young lady in charge just then happened not to know English, but I showed her my ticket, asked her to check whether everything was still in order, and the transaction was conducted amiably and efficiently. When it was over, I thought I would take advantage of the occasion and learn how to say "to reconfirm" in Hungarian. I asked: "tell me please—what I just did, what you helped me to do with my reservation—how do you say it, what's the word?" "Oh that", she said; "it's rekonfirmálni".

My main function in Hungary was not that of an expatriate returning for a visit, but that of a mathematician from far away establishing and cementing scientific relations. I worked at it. I re-met and met dozens of people—big shots such as Rényi, Vincze, and Alexits, and young beginners such as Májusz, who studied with Gelfand in Moscow, and Durszt, a disciple of Nagy from Szeged. Some of it was hard work indeed; the hardest was the "polylogue" I unexpectedly found myself in. As far as I knew I was just going to have coffee at the Research Institute that afternoon and shake some hands. My hosts planned something else. I was indeed given a demitasse of black liquid that tasted like hydrofluoric acid, but once that was at hand we all trooped up to a large room, where some 30 chairs were arranged amphitheater style with three other chairs facing them. I was told to take the middle one of the three, Vincze and Alexits surrounded me, and the rest of the party filled the thirty seats and stood behind them by the wall.

The grilling began. What is the Ph.D. training like at Michigan? What do I think about the role of applied mathematics? How much do American mathematicians earn? Do we get Russian books and journals? I did the best I could, about 80 percent of it in Hungarian, but when something got complicated and precision was more necessary than usual I lapsed into English.

The business of the Research Institute versus the University puzzled me then, and I never got it completely straight. Roughly speaking, research is done at the Institute and teaching at the University, and the two are administratively and financially (as well as geographically) quite separate—but I was told that this is quite a rough approximation indeed. The two cooperate—but not always, and not wholeheartedly. Some members of the Institute lecture at the University sometimes, and some professors at the University have part time appointments at the Institute—but there is definitely rivalry between the two and relations are more often polite than cordial.

The principal ceremony that a mathematical visit always involves is the lecture. My lecture came on my last full day in Hungary. The day was Good Friday, which officially doesn't mean a thing behind the Iron Curtain. I explained in advance that I couldn't possibly lecture in Hungarian. My control of that language is more than adequate to talk about movies and travel and schools and food and family and gossip; it is just barely adequate to understand and occasionally chip in when the subject is politics and academic organization and philosophy; but it is completely hopeless when it comes to talking about $\sqrt{2}$, $\int_a^b f(y)\,dy$, the dominated convergence theorem, and the principle of uniform boundedness. I never learned the technical words, and, consequently, both my lexicon and my syntax abandon me exactly when they are the most needed. Nothing doing, I said; I am an American mathematician, and I lecture in English.

Just for fun I wrote out an introductory paragraph (expressing my gratitude for having been invited and apologizing for not being able to lecture in my mother tongue), I asked Rényi to translate it into good Hungarian, and I memorized it. When I stood up to talk, I began with that paragraph, and it went all right except for one thing. It just happened that on that very day a delegation of about a half dozen mathematicians from Ethiopia was visiting Budapest, and, of course, as a matter of courtesy, they were invited to attend the meeting of the Bólyai János Mathematical Society. Their English was just fine, but when they heard my initial paragraph they became justifiably frightened—they blanched, if that's the right word.

The lecture went as lectures go, for a while. When I was about half through, the rear doors opened and four waiters entered balancing trays of coffee, which they started distributing with a clatter of saucers and spoons. Oh, yes, they brought mine right up to the podium for me. I managed to shout over it all, but I did wish that it hadn't happened.

All is well that ends, and my visit ended. It was a small but pleasant surprise to learn that the Bólyai János Society was giving me an honorarium of 500 forints (about $20 or $30); I was expecting zero. The newspaper I bought at the airport next morning carried the legend "Proletarians of the world, unite!" on its front page. The man who looked like a brigadier general and who examined my passport thought I was the one who left a boarding card on his desk, and he called after me: "Comrade Halmos!" When I boarded my Sabena flight (next stop Brussels) I felt lighter—I had re-crossed The Curtain—I was back home.

That was twenty years ago, and many things are different now. There is more soap at the Gellért, there is more light on the streets, there is more plaster on the walls, and there are more English books and newspapers (including the New York Times). The customs and immigration inspectors are just as high and mighty as ever.

Scotland 1965

The British Mathematical Colloquium, BMC for short, was to meet at Dundee early in April 1965. Professor W. N. Everitt (who became Norrie Everitt soon after I met him) was in charge, and he invited three "survey" lecturers, one for each of the three days of the meeting, namely Claude Chevalley, Arthur Erdélyi, and me. His letter reached me in May 1964, just six weeks after I arrived home from Hungary, and by then I was rested up and eager to re-join the mathematical jet set. Yes, thank you very much, I answered, and thus started eleven months of complicated negotiations.

There was no special complication about the BMC. They have some funds; not very much, but enough to pay domestic travel and subsistence expenses. Transatlantic travel was to come from the NSF—which meant, of course, application forms, carbon copies, abstracts, and other such paraphernalia—but I was too energetic to be satisfied with only that much activity. As long as I am going to Europe, I might as well do it up brown, I said to myself. Doob (who was president of the AMS then) and Albert (who was president-elect) told me about a recent exchange agreement between the academies of the U.S.A. and the U.S.S.R. The idea was that each country was to send to the other 20 prominent scientists some time during the years 1964–1965, "at least half of whom shall be members of the respective Academies, for a period of up to one month each, to deliver lectures, conduct seminars, and to study scientific research on various problems of science." I was 48 years old; I was as "prominent" as I was ever going to get; and I applied to be one of the 20. And, I said to myself, as long as I was in Scotland and Russia, I might as well go to Hungary too, and I initiated personal correspondence with friends and official correspondence with the Hungarian Academy of Sciences, to see what could be done.

Correspondence at or near the academy level can get complicated, and I was doing it with four countries simultaneously: it became quite complicated. Added complications: in this country I was dealing with both the NSF and the NAS; in Scotland there was to be not only the BMC, but, back to back with it, a joint meeting of the Edinburgh Mathematical Society and the London Mathematical Society; in Russia I was supposed to visit both Moscow and Leningrad, and, on the side,

I was negotiating about royalty payments for my books that had been translated into Russian; and in Hungary lectures were arranged in both Budapest and Szeged. The eleven months I had were barely enough to administer the two months of travel. There were some successes and some bureaucratic setbacks, but in the end all the documents and photographs and signatures and seals (not to mention dollars, pounds, rubles, and forints, as well as airplanes, cars, and trains) were where they were supposed to be, and I was off—ready for a long spring of touring and mathematicking.

The British part of the trip was three weeks, but only the first of them, in Scotland, was mathematicking; my wife joined me for the rest and we had a holiday in England. Avoiding the noisy chaos of Heathrow, I flew direct to the civilized, small, and efficient airport at Prestwick, spent a night trying to find those five hours that I misplaced somewhere in the mid-Atlantic, and then leisurely drove a rented Ford Zephyr to Dundee. It's easy: all you do is follow the signs to Glasgow, then get hopelessly turned around on the narrow curvy streets, find the other end of town somehow, follow the signs to Stirling, get lost in town, and then, eventually, follow the signs to Dundee, and ask a traffic warden where the Angus Hotel is. Along the way you listen to Mozart on the Third Programme, which presently turns into the news in Gaelic.

Everitt and a colleague of his, R. P. Pearce, came to welcome me and we had dinner at the Angus. Everitt had it charged to my room, and then hastened to reassure me. "Fear not", he said; "we've arranged for the bill to be sent to the Colloquium. We really are who we claim to be."

The Colloquium was very well attended: over 200 people were there for the full three days, I shook hands with half of them, and tried to remember their names, but by now, almost 20 years later, my memory can attach faces to only 20 or 25 of the ones that appear on the official list.

I met Frank Bonsall for the first time. He was still at Newcastle then; later he moved to Edinburgh, where I got to know him well. He is very English (as contrasted with Scottish), but he doesn't make an issue out of it. He learned all about Scotland and its ways, and when I saw him in Budapest, or in Crawfordsville, Indiana, he proved himself to be equally adaptable there too. When my lecture came around, Bonsall was the one who introduced me, and I can't resist quoting a sentence. "Professor Halmos", he said, "may look like one mathematician, but in reality he is an equivalence class and has worked in several fields, including algebraic logic and ergodic theory; this afternoon his representative from Hilbert space will speak to us."

Sir Edward Collingwood was there—I had heard a lot of nice things about him, we had friends in common, and I dared to introduce myself and ask whether I could buy him a drink. He said I could. His

E. Collingwood, 1965

hair was white, what there was of it; he was on the short side of average height; and he had beautiful manners. He listened to what people said to him. He was a complex function theorist of good reputation, but a wealthy man who didn't need a university job to make a living. He had been knighted for his services to the British hospital system.

Rankin was there, Robert Alexander Rankin, whose Gaelic name is MacFhraing (pronounced something like Mackt Rhankh!). He is a well known analytic number theorist, and, just for fun, he wrote one of his papers in Gaelic. Boas, as editor of Math. Reviews, not realizing the connection, and not having a large supply of people who could read Gaelic, sent the paper to Rankin to review. Rankin wrote the review all right, but his conscience wouldn't let him keep his real identity a secret from Boas. The review was published anyway.

John Erdos was there, a young functional analyst, who inevitably, but only temporarily, came to be known as "the wrong Erdös". He was born in Hungary but left it at the age of 8 and was brought up in Australia and England. He speaks both Hungarian and English better than I do, and he insists that people pronounce his name in the anglicized way. He doesn't want to be "Air-daish"; he demands "Erdoss". As long as Erdös is as famous and ubiquitous as he is, Erdos is bound to have a hard time with his linguistic reform. (Do all readers know that I reject "Hal-mush", some people's notion of the "right" way to pronounce me? Please, please, say "Hal-moss".)

Mathematical gems and paramathematical anecdotes were being bandied back and forth. One of the latter concerned the mathematician Lord Cherwell, formerly Frederick Alexander Lindemann, and his lecturing style. Most of the time he was inaudible, we were told; when he was not inaudible, he was unintelligible; and the few times when he was neither inaudible nor unintelligible he was wrong. This was the same man, by the way, that Littlewood told his Freudian story about. Littlewood didn't like the man, and his emotion was strong enough that his memory blocked out the name, the pre-peerage name. They had been colleagues at Cambridge, Lindemann and Littlewood, and Littlewood found it embarrassing at faculty meetings to talk against the deplorable proposal just made by Professor... uh.... Littlewood thought he could solve the problem by a mnemonic device: every time he needed his colleague's name, he would think of π, and then of Ferdinand Lindemann, who proved the transcendence of π, and there he would be. It worked, in a way, as Littlewood later related. The next time Littlewood lectured on π, and on its transcendence, and wanted to give credit for the proof, all he could say was that this profound fact was first proved by... uh...!

Wendy Robertson drove over from Glasgow for a part of the Colloquium with 4.8 children (the 4 were well-behaved, the .8 was the most visible). She and her husband Alex (perversely pronounced Alec) are both mathematicians and have even collaborated on topological vector spaces. An ebullient member of the crowd was that irrepressible redheaded Irishman, Trevor West, who frequently says "let haitch be a Hilbert space". He is an almost total tee-totaller who talks like a lush, a devoted rugby player and referee, and (later) a senator, representing Trinity College, Dublin, in the Irish parliament.

Not every minute of my time at the Colloquium was spent at the Colloquium. I did a lot of walking in Dundee and some shopping. One of the best walks was up the Dundee Law, a hillock of some 500 feet from which you can inspect most of the town. Another pleasant walking and shopping expedition was with Chevalley. He and I happened to be sitting next to one another at a couple of talks, and at the coffee break we had very little difficulty agreeing that two was enough. We skipped out on the third talk of the morning and went to buy sweaters for Mme. and Mlle. Chevalley. We (Chevalley and I) had known each other in Princeton and had often played Go (he was an addict for a while); thrown together in Dundee, far from our home bases, we found it easy to be friendly. Later that day Chevalley gave his lecture (about algebraic groups); I sat in the front row and tried to look intelligent. The talk was very well received. I wondered at the time: is it a British custom, or a general European academic one, that an audience signifies its approval not by applause only, but by a vigorous stomping of feet as well?

Erdélyi's talk was the next day. Erdélyi was Hungarian originally, but much of his university education and all of his professional career was away from Hungary: in Czechoslovakia, in Scotland, and in the U.S. He left Edinburgh for Pasadena, and was at Cal. Tech. for 15 years. His largest single job there, the original reason for his being invited, was the "Bateman project". The project was to organize, in effect to re-work and re-write, and then to publish, Harry Bateman's posthumous notes. The result, a collaborative effort, was five gigantic volumes. Erdélyi's main interest was in hard analysis, including the subject called "special functions". Eventually he returned to his beloved Edinburgh, and that's where he was based when I saw him in Dundee. His lecture was a work of expository art: polished, clear, informative, good fun, altogether beautiful—and rather easy. He talked about nonstandard analysis—not at all his usual sort of thing—and he spoke kindly of it. When it was over Chevalley snorted and sneered—trivial, awful talk, tripe, the nerve of the man saying that such stuff can do what distributions can!

That evening a group of us was invited, I might almost say ordered, to the Master's reception. The Master would be called the college president in the States; the reception was to be from 7:45 to 8:30. The guests were, mainly, the organizers and the speakers, and some of the important dignitaries from abroad—very elect. We were all in our best clothes, all forgathered, and, sharp at 7:45, we were guided by Everitt up to the Master's room, to be introduced, sherried, and instructed to sign the visitors' book. We never learned the Master's name. I chatted with him a few minutes and learned that his specialty was Scottish law. At 8:30 sharp we shook the Master's hand and left.

My talk was the last one on the program. Everitt confessed later that that was his doing; "a bit of conscious villainy" he called it. I was somewhat apprehensive of the odd rubber blackboard. It's a belt stretched around a frame and it goes round and round; after twenty minutes or so, as you push up the part you had just written on, the part that you started with comes up from the bottom. My subject was "Some recent progress in Hilbert space" (what else?), and it seemed to go over all right. The applause was polite, and the seventeenth BMC was declared concluded.

I am always nervous at conferences till after my talk; only then do I feel I can relax. The joint EMS-LMS meeting (held in nearby Edinburgh) the next day was a short one—as I remember the program consisted of just one lecture, by Rankin. My holiday began after that, and after *that* I had Russia to worry about.

The holiday got a little business mixed in with it, because we visited friends, and most of them were mathematicians, but it was a good holiday—Hull, London, Southampton, and several stops between. Hull was the most novel experience. The University was growing, and building, and, in the meantime, suffering. There were cranes in a sea of mud; Bill Cockcroft, Dean of Sciences at the time, had his desk next to exposed water pipes in a tiny dark cubicle.

He took me to meet the librarian, who was Philip Larkin, the famous poet. I was worried beforehand. I am not tuned to poetry, I rarely read it, and I don't enjoy it—but I had heard of the famous Philip Larkin, and what was I ever going to talk to him about? I needn't have worried. Larkin is a poet, and a poised and cultured gentleman. He told me, with deep professional interest, how much money his library spent on mathematics last year (£1606/3/7, including journals). At lunch we talked about bifocals (we were both beginners) and about copyright laws—and I think we both enjoyed the conversation.

Holidays end, and duty keeps calling. Friday night (April 23 it was) I saw my wife to her ship in Southampton; Saturday morning I drove up to London to return the rented car and attend to a few other such details; Sunday I rested, I worried, and I kept trying to go on learning some more Russian; and early Monday morning I checked in at the West London Air Terminal and put myself in the hands of Aeroflot.

Tourist in Moscow and Leningrad

I didn't want the tiny camera I was carrying in my pocket to brand me as a spy, so I pulled it out and waved it before the customs man and explained, in my most fluent Russian: "fotograf apparat". He grinned and waved me along: "Da, da; Minox." Nicky Kazarinoff, my Michigan colleague who was spending the year in Moscow, was at the airport to meet me, with S. V. Fomin, a disciple of Gelfand and Kolmogorov, and with R. V. Gamkrelidze and E. F. Mishchenko, a

R. V. Gamkrelidze, 1962

E. F. Mishchenko, 1962

couple of Pontrjagin-type control-theorists whom I had known when
they visited Ann Arbor. (I took them to dinner at Win Schuler's restaurant in Marshall, Michigan, about 42 miles from Ann Arbor. The
U.S. State Department had issued strict orders, in retaliation for some
similar orders that the Russians had been issuing: G. and M. were to
stay within a circle of radius 40 miles centered at Ann Arbor. Hay,
our chairman, wanted to avoid any possible charge of laxity; I had to
call Washington and get permission to show our visitors how a genuine American salad bar is operated.)

Nicky had imported his flaming red spacious station wagon, and we
rode into town in it. Owning a car in Moscow is not always fun. Nicky
said that the radio antenna of his car was broken off and removed
their first night there. We saw a car parked before a restaurant with
both front tires slashed. The locals with me offered two possible explanations: just plain fun, or resentment against the wanton display of
wealth that a large car represents. On another occasion, a week or so
later, when I was driven to a museum a few miles out of town, my
host removed three essential parts from under the hood and pocketed
them before he dared go inside.

The Hotel Budapest was different from the Kensington Palace Hotel,
where I spent the last two nights. In London a doorman opened the
taxi door and carried my luggage in, a young assistant manager in a
dinner jacket showed me to the room, the bed was luxurious, and the

bathroom had a generous supply of fuzzy towels on a heated towel rack. In Moscow the doorman had a two-day beard, a half-smoked cigarette dangled from his lips, and he paid no attention to us; he was in an engrossing conversation with the mailman. The hotel cashier operated an abacus (gracefully and effectively); they are common all over Russia. The elevator had a bad reputation; usually, I was told, you had to wait 10 minutes before it came. The room was small, cold, and dark (one 40-watt bulb hanging from the ceiling and another in a floor lamp), the bed was a cot, the narrow window looked out on the back court, and the radio was permanently tuned to one station—the choice I had was silence or the news in Russian, interspersed (rarely) with soupy ballads such as "Sous les ponts de Paris". On May 1 the program was different; emotional voices with a military band in the background made speeches in which I could catch one word in twenty— American imperialism, Viet Nam, peace, aggressor, solidarity, glory, Communist Party of the U.S.S.R. I had no reason to think that the room was bugged—I couldn't possibly have been important enough for that honor—but the idea passed through my mind because of how the radio operated. There was no on-off switch; the way I got silence was by turning the volume control all the way to the left. The receiver was always on; when I held my ear against it I could hear it hum and I could feel its warmth.

The only thing like Budapest about the Hotel Budapest was that one of the four languages in which the menu was printed was Hungarian. (The others were Russian, German, and English.) My breakfast cost 1.22 rubles. (That was a little more than $1.22; for purposes of approximate calculation the ruble and the dollar were equal.) I offered a ten-ruble note, which caused the waitress to go into a tizzy—no, no, she could never change *that*! I shrugged helplessly, so she took it and (judging from the time she took to get the change) donned her street clothes, took a bus to the bank, and filled out the application forms appropriate to 10 rubles before she returned. I left a 20 kopek tip = (.20 rubles), but she protested—no, no, that was too much, I should leave less! I encountered a similar hospitable (motherly?) attitude another time, at another hotel, when my order was interpreted to the waitress through the friend I was eating with. She agreed to bring me a glass of milk, but she didn't. Everything else came all right, but she explained to my friend that milk would be bad for me after herring.

Very few service people (such as waiters, shopkeepers, or taxi drivers) spoke any language other than Russian. One waitress at the Hotel Budapest did, but she made it plain that she'd rather take my order in Russian. The first time I met her (I must have arrived in the dining room too early) she came to my table and, pointing at me, presumably to make sure I knew that I was being addressed, she said: "You most vet ten minoot pliss".

I waited a lot more than ten minutes in the dining room of the Hotel Budapest. Service was always slow. I clocked it several times, and I found that I usually spent $3\frac{1}{2}$ hours a day in the dining room, mostly just sitting and waiting. Service isn't just accidentally bad, it is systematically bad. If you forget to order butter, you don't just casually mention it to the waiter the next time you catch his eye, and, if you do, he doesn't just go and pick some up for you—the operation is a major one. The order has to be written down, and the waiter has to go back to stand in the long line that he had to stand in for your main dish. The same sort of awkwardness is built into how you pay. There is no central cashier. Each waiter has his own purse, and if he can't make change out of that, the transaction breaks down.

There seemed to be enough waiters and waitresses for the number of guests, but perhaps their hearts were not in the job. One poor black guest (from Nigeria, I later learned) had a hard time. He and I had smiled and nodded at each other but never yet spoken, when at lunch one day he came over to ask me something. He asked in French, and that took a moment to straighten out; he then went on in slightly hesitant but perfectly controlled, educated English. Could I please explain to the waiters that he would always like water with his meal? It wasn't so much a matter of right now, but always, at lunch and at dinner too. He realized that it might be a nuisance to prepare (that's the word he used), so it might be easier if I explained it to them in advance, please? All he wanted, he said, was just plain water, natural water, not mineral water. Poor man. I did my best, I used my feeble Russian, and, sure enough, someone brought him a glass of water; but I saw him across the room at dinner that night, and he didn't have any then.

There were other ways in which life in Moscow was not always pleasant for the ignorant foreigner. There were no maps available, for one thing. The State Department had given me an American-made map of Moscow, with the names all transliterated into our alphabet, but it was not completely helpful. It was out of date, it was inaccurate, many street names had been changed since it was made, and the street of the Hotel Budapest was not on it. It was better than nothing just the same, and I used it a lot as I walked about the town.

I had been told that one of the good aspects of rigid state control is that even small statutes could be enforced, so that, for instance, there was no litter in Moscow. Not true. There is, quite a lot. It isn't as bad as downtown Chicago, but it's there. I saw a taxi driver crumple up some newspaper to wipe the inside of his windshield and then, as he was driving along, just throw the paper out the window.

The river Moskva runs through Moscow and provided me with many good walks along its shore and back and forth across its bridges. The city itself, all that I saw of it, seemed slummy to me. From far away a group of buildings would look shiny and attractive in the sun (the very rare sun), but when I got to them I could see the windows streaked, the dirt caked, the paint peeled, the sidewalk cracked. There was a permanent, unpleasant, peculiar musty smell in the city; one Muscovite said it's caused by the use of low-quality low-octane gasoline. During my stay in Russia I saw very few dogs and not a single cat.

Leningrad is touted as much more beautiful than Moscow, and even the detractors of Russian tourism must admit that it has its points. It is on the river Neva, which keeps splitting and, together with some canals, partitions the city into "islands". I took a long walk down the Nevsky Prospekt, a wide boulevard guarded by several elegant looking marble generals on horseback, and found it very satisfactory indeed.

People in the U.S.S.R. do not have some of the small freedoms that are taken for granted in many other countries. The most trivial example I can remember is that the hotels have exactly one door to the outside world. That was true of the Hotel Budapest and a couple of others that I saw in Moscow, and it was true of the elegant Hotel Astoria in Leningrad. The Astoria faces on three streets, but if you're not on the street where the unlocked door is, you have to go around a corner or two to get inside. A less trivial example is the difficulty (impossibility?) of getting English books, magazines, and newspapers. Perhaps the least trivial example is the use of domestic passports. They are not like full international passports, but they are more elaborate and informative than just an I.D. card. They come complete with photographs and seals, and a native Russian must produce his before he is allowed to register in a Russian hotel.

In both Moscow and Leningrad I took in some of the obvious tourist attractions and they helped a little to counteract the unfavorable impressions I've been mentioning. In Moscow, for instance, I went to GUM and to the circus and to the Kremlin theater.

GUM is an immense department store, or, perhaps, it is more fairly described as a shopping center. It is in a six-story building just off Red Square, but most of it is on the first two levels. As you walk along inside you are covered by an opaque glass roof, and you pass hundreds of little shops. Each shop is a little room, a stall, a booth, opening off the wide aisle, dedicated to a specialty—women's hats, or soap, or stationery. People jam past each other carrying shopping bags with fish tails sticking out of them. A female voice is steadily yammering away on a public address system.

As for the circus, it was like a circus anywhere, but I liked it a lot; it was a high-quality circus, very well done. There was only one ring, which I approve of, and the jugglers and the acrobats and the clowns were good. The stand-up comic was presumably funny (in Russian), and some of the animal acts, such as the boxing and motorcycling bears, made me sad, but all in all I was grateful to Sinai (more about him later) for taking me there.

The theater is inside the Kremlin walls and is extravagantly out of harmony with the old baroque of the rest. It is modern and magnificent —all glass, indirect lighting, spacious corridors, marble columns, and infinitely reflecting mirrors. The auditorium is huge, beautiful looking, well arranged. Each seat has a little loudspeaker of its own, and a plug for earphones (to get instant translations at international conferences). These devices were not used the night I was there; the performance was Verdi's opera Don Carlos. By local custom my hosts, Gamkrelidze and Mishchenko, took me first to the buffet—a very large room with dozens of counters where you buy food, and with dozens of tables (no chairs) where you go to eat what you bought (standing up). After we finished our meal, ten minutes before the scheduled start of the show, Mishchenko said we can go home now—this was the best part of the party. He was joking, but he wished he hadn't been. Listening to the opera was a sacrifice he had to make toward entertaining me. After the second act he told me, as if granting a boon, that we didn't have to stay any longer. I was enjoying the performance; in addition to the music and the singing, the scenery, the costumes, and the staging were all splendid; but we went home after the second act.

In Leningrad I went to the Hermitage, the art gallery, and I was impressed. It had been the Tsar's winter palace. It is gigantic. It has broad marble staircases, intricately beautifully inlaid parquet floors, spectacular green malachite tables, and literally hundreds of rooms. The walls are two feet thick, the ceilings are proportionately high, and the doors between the rooms were made big enough for elephants. How was it to live there?

I tried to see too much in one short visit; there was an awful lot to be seen. I saw Rembrandt, Rubens, two small Madonnas by da Vinci, Titian, Gaugin, Matisse, some blue Picasso people, and many many others. (A small added part of the fun was to decipher the Cyrillic versions of the names.) For such good stuff the display was incredibly bad. The rooms were dark, most of the time the electric lights were not turned on, and when they were they were so weak they hardly helped. Where it was light enough, directly by a window, all one could see was glare and reflection in the glass.

A small group of us was taken on a tour through the Golden Treasure Room. A well-trained guide lectured at us (mainly at me) in very good English, full of semi-technical words such as diadem, pendant, stylized stag, scabbard, and snuff box. She went in historical order, and we saw jewels made in 700 B.C. to A.D. 700 by Greeks and by barbarians, and then modern things made from the 17th century through the 19th. It was wonderful to see solid gold stags (about the size of a desk telephone) made 2500 years ago, along with solid gold helmets (lifesized) and laurel wreaths. I found it even more interesting to see diamond encrusted snuff boxes, and watches, and miniature portraits, and to marvel at the time they took to make, the luxury they represented, the style of life they symbolized. The last thing the guide showed us was "the supreme order of Denmark, the white elephant", and I don't think she knew why I grinned.

The summer palace of Catherine The Great, in Pushkin, near Leningrad, was less breathtaking, but it had its sumptuous aspects. All the rooms, from the large throne room to the very small duty officer's guard room were gaudy: gold, statues, bas-relief, wood carving, almost oriental in their exaggeration. One long series of rooms stretched 300 meters (one and a half Chicago city blocks) so arranged that all the doors were in a straight line. When they are all open they form a beautiful and apparently infinite vista of glittering gold and white.

St. Isaac's Cathedral is just across the street from the Hotel Astoria in Leningrad; it is the tallest building in the city. From the outside it looks like a church—nothing too special. Inside it is all marble and malachite—pink and gold and green. It is a museum now, but with very few exceptions all that the museum exhibits is the cathedral. The exceptions are a couple of statues of the architect, a few models of the cathedral itself, a desk where you can buy souvenirs and postcards, and, suspended from the very top tip of the cupola down to the very bottom of the interior, a Foucault pendulum. The rest is St. Isaac's Cathedral.

There are signs requesting people to speak softly and men to remove their hats. The hat protocol is unheard of in modern Russian life; most people have never heard that such an act used to be regarded as a courteous sign of respect. Gamkrelidze, who had read widely and had travelled in the West, had to explain to his wife that a man removes his hat for a lady.

What you see in the cathedral, in addition to the marble and malachite columns, are pictures, mostly shiny mosaics. There are saints, there are figures of Christ in all situations, and there are modern patriarchs—they are on the doors, on the walls, on the ceilings. They were made, they must have been, with devotion as well as art. Somebody really cared about the details. St. Isaac's Cathedral was the most beautiful thing I saw in Russia.

Life with Anosov

Dmitrii Anosov was assigned to be my host, guide, translator, and companion during my Russian stay. He has made a name for himself in ergodic theory since then, but he was just at the beginning of his career when I knew him (he was not yet 30, I guessed), and I think he liked the idea of being in charge of me. He had the Russian equivalent

of a Ph.D. and he was a very smart young man. His English was quite
good and his personality and mine were not so far out of phase that we
had any trouble getting along. We ate many meals together, attended
many lectures together (often as performers: I lectured, he translated),
we walked and talked and travelled together, and when a pair of shoes
I brought gave up the ghost, he helped me shop for a new pair. (The
latter was a traumatic experience with a poor stopgap result; I gave my
Soviet shoes to the Salvation Army as soon as I got home.)

D. V. Anosov, 1965

The three occasions when I was most thankful for Anosov's existence
were my official meeting with Vinogradov near the beginning of my
Russian stay, the trip to Leningrad near the end, and the publisher's
ceremony in the middle.

Ivan Matveevich Vinogradov, the powerful analytic number theorist,
was in his middle 70's when I met him; he was a short, fat, egg-bald
man, not quite senile, but tending that way. He had been director of the
Steklov Institute since 1932, and he had become as powerful in mathematical politics as he once was in mathematics. An audience was ar-

ranged for me just so I might pay my respects; it was scheduled to begin at 2:30 and end at 2:45. In fact, I was ushered into the presence at 3:15, and I wasn't given much chance to say anything. Vinogradov pontificated—how to build an institute, the young men are the mainstay, international cooperation is important—and he was inclined to ramble and to repeat himself. He spoke in mumbled Russian; poor Anosov tried to translate from time to time, but then Vinogradov shouted over him. When the clock rolled around to 4:45 I could no longer repress the worry about my 5 o'clock appointment; I made the polite speech that the occasion demanded and took my leave. Diplomatic worry: was it up to me, the visitor, to terminate the conversation, and, if it was, shouldn't I have done it long before?

The difficult parts of the Leningrad trip were the train (the famous Red Arrow) and Rohlin. I was used to trains. In the old days I often went from Chicago to both the east and the west coasts by train, and I knew how to sleep sitting up, or else curled up with my area minimized in an upper berth. The basic problems were the same in Russia, but the details were sufficiently different to be baffling—especially when there was no language to solve them in. (Anosov was in another car, two removed from mine.) There were four of us in my compartment, an old man, a young married couple, and I. The compartment had four bunks, two upper and two lower, and the procedure for retiring was complicated. We opened doors, turned out lights, closed doors, changed places, and in general did our best to observe amenities under approximately impossible conditions. Where do I put my suitcase? (Under the lower bunk.) Where is the top sheet? (Folded under the pillow.) How do you operate the faucet in the washroom? (Push up on a hidden stump.)

The train arrived on time, to the second, and Rohlin was there to meet us. His German was weak; his English was weaker, and he didn't want to use it at all. The result was not totally satisfactory; we spoke German mostly, except when I thought it was important that I make myself clear, when I lapsed into English. He understood just fine. As far as mathematics goes, Rohlin and I were both changing (or had already changed) away from ergodic theory, he to topology and I to operators; what we had in common were memories of work ten years old or more. The visit was neither profitable nor pleasurable to either of us.

Three of my books had been translated into Russian, and I had several times heard the rumor that when that's done the author gets royalties for the Russian edition. The money is not, cannot be, sent out of the country, but instead it is paid into a royalty account which the author can draw on whenever he is in Russia. In my experience that is true in spirit, but not in detail. There was no "royalty account" for me, and my preliminary correspondence with Mishchenko seemed to accomplish one thing only: it gave the pertinent authorities advance warning that I was coming. Once I arrived, the royalty was calculated retroactively.

The account was a legal fiction, but the money was hard cash; all I had to do was come and get it.

The Publisher of Foreign Literature was housed in a dilapidating building with dark corridors and creaky floors. The room Anosov and I ended up in was clean and pleasant; it was carpeted and had a large conference table in the middle. There we met three people: *the* publisher, one of his subordinates, and a mathematician. The latter was Boris Vladimirovich Shabat; I never learned the names of the others. The five of us sat at the table and made polite formal conversation about the weather and about the extent to which book publishing in the U.S.A. and the U.S.S.R. are alike and different. After ten minutes of that a clerk appeared (having been telephoned for), and put before me a receipt I must sign. Since it is an official document, I am not allowed to use a ball-point pen! I must write not only my name but the amount whose receipt I was acknowledging, in words, not in numerals—yes, it's all right to write it in English. I dutifully wrote nine hundred fifty four rubles and sixty four kopeks—which came to about $1000.00—and I was rewarded by being handed a large stack of cash and a few coins. I was told, moreover, to count it. I did; it checked. That concluded the business, but not quite the ceremony. Every major transaction such as this must be celebrated with a drink. It was the middle of the morning of a weekday, and we all had to work through the rest of the day; the drink that appeared was mineral water. We washed the deal down, made polite conversation for five minutes more—and that was it. Goodbye, shake hands, thank you very much.

One of life's problems I now had to face (it would be fine if they were all like this one) was what to do with my ruble wealth. I could not legally change it to dollars, I could not take it out of the country. My expenses for daily needs—room and board and transportation—were reimbursed by the Soviet academy. There is only so much caviar and there are only so many fur hats that I could use—and, besides, I had no desire to weigh myself down with extra luggage for several thousand miles. Solution: Kazarinoff and I haggled a while and came to an agreement. I gave him all my rubles, and he promised to write me a check as soon as he got back to the States. (He did.) The rate of exchange we agreed on was $2.00 for 3 rubles; since it was all pure velvet for me, I did not grumble.

Fomin and Gelfand

Who was the better mathematician, Gauss or Euler? Which state university has the better mathematics department, Illinois or Michigan? Which name will live longer in mathematics, Weil or von Neumann?

Mathematicians enjoy asking such questions, at all levels, and they enjoy offering and defending answers. It's fun to establish a rating system and then to fit (or cut or stretch) people to it, and, of course, it can be fun (or depressing) for each of us to learn how high or low we stand in the rating.

One possible rating system that I learned long ago goes as follows. Mathematicians of the first rank are the obvious immortals, the unarguable greats, the ones like Archimedes and Gauss. The second rank consists of people who have a great effect on the mathematics of their time, the important panjandrums like Felix Klein and Saunders Mac Lane, about whose permanent worth, however, it is difficult to make a prediction. A mathematician of the third rank is a steady contributor of deservedly respected additions to the literature, possibly the acknowledged leader of a "school", but one whose discoveries are not so deep or so original as to make it likely that his name will be remembered a hundred years after his death. Examples are dangerous, but necessary to make the idea clearer than adjectives can: names such as George Mackey, Alfred Tarski, and Antoni Zygmund will help to prove that I am still talking about definitely non-trivial people. I am inclined to put Garrett Birkhoff and Chuck Rickart in the fourth rank, along with Dieudonné and Kuratowski. As for myself, of course I cannot be sure where I belong, but if there were an election, I would like to be a candidate for rank four. Perhaps the distances between these ranks give a vague idea of how to continue; precise definitions are, of course, impossible.

Is this a good game? Perhaps so, but it can become confusing very quickly. Where should we put Riemann, and Poincaré, and Hilbert? It may be obvious that they are of rank one, but is it really obvious that they are in some sense equal to Archimedes and Gauss? And what about Paul Cohen and Charlie Fefferman and Kurt Gödel? Are they lesser than Hilbert but greater than Klein—and, if so, should we introduce ranks such as 1.5?

All this came to mind as I was preparing to write about one of the friends I made in Moscow. Just one year younger than I, Fomin was a gentle man, an intelligent man, a kind man. As a mathematician he was perhaps of rank five. He was a great friend and collaborator of Gelfand, and he worked with Kolmogorov also. He wrote the preface to the Russian translation of *Measure Theory*, and later he himself translated my *Lectures on Ergodic Theory*. I read the preface he wrote, slowly, painfully, with a dictionary, and I found that I had written a pretty good book, with some pretty serious faults. It had to be good—else why would the U.S.S.R. have permitted time and energy to be spent on translating it? Conspicuous among the faults was the regrettable American tendency of not giving adequate credit to the Soviet mathematicians on whose work the book was based. Example: for Egorov's

S. V. Fomin and I. M. Gelfand, 1965

theorem and Luzin's theorem, instead of referring to articles by Egorov and Luzin, the book refers to an earlier text (Saks) in which the bibliographic details of the Russian work can be found. (Guilty.)

Fomin's English was weak but adequate. I saw him relatively frequently, and we had a good time together. (A few years later I was able to arrange for him to visit me in Hawaii, and he had an even better time there.) He invited me to a dinner party at his house. That was rare. I saw only two or three of my Russian acquaintances at home; mostly we met in public places, such as classrooms and restaurants.

Fomin's home was the second floor of what looked like a hovel from the outside. It was a wooden house built 200 years ago. I was told when it was built, but no one could remember when it was last painted. The walls seemed to be leaning this way and that. Fomin's quarters consisted of two rooms (plus kitchen and bath): a small room that was combination living room, bedroom, and study, and an even smaller dining room. His university-age daughter didn't live there. Mrs. Fomin was nearly the same height as her husband and twice the weight—immense, bubbly, good-natured, and smart. She told me that she had never studied English, but she spoke it just fine, much better than he. She worked for a publishing company as its expert on biology, and she cooked for me the best meal, from borscht to blini, that I had in Russia.

Gelfand and his wife arrived an hour late. That was very much par for the course: Gelfand was always late. He told me, half ashamed, half proud, that although "Don Giovanni" was his favorite opera, the only way he ever heard the overture was on records—he never got to the opera house on time.

I had met Gelfand before the Fomin party, and I loved him immediately. He is a little man, spry, alert, sharp, pixyish. I thought at first that he was quite a bit older than I, a little Jewish grandfather, but I learned that he is only three years older—he was 52 in 1965. He worried about the cold I was catching; he wanted to lend me a sweater, and he insisted that I telephone him during the May 1 weekend if I didn't get better. He took me on a sightseeing tour of the Kremlin, and he showed me, among other things, the old Tsar Pushka—a large lethal looking cannon made during the Napoleonic war to help fight off the invader. It wasn't well made—something went wrong with the design and it never fired a single shot. It is impressive looking though, and I wanted to photograph it—but Gelfand wouldn't let me. No, no—we're in the Kremlin—this is, after all, a piece of military equipment—it won't do. He asked a guard—and the guard laughed: "Sure, take all the photographs you want of it."

Gelfand speaks English, in a way; he reads many English books, but he was learning the language, re-learning it, during that long afternoon at Fomin's house. At 2 o'clock his English was slow and hesitant, but it kept improving minute by minute; at 6 o'clock he spoke fast and almost as authoritatively as in his staccato Russian. He asked what I was interested in, and then he talked mathematics; he bubbled over with his ideas on my subject, he explained its connections with other subjects, and he told me which parts of it were important.

At 6 the dinner party organized itself into taxis and rode over to the university. The famous Gelfand seminar was scheduled to start at 6:30, and, unprecedented!, Gelfand was on time. That was easy to fix: we milled around for half an hour, I was introduced to a dozen people, and we chatted. The apparently desired effect was achieved: the seminar could start, as always, a half hour late.

I was scheduled to speak at the seminar that evening, and I had been forewarned. I was told to prepare a two-hour talk, but not to be surprised if it stretched to three hours; the translation would take time, the audience would ask questions, and Gelfand would keep making comments. What I was not prepared for was the last minute program change. During the half hour of milling around an unexpected visitor appeared: Ciprian Foiaş, from Rumania. I had known of Foiaş before, we had had some editorial correspondence, and I had read the papers he and Nagy collaborated on. I was glad to be able to attach a face and a person to the name. He just happened to be in Moscow and dropped in to attend the Gelfand seminar. Gelfand was happy to welcome him, and on the spur of the moment invited him to give a talk—right then, that same evening, splitting the time with me.

That upset me. I had five minutes to re-plan the whole show. My carefully organized timing had to be discarded, and difficult decisions (what must I keep?, what can I do without?) had to be made

A. A. Kirillov, 1965

instantaneously. It worked out, somehow, but I wouldn't want to do it again. Sasha Kirillov, a rosy-cheeked baby-faced 28 at the time, was the translator, and he was a big help: in a couple of minutes we both caught on how to bounce the ball back and forth between us. Once every paragraph or so Gelfand interrupted and, like a Greek chorus, explained the plot to the rest of the audience. I was stretched to the limit of my tensile strength and at the end of my 90 minutes I was ready to collapse.

When I stopped, Foiaş began, and all hell broke loose. Foiaş's English was nowhere near what it has become by now; then it was fractured, to say the least. Each time when Kirillov started to translate, Foiaş would break in and shout him down, whereupon Gelfand would use *his* broken English to drown Foiaş out and, more or less simultaneously, use Russian to tell Kirillov and the rest what was going on. It was a circus, a three-ring circus at least.

Mathematicians in Moscow

Some of the younger mathematicians I met in Russia told me that everyone finds Kolmogorov uncomfortable to be with, and when I was introduced to him I saw what they meant. He was stiff and formal and very much aware of his own importance. He was faultlessly polite to me, but I couldn't relax, I felt awkward. At our first meeting we agreed to speak German; his was no better than mine. We arranged to meet a

A. N. Kolmogorov, 1965

couple of days later, and Kolmogorov came to the Hotel Budapest to pick me up. The telephone in my room rang exactly when he said he would arrive: "Hier spricht Kolmogorov—ich bin hinunten", and we were off.

I wish I knew why I was so ill at ease with him. Partly, of course, I was overwhelmed by the great presence. I knew that we all love the friendly relaxed peasant, even if he is uncouth, more than the awkward, ill-at-ease gentleman—but I couldn't help being awkward. I felt that I had to perform and I didn't know my lines. I had to speak German grammar (ludicrous), I didn't remember that Kolmogorov had worked on mathematical logic (a faux pas), I had to understand his new views on entropy in broken German (almost), and I had somehow to resonate with his soul (couldn't).

We rode in an academy car (the usual five-year-old Chevrolet-like government issue) driven by a chauffeur bundled up in a ragged overcoat, wearing a soft cap, and dangling a cigarette from his lips; at the end of the ride Kolmogorov signed a chit on the pad that the chauffeur handed back. The wing of the university we entered contains apartments. Kolmogorov's is dark, cold, and relatively large. I saw five rooms. They were not tiny, but none was as large as my bedroom back home in Ann Arbor, and all were crowded—stacks of reprints in one corner, a collection of theatrical masks somewhere, and a couple of skis somewhere else. "Is this where you work?", I asked. "No, no", he said; "I work out at the dacha; I am here only three days a week."

The rest of the party consisted of Fomin, Kolmogorov's wife, and Pavel Sergeevich Alexandrov (the topologist of Alexandrov–Hopf fame). Kolmogorov made conversation; he showed us color reproductions of the work of a Lithuanian artist, and I kept saying sehr interessant, sehr hübsch, das hab' ich gern. Presently we sat down to a huge and elegant dinner, from caviar to coffee, served by a maid. Alexandrov was a gentleman of the old style, and during dinner he entertained us by reminiscences of pre-war Germany. He spoke German almost as a mother tongue, and Mrs. Kolmogorov and Fomin spoke it almost not at all, but Alexandrov's charm and his personality carried the day. It was all more cosmopolitan than homey, and it was for me a memorable occasion. At the same time, for both Kolmogorov and for me, it was a duty conscientiously attended to; we didn't hate doing it, but we were glad when it was over.

M. A. Naimark, 1965

Naimark, everybody's uncle Mark Aronovich, was a famous mathematician and a very good one. He was not as great as Kolmogorov, but everybody seemed to love him, and I happily joined the crowd. He had been quite ill (a stroke, I was told). His right side, which had been paralyzed, had almost completely recovered, but recovery took five months first in a hospital, then in a sanatorium, and then at home, and he still looked and acted weak. He went to a lot of trouble to be hospitable to me, and I found him a pleasure to talk to. He wangled an invitation to Sergei Mihailovich Nikolskii's birthday banquet for me, and we had a chance to sit together and chat for several hours.

The banquet was of a sort apparently not at all uncommon in Russia but new to me. Nikolskii had turned 60 just three days before, and he was simultaneously the guest of honor and the host. He rented one of the dining rooms at the Grand Hotel (officially re-named the Moskva, but the old-timers insisted on using the old name), and he invited about 100 people to help him celebrate. (Rumored cost: 2000 rubles.) When Naimark and I arrived, about half the crowd was already there, standing around and chatting; the tables were set and laden with food and drink. There were two or three bottles within easy reach of every seat, in a periodic pattern of cognac, lemonade, wine, mineral water, vodka, etc., and a similar and equally generous cold buffet of caviar, pâté, tongue, tomatoes, fish, and other such goodies. When we sat down, we set to; the rule seemed to be to talk, eat, and drink at the same time, all the time, with two kinds of interruptions.

The most frequent interruptions were the toasts. Whenever the spirit moved someone, he would rap on his glass to call for attention, and then propose a toast—praising Nikolskii, Nikolskii's work, Nikolskii's skiing, Nikolskii's teacher (Kolmogorov), Nikolskii's institute—whereupon everyone would clink glasses, drink, and applaud. Some of the time you would clink with one or two of your immediate neighbors only, but sometimes, after a specially rousing speech in favor of a good cause, you would clink around, each with each, in groups of eight or ten. At the end of the really fine toasts half the crowd would get up, crowd around the head table, and clink furiously with each other and with Nikolskii. From time to time Nikolskii himself would get up and, going to one or another part of the room, clink many times.

The interruptions that did not involve drink involved food. I thought that supper was to be equal to the buffet that had awaited us, but nothing of the sort—that was just the hors d'oeuvres. Presently waiters brought each of us a hot crab cocktail, and later a huge plate of chicken and trimmings, followed by several dessert courses.

Between bites and gulps, people wandered around in an informal way, some left the party early, a few late-comers arrived to replace them, and many went table-hopping. I was taken around to shake some hands. One of them belonged to Andrei Nikolaevich Tihonov (products of compact spaces are compact), a roly-poly Santa Claus with a round body, round head, white hair, and white goatee. Another was Yuri Vasilevich Prohorov, an excellent young (35?) probabilist, looking like a rusty-haired Spencer Tracy. Prohorov seemed to be as impressed to meet me as I was to meet him. He was astonished that I was in Moscow—no one had ever told him that I was expected. He grumbled about the non-organization that would permit such things. Lev Semenovich Pontrjagin (*the* Pontrjagin) was at the head table, and I went to say hello to him—we met when he visited Ann Arbor. I intended to reintroduce myself: "Professor Pontrjagin, I am Paul Halmos; we met at the University of Michigan" (since he is blind I assumed he would need some extra auditory recognition clues), but he didn't let me get it all out. He remembered immediately, he stuck his hand out in my direction, and he welcomed me to Moscow. His control of English is curiously deceptive—he pronounces the few words he knows so nearly perfectly that you tend to think he speaks the language much better than in fact he can.

Almost not meeting Prohorov reminds me of some others whom I wish I had met but did not. One is a man named F. V. Shirokov who solved a problem I posed. He sent me a preprint, we exchanged a few letters, and I became quite fond of him. I was looking forward to meeting him and I was disappointed not to. Another was one of the founding fathers of modern harmonic analysis, Georgii Evgenevich Shilov, and how I managed not to meet him is quite strange. I was invited to lecture to the Moscow Mathematical Society one day. After we had all trooped into the lecture hall, and sat down, the chairman of the session, whom I didn't know, called the meeting to order and introduced the speaker (me, that is). Since he spoke Russian I could only guess what he was saying, and I whispered to Anosov, sitting next to me: "What's he saying and who is he?" Anosov was surprised. "Haven't you met him yet? That's Shilov, and he is introducing you." Good, I thought; I supposed we would meet after the lecture—but we never did. When the lecture was over, and I answered the customary courteous question or two, everyone got up and left, and Shilov had disappeared. Strange.

There were many other Soviet mathematicians I could have met but did not, and there were many I did meet but have not mentioned here. I'll conclude this selective roll call by telling about two more whom I did meet, two important and famous mathematicians, Sinai and Krein.

Yakov Grigorievich Sinai was in on the ground floor when entropy became a concept of ergodic theory; I knew his work, and I had quoted it and worked on it. He was still young when I met him (30), and impressive. We took to each other at once, and had much to talk about —mathematics, of course, and the way mathematicians are trained in the U.S.A. and the U.S.S.R., and much else, both professional and mundane. (He is the one who took me to the circus.) He invited me to attend his seminar one evening (to listen, not to talk), and I was impressed by the seminar and the students attending it almost as much as by him. What happened was this. Sinai and I walked in, I sat down, and Sinai spoke in Russian for about three minutes. I could catch some of the drift: he pointed to me and said my name, and suggested that the rest of the session be conducted in English. And it was, and it went smoothly, and I tried to imagine something like that happening back home. Several students reported on the theorems and proofs that they had been learning and thinking about (one who has acquired a power-ful reputation since then was Katok), all in English; the listeners asked questions and discussed the answers, all in English.

My meeting with Mark Grigorievich Krein (pronounced "Crane"), the more famous of two brothers, was more strenuous. Naimark brought us together at the Steklov Institute and then left. The session that ensued had two main participants, Krein and Foiaş, and two minor characters, Krein's daughter and me. I was the original reason for the get-together, but I don't have the right personality to have been able to keep my end up in that company. The daughter was there as a trans-lator: she spoke imperfect but serviceable English and, so far as I could tell, the same kind of French, and she had no personality problem about interrupting and outshouting people.

There was something a little strange about my meeting Krein in Moscow. Krein's home base was in Odessa, and, mainly for that reason, I originally put Odessa down as one of the places in the Soviet Union I wanted to visit. Through Mishchenko I got the message from Krein: he will come to Moscow, or Leningrad, or anywhere else to meet me, but he will not meet me in Odessa. I never figured that one out.

M. G. Krein, 1965

Krein was sixtyish, stocky, round-faced, and nearly bald; he had a dark moustache and a bad tic—his right eye kept winking out of con-trol. He had the nervous habit of saying "tak" and "da" ("so" and "yes") several times in each sentence. "A necessary and sufficient tak condition tak is da that tak...". The daughter would chime in often, sometimes to change Krein's incorrect (but understandable) English to something defensible but unidiomatic, and sometimes even when he was doing fine. Krein would say something about "this estimate" (perfectly correct mathematical English), and she would say "estimation, Papa". When Krein said "this term Brodsky proposed tak... suggested...", the daughter said "introduced, Papa". When he stressed the wrong syllable, as in "arTICle", she would correct in mid-sentence. When he said "this is point I wish to stress", she said "emphasize, Papa, emphasize".

Most of our three hours together was a decibel contest between Krein and Foiaş. They were good-natured some of the time and not the rest, and they were trying to out-point each other all the time. Krein would begin to explain something at the blackboard, and, in mid-formula, Foiaş would say "I have a question", and proceed to explain something of his own. Krein said "I did that in 1940", and Foiaş countered "ah, but not this general—let me show you the proper method...". It was back and forth: I did it first, no, I did it better, no my student did it more general.

Most of the time I was just sitting there, both amused and depressed. I felt guilty too, but I am not sure why. Could it be that I don't love mathematics enough, or that I don't like people enough? I wouldn't dream of buttonholing a stranger, not even one whose books I have read, to tell him "here's something important I proved last week". In the first place I don't think what I proved last week is that important (it never is), in the second place I would feel foolish, and in the third place I don't think he would care.

At the end of the afternoon Krein proudly told me how lazy and inefficient he was about answering letters—now that he had this wonderful opportunity of personal contact, he would take advantage of it. He handed me a large stack of reprints (about seven inches thick) and instructed me to give them to the various American mathematicians he indicated. I was too cowardly to refuse. The next day I took the whole stack to the post office, had them wrapped there (a service not available in the U.S.), and shipped them to their intended recipients. It would have been easier for Krein than it was for me to explain to the young woman who did the wrapping what he wanted done, but it would have cost him the same number of rubles.

I met a few visitors to Moscow in situations similar to my own; we started conversations as we were waiting for our dinners to appear. I remember two in particular: an Italian historian and a Canadian geologist. The Italian was a mild man, but he was emphatic on one point; this was his "prima" time in Russia and he would see to it that it was the "ultima". The Canadian expressed himself in stronger language. He wasn't even sure whether he'd stay his full month, he said. There wasn't anything for him to do really; the whole thing was no bloody good.

What were they complaining about? Culture shock, in part, inability to communicate, neglect, non-organization and disorganization, lack of interest in their visit and frequently even lack of knowledge of it—those are some of the answers. Instead of being at home in familiar friendly surroundings, here they were in a strange, cold, dark city, alone, wasting time.

Somewhat to my surprise I found myself trying to cheer them and taking the other side. Maybe there is some use in seeing the faces of our Soviet colleagues, I said, maybe it is good to establish direct contact, maybe it is good to learn what they are doing before they publish it—these are some of the standard obvious reasons for personalizing professional communication—but my heart wasn't much in what I was saying. In truth I shared their view and sympathized with their emotion. An exchange program such as we were participating in could and should be an instrument for effective international collaboration, but for many petty reasons of human and bureaucratic imperfection we found that it wasn't so. Sad. On balance I didn't enjoy myself in Russia. I don't think I accomplished much by way of either the transmission or the reception of knowledge. All I can say is that the trip was an experience and I learned something from it, but I cannot clearly and convincingly specify what that something is.

As others see us

In 1964 I went to Budapest directly from home, and I called it shabby, dingy, dark, and depressing. In 1965 I went to Budapest from Moscow and, in comparison, I found it an elegant city, full of light, color, and gaiety. I spent 10 days in Hungary that second time, including a few in Szeged, but they were more of personal and touristic interest than professional and mathematical.

I did, to be sure, give a couple of lectures and see some mathematicians. By a pleasant coincidence I ran into Jerzy Łoś (pronounced Wash), who was driving around Europe on holiday, and who just happened to be passing through Szeged while I was there. He is one of the main discoverers and developers of the theory of ultraproducts, a mathematician and logician of broad culture, and a fascinating human being. We spent an evening together, dining and chatting. I admired, among other things, his linguistic courage. He knew not a word of Rumanian or Hungarian, but just by force of personality, remaining quiet and courteous at all times, he could make himself liked and understood in the bookstores, filling stations, and restaurants of Bucharest and Budapest. He quickly made it known to everyone that he is from

J. Łoś, 1965

Poland; since Poles are by and large popular in Eastern Europe, including Hungary, that helped.

Another mathematician I met in Szeged is Attila Máté. He was very young at the time (16?), and his interest in those days was set theory and, in particular, the theory of frighteningly large cardinal numbers. He was bright enough that, almost unprecedented, the authorities allowed him to finish Gymnasium two years early and accepted him in the university. Since then he has shown that his talent can encompass several totally different parts of mathematics. When, for instance, Carleson settled the Luzin conjecture (the one about the pointwise convergence of functions in L^2), Máté wrote up an exegesis that helped the rest of us to understand what the proof was all about. A few years later he emigrated and seems by now to be settled in Brooklyn as well as he once was in Szeged.

Instead of mathematics, what I learned during my second return visit was a collection of curious facts about Hungary and that country's attitudes toward America.

Consider, for instance, the grassroots meaning of democracy. The car that brought several of us, including Nagy and me, back from Szeged to Budapest was driven by a university employee. Some of the conversation included him (although somewhat condescendingly, I thought); when we reached Budapest and dropped Nagy off he shook hands with the driver. Before we reached Budapest, however, we stopped for coffee, and we met another university chauffeur; he put his hands on the sides of his trousers, in a posture of military attention, and bowed to Nagy. Our driver offered to take our coffee coupons (you pay before you drink and get coffee with the coupon you bought) and do our standing in line for us; it was taken for granted that the offer would be accepted. We, the aristocrats, took our coffee to one part of the "Presso" place; the drivers took theirs to another. Having asked about the protocol, I offered our driver a tip when I said goodbye to him. First he said no, but presently he accepted it. Socialized monarchy makes for a strange sort of democracy.

A social-political sort of gallows humor (often tinged with vestigial ethnic dirty cracks) has always been popular in Hungary. I was told, for instance, that in Russia if you can read and write, you are called intelligent; if you can read or write but not both, you are called a specialist. In Hungary, the literacy joke went on, cops cruise in twos: one who can read and one who can write. In Rumania, however, they go in threes: one who can read, one who can write, and one to keep an eye on those untrustworthy members of the intelligentsia.

Some of the good aspects of monarchic culture are still in effect in Hungary. Service, for instance, in restaurants and hotels is excellent— totally unlike the Soviet version. The breakfast waitress in Budapest took orders, and was even able to kid the customers, in French, German, and Hungarian; she brought my breakfast in five minutes, not 35. The doorman was an elderly man in a neat uniform who apologized to me, in bad English, for not being able to speak English; besides Hungarian, he said, he can speak only French, German, Italian, and Russian.

The minor services in Hungary (such as the ones provided by small shops and the streetcar lines) are good; the major ones (such as the post office and the airline) are dreadful. Révész told me that once he telephoned Budapest from Debrecen (the distance is approximately the same as from Champaign, Illinois to Chicago); he was not in the least surprised that it took $2\frac{1}{2}$ hours for the call to go through. Another time he phoned Budapest from East Lansing, Michigan. The long-distance operator apologized before she went to work on the call. She explained that the Hungarians are sometimes a little slow to answer. Perhaps it would be better if he hung up; she would ring him as soon as she got through. She rang in 15 minutes, and apologized again for being so slow.

I saw Tenessee Williams' "Summer and smoke" in Szeged (in Hungarian, of course), and "The Oppenheimer affair" in Budapest. The latter was based quite closely on the facts, with some curiously off-beat details. Example: there was a soda-water dispenser on a refreshment table in the presumably American courtroom; "Evans" was pronounced "Eevans"; J. Robert Oppenheimer's first initial was pronounced gee; and the American flag had 50 stars on it (an anachronism). The actor who played Oppenheimer didn't resemble him at all, but he acted him very well—he must have studied films showing

Oppy, and he moved and spoke true to life. I wrote Oppy afterward and told him that it was odd to hear him speak fluent Hungarian.

Despite such cultural exports as Tennessee Williams and J. Robert Oppenheimer, there was a lot of misinformation about life in the United States that everyone believed to be gospel truth. It was well known, for instance, that shoes are never repaired in America; when they wear down, they are just thrown away. (On the other side of the picture, I was surprised to learn that in Hungary there were run-repair shops; when a woman's stocking—silk stocking?—develops a run, that's where you take it for repair.) I was also told, authoritatively, that (a) in America everyone is always breathlessly running, and that, therefore, the quiet life of small, provincial Hungary is much better, and (b) in Hungary everyone is always breathlessly running to make a living and fend off starvation, so that the more leisurely pace of the new world is enviable.

Cultural differences on top of linguistic ones make communication extra hard. Speaking a language that you don't feel to be your own you have to concentrate on the way you express yourself—on strange idioms as well as strange ideas. Even if you are asked to explain something you know perfectly well, and even if you have the words to do it with, the culture can get in the way. To teach calculus to a Hottentot, you have to know more than calculus and Hottentot—you have to know Hottentot culture, you have to know where in the experience of your listener straight lines occur, what in his experience moves, and what the units by which he measures things are. To explain American universities to my sister-in-law I had to understand Hungarian schools, Hungarian dentistry, Hungarian railroad slang—Hungarian culture, in a word—and because I didn't, I couldn't. It was good to get back home where I could think without even thinking about it.

•

CHAPTER 14

How to do almost everything

Rejections

I was at the height of my powers in the 1960's—which may be interpreted as saying that, having reached and passed the maximum I was capable of, I was over the hill. The height of your powers takes an instant to attain; from then on it's down the rest of the way. I have spent all my life with one value system—research is all—and height therefore refers to theorems, or, in a better word, insights.

The three most visible concepts on my mathematical horizon in the 60's are identified by the words Toeplitz, Cesàro, and quasitriangular.

Toeplitz operators played a big role at the farewell conference in Chicago just before I went to Michigan. I don't know why Toeplitz matrices are attractive, but it is a fact that every mathematician who is introduced to them finds them natural and pleasant. (They are the matrices with constant diagonals: $a_{i,j} = a_{i+k,j+k}$.) Arlen Brown and I became fascinated by them, read about them (principally in the papers of Philip Hartman and Aurel Wintner), and started asking questions about them. We got some satisfying insights—we thought we really understood what made the Hartman–Wintner theorems tick, we gave a characterization of Toeplitz operators, and we laid the foundations for their algebraic study. Our paper is old stuff by now (it appeared in 1963), the theory has made much progress since then, but I am proud to have had a hand in it.

The work on Cesàro operators was another collaborative effort. It was written in a year when my two-man seminar with Allan Shields became a three-man seminar by the addition of Arlen Brown. We were all at Michigan then, and we got together Tuesday afternoons for a couple of hours to drink coffee and talk Cesàro. (Surely everyone knows the Cesàro matrix? It is the one that converts a sequence into the sequence of averages; its n-th row has the first n entries equal to $\frac{1}{n}$ and the rest equal to 0.) We looked at boundedness problems and spectral problems for that matrix and for its "continuous" analogues, both finite and infinite. We solved a lot of them, but we were stumped by one: we couldn't prove that the classical Cesàro operator is subnormal. The proof was found six years later by Tom Kriete and David Trutt; it is a difficult and deep achievement.

The polynomially compact business was one of the exciting mathematical events of the 1960's. The first published proof that compact operators always have non-trivial invariant subspaces is due to Nachman Aronszajn and Kennan Smith; it appeared in 1954. Smith pointed out, I might almost say complained, that the proof was "tight". It left no room for modifications and generalizations; it proved exactly what it was designed to prove, no more. It did not, for instance, help to prove that a square root of a compact operator also had to have non-trivial invariant subspaces.

Aronszajn taught me the proof on a restaurant napkin several months before the paper appeared. I understood it, I cherished it, and along with many others I kept trying to "loosen" it so as to be able to apply it more broadly—but all to no avail. It was a surprise, a jolt, when, early in 1966, Abby Robinson sent me a preprint written in collaboration with his young disciple Allen Bernstein: yes, square roots of compact operators did have invariant subspaces. Bernstein and Robinson found the right algebraic context to set the result in. They replaced x^2 by $p(x)$, where p could be an arbitrary non-zero polynomial; they proved that if an operator A was such that $p(A)$ was compact, then A had invariant subspaces. (I think I invented the phrase "polynomially compact" to describe such operators.) Their main accomplishment was the sought-for "loosening"; they showed that the Aronszajn–Smith technique was not as narrow as we feared.

The Bernstein–Robinson result inspired me to re-study that technique, and I thought I saw what made it work. I abstracted the principal property that did the trick and called it quasitriangularity. In the paper I wrote on the subject I took the first few steps in the theory of quasitriangular operators, and conjectured that they are indeed our friends, i.e., that they all have invariant subspaces. The notion caught on, by now there are many papers on it, and it has become a standard part of operator theory. The Rumanian school worked especially hard on it, and some of them (namely Apostol, Foiaş, and Voiculescu) proved the deepest and most spectacular theorem: they proved that the *non*-quasitriangular operators are our friends for sure. That is: non-quasi-triangularity implies invariant subspaces; if my conjecture turns out to be right, then the invariant subspace problem is solved affirmatively. I admit that that rocked me. Since I am convinced that the solution of the invariant subspace problem is negative, I abandon my conjecture, and shall proceed instead to search for a quasitriangular counterexample.

I liked both the Toeplitz paper and the quasitriangular one, and I am no end pleased that some of the operator theory world has come to share my opinion. Both papers have been called classics (not by me) and both seem to have become compulsory entries in all the bibliographies of their subjects. The reason I immodestly mention these facts is to give courage to the young people who read these lines and are having trouble with persnickety referees and editors. The point is that both papers were rejected when I first submitted them, Toeplitz by Acta Mathematica and Quasitriangular by the American Journal of Mathematics. Writing to Ambrose about the latter one I said: "... just had a paper rejected last week ... I hated that ... it injured my pride .. but all I actually did was put the manuscript in a different envelope and send it to another journal ... it's a pretty good little paper actually, with a small but definitely new idea, and I think it should and will get published." Cheer up; the judgments of referees and editors are not always right.

How to do research

Can anyone tell anyone else how to do research, how to be creative, how to discover something new? Almost certainly not. I have been trying for a long time to learn mathematics, to understand it, to find the truth, to prove a theorem, to solve a problem—and now I am going to try to describe just how I went about it. The important part of the process is mental, and that is indescribable—but I can at least take a stab at the physical part.

Mathematics is not a deductive science—that's a cliché. When you try to prove a theorem, you don't just list the hypotheses, and then start to reason. What you do is trial and error, experimentation, guesswork. You want to find out what the facts are, and what you do is in that respect similar to what a laboratory technician does, but it is different in its degree of precision and information. Possibly philosophers would look on us mathematicians the same way as we look on the technicians, if they dared.

I love to do research, I want to do research, I have to do research, and I hate to sit down and begin to do research—I always try to put it off just as long as I can.

It is important to me to have something big and external, not inside myself, that I can devote my life to. Gauss and Goya and Shakespeare and Paganini are excellent, their excellence gives me pleasure, and I admire and envy them. They were also dedicated human beings. Excellence is for the few but dedication is something everybody can have—should have—and without it life is not worth living.

Despite my great emotional involvement in work, I just hate to start doing it; it's a battle and a wrench every time. Isn't there something I can (must?) do first? Shouldn't I sharpen my pencils perhaps? In fact I never use pencils, but "pencil sharpening" has become the code phrase for anything that helps to postpone the pain of concentrated creative attention. It stands for reference searching in the library, systematizing old notes, or even preparing tomorrow's class lecture, with the excuse that once those things are out of the way I'll really be able to concentrate without interruption.

When Carmichael complained that as dean he didn't have more than 20 hours a week for research I marvelled, and I marvel still. During my productive years I probably averaged 20 hours of concentrated mathematical thinking a week, but much more than that was extremely rare. The rare exception came, two or three times in my life, when long ladders of thought were approaching their climax. Even though I never was dean of a graduate school, I seemed to have psychic energy for only three or four hours of work, "real work", each day; the rest of the time I wrote, taught, reviewed, conferred, refereed, lectured, edited, travelled, and generally sharpened pencils all the ways I could think of. Everybody who does research runs into fallow periods. During mine the other professional activities, down to and including teaching trigonometry, served as a sort of excuse for living. Yes, yes, I may not have proved any new theorems today, but at least I explained the law of sines pretty well, and I have earned my keep.

Why do mathematicians do research? There are several answers. The one I like best is that we are curious—we need to know. That is almost the same as "because we want to", and I accept that—that's a good answer too. There are, however, other answers, ones that are more practical.

We teach mathematics to the engineers, physicists, biologists, psychologists, economists—and mathematicians—of the future. If we teach them to solve only the problems in the book, their education will be out of date before they graduate. Even from the crude, mundane, industrial, commercial point of view, our students must be prepared to answer future questions that weren't even asked when they were in our classes. It is not enough to teach them everything that's known—they must know also how to find out what has not yet been found. They must, in other words, be trained to solve problems—to do research. A teacher who is not always thinking about solving problems—ones he does not know the answer to—is psychologically simply not prepared to teach problem solving to his students.

One part of doing research that I am no good at, and therefore never liked, is competition. I am not sufficiently quick to win kudos by scooping people. My substitute for trying to be the first was to go off in a direction orthogonal to the mainstream and hope that I could find a small but deep backwater of my own. Loath to waste time trying to prove the outstanding conjecture and then fail, I have tried instead to isolate the missing concept and to formulate the fruitful question. You can't do that often in one lifetime, and if the concept and questions are indeed the "right" ones, they get widely adopted and you're likely to find yourself outdistanced in the development of your own subject by the people with the powerful techniques and the deep insights. Fair enough, I can live with that; it's a fair division of labor. Sure I wish I had proved the subnormal invariant subspace theorem, but at least I did something by introducing the concept and pointing the way.

One aspect of staying outside the competition was my de-emphasis of hurry. What's wrong, I asked myself, with living a year or two behind the most recent elaborations? There is nothing wrong with it, I told myself, but even for me that answer was sometimes unworkable, and for some people, psychologically differently constituted, it is always wrong. When the Lomonosov news broke (about simultaneous invariant subspaces of commuting compact operators) and the Scott Brown news (about subnormal operators) I was as excited as the next operator theorist, and I was eager to learn the details quickly. That sort of breakthrough is, however, rare enough that I could still live most of my life happily behind the times.

Very well then—out of the competition, orthogonal to the mainstream, and behind the times—what do I actually do? The answer is that I write. I sit down at my desk, pick up a black ball-point pen, and start writing on a sheet of $8\frac{1}{2} \times 11$ ruled paper. I put "1" in the top right corner, and then begin: "The purpose of these notes is to study the effect that a perturbation of rank 1 can have on the lattice structure of ...". When the paragraph is finished, I label it with a big black bold "A" on the margin, and go on to paragraph B. The page numbers and the paragraph letters constitute the reference system, often for several hundred pages: 87C means paragraph C on p. 87. I put the pages in three-ring notebooks, labelled on the spine: Approximation, Lattice, Integral Operators, etc. If a project succeeds, a notebook becomes an article, but success or no, the notebook is hard to throw away. I always have dozens of them on the bookshelves next to my desk; I keep hoping that the incomplete ones will grow and that the ones that led to a publication will turn out to contain that overlooked gem of an insight that's needed to solve the main open problem.

I keep sitting at my desk for as long as I can—which might mean for as long as I have the energy or for as long as I have the time. I try to arrange matters so as to stop at an upbeat, such as a lemma settled or, in the worst case, an unexamined and not obviously hopeless question raised. That way my subconscious can go to work and, in the best case, make progress while I walk to the office, or teach a class, or even sleep at night. The elusive answer tries to keep me awake sometimes, but I seem to have developed a way of fooling myself. After tossing and turning a while, not long—just a few minutes usually—I "solve" the problem; the proof or the counterexample arrives in a flash of insight and, content, I roll over and fall asleep. Almost always the flash turns out to be spurious; the proof has a gigantic hole in it or the counterexample is not counter anything. No matter; I believe the solution long enough to lose consciousness. The curious thing is that at night, in bed, in the dark, I never remember to distrust the "insight"; it is so welcome that I accept it without question. On a few occasions it even turned out to be right.

I don't mind working by the clock. When I must stop thinking because it's time to go to class or time to go out to dinner, I cheerfully put my notes away. I might keep mulling over the problem as I go downstairs to the classroom or get the car started and close the garage door, but I don't resent the interruption (as some of my friends say they do). It's all part of the pattern of life, and I am comfortable in the knowledge that in a few hours we—my work and I—will be together again.

Where do the good questions, the research problems, come from? They probably come from the same hidden cave where authors find their plots and composers their tunes—and no one knows where that is or can even remember where it was after luckily stumbling into it once or twice. One thing is sure: they do not come from a vague desire to generalize. Almost the opposite is true: the source of all great mathematics is the special case, the concrete example. It is frequent in mathematics that every instance of a concept of seemingly great generality is in essence the same as a small and concrete special case. Usually it was the special case that suggested the generalization in the first place. One precise way to formulate "in essence the same" is as a representation theorem. The Riesz theorem about linear functionals is typical. Fixing one vector in an inner product defines a bounded linear functional; the abstract concept of a bounded linear functional is a seemingly great generalization; the theorem is that, in fact, every instance of the abstract concept arises in the concrete special way.

This is a subject, one of many, on which Dieudonné and I seem to disagree. I gave a colloquium talk at Maryland once, one of the times when Dieudonné was visiting there. The subject of the talk was positive approximation. The problem I had set myself was, given an arbitrary operator A on a Hilbert space, to find a positive (non-negative semi-definite) operator P that minimizes $\|A - P\|$. I was lucky: it turned out that there was a small concrete special case that contained within itself all the concepts, all the difficulties, and all the steps needed to understand and to overcome them. I focused my talk on that special case, the operator on $\mathbb{C}^2$ defined by the matrix $\begin{pmatrix} 0 & 1 \\ 0 & 0 \end{pmatrix}$, and I felt proud: I thought I succeeded in communicating a pretty problem and its satisfactory solution without getting bogged down in irrelevant analytic technicalities. Dieudonné was polite and friendly, but clearly condescending afterward; I don't remember his exact words but, in effect, he congratulated me on a droll performance. It seems to have struck him as "recreational mathematics", which is a sneer in his vocabulary; pleasant enough, but contrived and superficial. I thought (and continue to think) that it was more than that; the difference in our values was caused by the difference in our points of view. For Dieudonné the important result is, I think, the powerful general theorem, from which it is easy to infer all the special cases you want; for me the greatest kind of step forward is the illuminating central example from which it is easy to get insight into all the surrounding sweeping generalities.

My greatest strength as a mathematician is the ability to see when two things are the "same". When, for instance, I kept puzzling over David Berg's theorem (normal equals diagonal plus compact), the insight came when I noticed that his mess resembled the proof that every compactum is a continuous image of the Cantor set. From then on it did not need great inspiration to use the classical statement, rather than its proof, and the outcome was a new and perspicuous way of getting Berg's result.

I could cite many instances of the same kind of thing. Some of the most striking ones occur in duality theories. Examples: the study of compact abelian groups is the same as the study of Fourier series, and the study of Boolean algebras is the same as the study of totally disconnected compact Hausdorff spaces. Other examples, not of the duality kind: the classical method of successive approximations is the same as Banach's fixed point theorem, and probability is the same as measure theory.

That kind of insight makes mathematics clean; it trims off the fat and shows what's really going on. Does it advance mathematics? Are the great new ideas really just recognitions that two things are the same? I often think so—but I am not always sure.

There; does that answer the question of how to do research?

The invariant subspace problem

To be a research mathematician you have to do more than pose and solve research problems; you must stay in touch with the steady and inexorable rise in the number of pages that are daily added to the literature, and you must certainly be aware of the occasional piece of spectacular news that surfaces, be it a collapsed conjecture or a glorious

L. de Branges, 1965

victory. One of the most spectacular mathematical news items of the 1960's was the de Branges–Rovnyak announcement of their solution of the invariant subspace problem.

Louis de Branges and I started corresponding and exchanging reprints in 1962. A year later (he was at Purdue by then) he asked me to communicate a Research Announcement of his to the Bulletin, and I was glad to do so. Our relation was friendly but formal. I saluted him as "Dear Professor de Branges", and his letter to me contained sentences such as these: "The interest which you have expressed in my work is much appreciated. Not everyone is in a position to understand it."

I met him and his coterie of worshipful students one April afternoon in 1964 when I paid a colloquium visit to Lafayette. I went to his office at the time we had agreed on, but he wasn't there; David Trutt (one of the coterie) met me and, looking mysterious, said that de Branges had been delayed. The delay wasn't long; de Branges arrived presently, with his wife, with Jim Rovnyak, and with two or three others of his current students. He greeted me, and, handing me a manuscript, he explained the delay: we were up all night, he said, putting the finishing touches on this and getting it all typed and ready for you. "This" was just a few pages, an announcement, not a full paper; what it said was that every bounded linear operator on a complex Hilbert space of dimension greater than 1 has a non-trivial closed invariant subspace!

I was flabbergasted, of course, and impressed, and there wasn't much time for anything else; the pre-colloquium coffee was to be served soon thereafter. Back in Ann Arbor next week, I wrote to Ed Spanier, who was the appropriate Bulletin editor for such matters, as follows.

"The enclosed paper by de Branges and Rovnyak titled "The existence of invariant subspaces" is hereby transmitted for publication as a Research Announcement in the Bulletin. I have read it twice and found it exciting. It is too condensed and proofless for me to be sure that it is correct, but there is no doubt in my mind that it deserves publication, and quick publication at that. The problem is important and has been outstanding for a long time. de Branges's reputation is good; even in the event that the paper has a mistake in it (and of course we all hope that it does not) the mistake is sure to be an important one. This sort of paper is what the Research Announcements were founded for."

A week after I sent that letter, the big manuscript arrived, the one that promised to supply all the details of all the proofs that the announcement didn't have room for. One copy was submitted for publication in the Transactions, and I was one of several people to whom de Branges and Rovnyak sent preprints. With that a round of correspondence began that grew rapidly in frequency and volume till it came to a climax six months later. A careful and not overly energetic referee can take much more than six months to reach a conclusion. This time, however, "everybody" was interested, many mathematicians were looking at and swearing at the preprint (it was very hard to read—the exposition was too condensed in some spots and too verbose in others, and it was unclear throughout), and several groups were in touch with each other, telling one another about gaps that they found and ways of filling them.

Ron Douglas was a junior member of the faculty at Michigan then, a young man of considerable mathematical talent, boundless ambition, and hyperactive energy. He thought fast and tried to talk almost as fast, with the result that half of what he said was nonsense, and the other half the expression of penetrating insight. He usually caught the nonsense himself before anyone else had time to notice and to react to it, and he covered it with more rapid-fire patter. His way of saying "whoops, that's not what I should have said" was "in other words" (pronounced 'n'zer w'rs, sometimes more than once in the same sentence); listening to him talk mathematics was an illuminating, awe-inspiring, and very tiring experience.

Ron and I started reading the de Branges–Rovnyak opus and getting together to discuss it. At the same time Peter Lax, one of the editors of the Transactions then, wrote and asked if I'd be a member of a refereeing committee of three—an unusual step. Committee or no, I could name a couple dozen mathematicians who were going through the paper, often line by line, word by word. That's a large number of readers, especially for a paper in its fetal stage. Arlen Brown was one of them, and Peter Fillmore, and so were Henry Helson, Shizuo Kakutani, Allen Shields, and Joe Stampfli—everybody wanted to be in on the act.

Early in July I wrote a circular letter to the editor and the other referees involved. "I have worked very hard on this stuff for two months now, and I am tired. I would like to get back to some of my own theorems. Douglas, on the other hand, is not tired yet. Proposed solution: I hereby resign from the referee committee, and I respectfully suggest that Peter put Douglas in my place." Peter's circular reply came promptly. "Coach to team: R. G. Douglas going in at left end substituting for Paul Halmos. He will be carrying Theorem 5 on the next play."

Theorem 5 was only the first of the trouble spots: Ron found a slip in the proof and a counterexample to the statement. de Branges fought a valiant battle for a while but he retreated before long and then showed how to get around the difficulty. Lemma 12 was the next one that gave Ron a problem, and, retroactively as it were, Theorem 3 sprung a leak. de Branges put another finger in the dike, and thus stopped Theorem 14 from collapsing. And so it went, and kept on going. Douglas and de Branges exchanged letters at a furious rate and I have copies of most of the correspondence; it makes fascinating reading.

Some time early that fall, perhaps in September, de Branges and Rovnyak prepared a revised preprint. That preparation cost them a lot of time, a lot of effort—and a lot of money. The financial outlay was great enough that they took an unusual step, but not an unprecedented one; they asked that people who wanted a copy of the corrected version help defray the expense by paying $3.00 for it. Many of us, myself included, were more than willing to do so.

Late in September Purdue University issued a press release, SOLUTION TO WELL KNOWN MATH PROBLEM ANNOUNCED BY TWO PURDUE SCIENTISTS, which made no effort to minimize the importance of the accomplishment. That didn't bother me. (Jack McLaughlin, looking at it later, commented: "that's a lot of words to eat.") What did bother me was a paragraph in the middle, which, I was pretty sure, represented the attitude of the authors. It read as follows.

"[T]heir proof is yet to be confirmed. The work is so complex—comprising 70-odd pages of closely knit mathematical reasoning—it may take as long as a year for other mathematicians to study the proof in order to validate it. Editors of Transactions of the American Mathematical Society, the journal to which the de Branges–Rovnyak work was submitted last May, are presently undertaking this task."

My bête noire again: authors exploiting editors and referees as assistants, as "validators", as proof checkers.

The dénouement came in December. Peter Fillmore found the unbridgeable gap and de Branges wrote me a letter that I found sad. Here is what he said, in part. "I apologize for asking your support of a false result, but I see no way in which the risk could have been lessened. It seems to be impossible to obtain a reliable confirmation of substantial discoveries before they are declared."

I asked him to prepare a short statement to be published in the

Bulletin (where the Research Announcement had already appeared), to set the record straight. He did. It consisted of a few sentences only, of which the second was: "We now withdraw the announcement and make no statement either for or against the existence of invariant subspaces." Research Announcements in those days appeared with indications of the names of the Council members who accepted them. The original de Branges–Rovnyak announcement, for instance, had the phrase "Communicated by P. R. Halmos" under the names of the authors. In connection with the retraction paragraph I flippantly suggested to the editors of the Bulletin (not to the authors!) that it be headed "Excommunicated by P. R. Halmos", but no one even cracked a smile; the circumstances were too serious for joking.

Except for two postscripts the story is finished.

P.S. The retraction came out in the March issue of the Bulletin in 1965. A few months later (in August) there was a de Branges–Rovnyak abstract in the Notices, as follows. "We previously announced that a bounded linear transformation in a Hilbert space always has invariant subspaces. This claim was withdrawn when a gap in the argument was discovered by Dr. P. A. Fillmore. We have filled the gap by using a new method of obtaining invariant subspaces from a factorization of the characteristic operator function. The result which we originally announced is true." This last gasp was a short one. In the November 1965 issue of the Notices there is an erratum. "The announcement is withdrawn. R. G. Douglas has found a gap in the proof of the theorem."

P.P.S. An envelope appeared in my mailbox one summer day containing a friendly rueful note from Jim Rovnyak. I no longer have the note, but in spirit it said something like this: the merchandise we sent you was defective, and the least we can do is refund your money. The note was accompanied by a check for $3.00.

No—on second or third thought, no story about mathematics is ever finished; theorems go on forever, in infinite progress, and sometimes individual mathematicians seem to go on forever, trying, failing, trying, failing, and—we hope—trying and gloriously succeeding in the end. Louis de Branges's missed target is not the end of his story. In 1984 his became a name that will be long remembered not for what he didn't do but for what he did. He proved Bieberbach's conjecture (first formulated in 1916) on the Taylor coefficients of univalent analytic functions, after dozens (hundreds?) of mathematicians have tried and failed to do so.

Friends can help

There was a shortage of mathematicians in this country in the post-depression and postwar 1940's, and by the end of the 1950's there were committees studying how to stem the over-production of Ph.D.'s. In the

1960's we were riding high again—there were not enough of us to go around. Anybody could get a job, a good job, and if he didn't like it, he could get another one. We tried to solve the problem of the number of mathematicians being too small by permuting them. Another attempt at a solution was to convert Normal Schools (established originally for training teachers) and cow colleges (sometimes called Agricultural and Mechanical Colleges or Institutes) into Ph.D. producing universities. The effect was that quality went down, salaries went up, and the mad scramble for staffing the newly created Ph.D. programs created more high-level jobs than it produced people needed to fill the low-level ones. In the 1970's the up became down again. At first the recession hit the fresh Ph.D.'s the hardest, and there were stories of people having to leave the academic life and go to work in daddy's hardware store. That loosened up after a while, and the next ones to have grounds for com-plaint were the "6-year-olds"—the ones who had received their doc-torates six years earlier and who, because of the up-or-out rules, had to be granted tenure or be fired. They were often fired. After a slightly slug-gish beginning, the 1980's seemed to promise upturn again—the custo-mers were clamoring at the gates and there were not enough warm bodies to put in front of them. Is there a pattern here?

One standard way of making job offers attractive in the 1960's, in addition to raising salaries and lowering teaching loads, was to make it easy to get sabbaticals and leaves of absence. You were always bidding farewell to your friends, either because they had just accepted jobs in Seattle or Stony Brook, or because they were going to spend a year in Berkeley or in Bonn. Nobody ever quite knew where anyone else was; all that it meant to say that someone was "at the University of Chicago" was that when he wasn't on leave of absence for a year, or taking his quarter out of residence, or attending a two-week conference in Tel Aviv, he probably spent some of his time in Eckhart Hall.

I did a lot of travelling in the 1960's, but there was only one academic year when I was on leave. In 1965–1966 I was seduced by the climate and went to teach at the University of Miami (irreverently known as Suntan U.). I made friends there, notably Jimmy McKnight and Bob Bagley, and I enjoyed the freshly picked strawberries in Janu-ary, but intellectually I was glad to get back to slushy snowy Ann Arbor. Jimmy and Bob were good friends of each other, and they allowed me to join their club. They were both dedicated mathemati-cians, trying to do the best they could with students whose intellectual quality, pre-university training, and anti-academic attitudes were hope-lessly, discouragingly bad. Bob is still there, still doing it, but Jimmy is not.

Jimmy and I used to correspond. He loved words, he loved to play with them, and he wrote the most involved letters imaginable; James Joyce must have taken lessons from him. The last letter I wrote him

J. D. McKnight, 1965

was a verbose six-month overdue update, in which, among other things, I analyzed myself and him and compared us. "The memorable and visible part of *your* life", I wrote, "is its complete dedication to truth and honesty. You tell your students the truth and you do your best to insist on honesty from them and from colleagues and friends and chairmen and deans, and you do all this at a great cost, completely unselfishly. What's more you *seem* to be able to do it naturally, without strain. Me, I huff and I puff, and I blow the house down, and make a lot of noise, and find myself making—having to make?—compromises anyway. I envy you."

Jimmy was chubby and lazy. He liked his beer, he was one of the warmest and wittiest human beings I have ever known, and he smoked himself to an early death.

Despite the Michigan climate, especially in comparison with Miami, I liked life in Ann Arbor. Washtenaw County is neither flat nor hilly; it is gently bumpy, wooded, and overlaid with a marvelous grid of county roads. They serve the country-dwellers, traffic on them is light, and they run mostly straight, north-south or east-west, with intersections coming regularly, once a mile. My bad foot makes me prefer smooth, paved terrain to gravel or grass or any other kind of irregularity—the county roads near Ann Arbor were the best I ever had for walking. I clicked off an intersection once every 15 minutes, up to about 12 or 14 miles worth of them without a stop. My record is 26 miles in one stretch, but that time I made stops. My wife drove along the selected route, and about

J. E. McLaughlin, 1963

three or four times along the way we'd meet; I'd rest a few minutes and sip some tomato juice. The whole trip took seven hours.

Another thing that made Ann Arbor pleasant was that in its large mathematics department I was able to find several colleagues who were personally just what I liked and professionally just what I needed. What I think I need is a logician, an analyst, and an algebraist; if, in addition, there are people around whom I can consult when I run into trouble with general (set-theoretic) topology, homological algebra, and the elements of category theory, then I think I am in good shape. In each of logic, analysis, and algebra what I hope for is someone who is willing to put up with the part of the subject that I can do, and is able to teach me some of the other part. My hopes were fulfilled at both Chicago and Michigan. At Chicago my advisors were Frank Bausch, Irving Segal, and Irving Kaplansky, with Norman Hamilton and Saunders Mac Lane covering the periphery; at Michigan the team consisted of Roger Lyndon, Allen Shields, and Jack McLaughlin, and I could use them as both infield and outfield.

Frank Bausch was a member of the College faculty. He never finished the Ph.D. thesis he started to write under Church, but he was smart, and he was an avid reader with a retentive memory; I learned a lot from him. Irving Segal was not a physicist yet in the days when I pumped him about harmonic analysis and multiplicity theory. I didn't find him easy to learn from; he often didn't understand why I was having trouble with such simple questions. His answers tended to be terse, but he almost always knew what I was wishing I knew, and eventually his answers made sense. As for Kaplansky, he was super, and there is nothing else I can say. He knew (and knows) ten times the algebra I could ever use and he knows how to make it accessible.

Norman Hamilton, whom I met in Princeton, became a graduate student at Chicago and one of André Weil's few Ph.D. students. As a learner and understander of mathematics he was in the genius class, but he never published any. Why not?, I kept asking myself. What is the quality he lacked? Ambition? Patience? Tenacity? Whatever its name, it is an indispensable secondary characteristic for creativity. A runner must have legs, yes, but legs alone without, say, eyesight, will not make him a champion. A mathematician must have talent, yes, but talent alone without the willingness and the ability to organize it will produce nothing but the empty set every time. Be that as it may, Hamilton was one of my Chicago crutches, and I loved having him around to lean on. He was my assistant one time when I taught general topology, and I am sure I learned more from the course, through him, than the students learned from me.

So much for the flashback to Chicago. At Michigan, Roger Lyndon had a name as both a logician and a group theorist, and was always ready to be helpful, and I have already told about Allen Shields and our two-man seminar. I had known them and thought of them as friends before I moved to Michigan. Jack McLaughlin I almost didn't know. We had met a couple of times, at an AMS meeting and at a lattice theory conference, but we never really got acquainted. When we became colleagues, Jack resented me at first, I think. I came in as a big-shot, a bit of a hullabaloo was made of my appointment, and he was suspicious of all the sound and fury. Ultimately he became my best friend at Michigan, and when I left he was the one I missed most.

Jack is a hard-working and broadly-informed algebraist, but the sort of algebra he writes papers about is not the sort I am fond of. Example: he discovered one of the notorious sporadic simple groups. I am a hard-working and quite knowledgeable operator-theorist, and the kind of operator properties that interest me leave Jack completely cold. It turned out, however, that there is a part of mathematics we both know and like. It is a small subject, not considered deep. It has, they say, no intrinsic importance; it is merely a useful tool and an occasional source of examples in other subjects. Both Jack and I tend to be shamefaced about admitting we like it; it is a little like admitting that you read westerns. The subject I am talking about is linear algebra. Jack and I enjoy finding and sharing puzzles about matrices. Once we found that out, we found other common interests and tastes, in food, in walking, in music, and in literature, and we enjoy each other's company. Jack has forgiven me the initial hullabaloo of my Michigan years; anybody who understands and likes the Jordan canonical form can't be all bad.

How to recommend

My life at the University of Michigan was the normal academic life —research, teaching, and service—with all the extra chores that that involves.

One of the major chores of academic life is correspondence: we are always getting demands for information from students, job searchers, cranks, colleagues and their chairmen and their deans, authors and their editors and their publishers, and, of course, government agencies. There is something odd about the epistolary distribution function: we can't all be doing more than our share of professional letter writing, but we all seem to think we do. In the course of years of grumbling about my share, I have developed a few basic rules that I live by, and I have filled many filing cabinets with copies of letters that I can't get myself to throw away. The best way to explain how I correspond is to set down the rules and show a few of the letters.

The rules are three in number. (1) Answer all bona fide letters without delay; be prompt. (2) Do not use flamboyant rhetoric; be plain. (3) If you have said what needed saying, stop; be brief.

Some explanations are in order.

Rather than define bona fide letters, I'll give some typical examples of ones that are not. The four highest on my list (I think I mean lowest) are (1) outright cranks pleading for official acceptance, (2) high school students demanding private tutoring, (3) rejected fellowship applicants with no academic credentials, usually from abroad, begging for professional supervision or financial support, and (4) doctoral candidates from the School of Education, with their letters attached to 24-page questionnaires soliciting statistical evidence. In all of these cases, and many similar ones, I not only don't answer promptly, I refuse to answer at all.

Mathematical cranks have been a source of amusement and vexation for as long as mathematics has existed. A fiendish suggestion for how to deal with them is attributed to Eugene Wigner. Too polite to leave letters unanswered, he recommends plugging them into each other. "Dear Mr. A: you might be interested in the views expressed in the enclosed letter I recently received from Mr. B...". I have heard that, ingenious as it is, the technique doesn't work; both Mr. A and Mr. B bounce right back and want you to continue the correspondence.

I have a thickish file of crank letters received in the course of the passing decades. Most of them concern the old chestnuts (such as angle trisection and Fermat's last theorem). The circle-squarer I am fondest of disposed of his task in a page and a half. He commented that his technique "will solve many mathematical headeques", which I thought was nice. His principal theorem was the statement

$$2\sqrt{\pi} = \frac{\pi}{2} + 2,$$

which I also enjoyed. I unscrambled the equation and found that it has exactly one (double) root, namely 4. That's a lot better than most circle-squarers can do; usually they get π to be something ugly, like 3.14.

The demanding high school students have been assigned a project by their homeroom teacher. Write a biography of von Neumann; what is topology?; explain the meaning and the use of Hilbert space—these are typical examples. One of the conditions of the project is that the student interview an expert, preferably in person, alternatively by telephone, or in the worst case by correspondence. My reaction is to wonder whether murdering the teacher would be called justifiable homicide.

The fellowship applicants, pre- or postdoctoral, are the most pitiful lot. They sometimes write in flowery English, full of mistakes, addressing *you* personally and praising you to the skies, in a way that makes it obvious that they have sent copies of the same letter to dozens of others. At other times they are more demanding and business-like. Here is a sample that I received not long ago.

"A large number of mathematicians have suggested me to reffer to and request any professor concerning "The Mathematical Association of America" to accept as a supervisor of my D.Sc. thesis. The topics of my D.Sc. thesis has been proposed to be "SOME CONTRIBUTIONS TO THE THEORY OF NUCLEAR LOCALLY CONVEX SPACES".

"In English kindly give your acceptance as a supervisor of my D.Sc. thesis and please furnish a copy of the synopsis of the above topics, duly signed and approved by you as a supervisor of a D.Sc. thesis so that I may sign therein as a candidate. If you fail to accept it, kindly give the name and address of the person who shall accept the same and who shall send a copy of the synopsis duly signed and approved by him, for which I shall be highly obliged."

No answer—right?

I cannot lay my hands on one of those 24-page questionnaires—they are fatuous enough and make me angry enough that I don't even file them—but I suspect that we have all seen examples. They are outrageous impositions on our time and patience, and they would be that even if they made sense, which they usually don't.

What is a prompt answer? Proposed definition: an answer sent off between zero days and seven. I answer many letters the same day I receive them, most of them one day later, and all of them, without exception, inside of a week. The exceptions to "without exception" occur only when I am out of town; in that case prompt comes to mean within seven days of the letter actually reaching me when I get back home.

The line between a bona fide letter and one that is not is sometimes very thin. Troubled applicants needing support are often pushy and obnoxious, but at other times they are sensible as well as sensitive human beings. If I have the slightest reason to think they are the latter kind, I answer them, of course, even if all I can say is that there is nothing I can do for them.

Here are a few excerpts from one begging letter, a touching one, that I couldn't just ignore.

"Kindly forgive me for taking the liberty of writing to you without any previous introduction." [The apology is unusual. The writers of most such letters seem to be sure that the world owes them something—or at least that I do.]...

"In 1963 I was awarded Ph.D. in Mathematics...". [Many others write long before they reach that stage, or even cheerfully admit that they are failed Ph.D. candidates.]

"I am enclosing herewith a list of my published papers...". [Extremely unusual; here is something solid that can be looked at and evaluated.]

"Kindly allow me to know if it will be possible for you to spare your valuable guidance to me and also if I will be able to get some fellowship or teaching assignment for the purpose.... I will try my best to prove worthy of it."

The subject of this man's papers was not one that I had much respect for, and, in any event, financially there was nothing I could do for him. He might have written the same letter to others, but internal evidence made me guess that he sent it to two or three of us at most. Here is somebody, I said to myself, who is serious about wanting to be a mathematician. I couldn't be sure how much talent he had, but I did feel sure that mathematically speaking he was a person of bad taste. My time was precious to me, and I didn't want to waste any of it, but at the same time I wanted to save him the long pain of false hope. Here, in toto, is what I wrote (on the same day that I received his letter).

"Dear Dr. N: Sorry—no. Sincerely, P. R. Halmos."

That, by the way, was not the shortest letter I ever wrote. Many years later, as editor, I asked Ed Hewitt to review a book. His answer said, in part: "How many pages, double-spaced typewriting, do you want? I can send you a telegram or write ad nauseam." My answer to that ran, in toto, as follows.

"Dear Ed: 8. Yours, Paul."

[P.S. He never wrote the review.]

Letters of recommendation constitute probably the most frequent kind of writing that we all have to do. My theory of them has two tenets: (1) always tell the truth, but sometimes not the whole truth, and (2) no matter how much good you want to say, be sure to say something bad.

If a candidate asks me to recommend him for two jobs at the same time, one at the University of Chicago and the other at the University of Miami, I don't necessarily send the same letter of recommendation to the two places; that's what I mean by not telling the whole truth. The question is just one of emphasis. Even at Chicago if a letter about a candidate says, in so many words, "he is a bad teacher", and goes into graphic detail about students from the School of Business leaving his calculus classes in droves, then a conscientious dean, or even the chairman, might decide to maintain the research level of the department with other candidates. And, obversely, telling a Miami dean that the quality of the candidate's research is not high enough to earn him tenure at Chicago, is almost certain to have the effect of a black ball. Tell Chicago that "in research X is as good as Y, or better—he keeps trying to improve his teaching, but he still has a way to go", and tell Miami that "Z is an inspiring and popular teacher, and he is almost certain to continue making sound and steady contributions to the research literature."

The purpose of saying something bad with the good is to establish your standards with the administrator who reads your letter. To convey information some contrast is necessary. Comparisons are the most helpful ("X is better than Z, but not as good as Y"); if they are not available, then "self-comparison" might do as well. ("His solution of the Ronda problem is deep mathematics, but his paper on cube roots struck me as a potboiler.") Nobody is perfect, and everybody knows it; unstinted praise (or its opposite) is more likely to arouse suspicion than carry conviction.

Here are three samples from my collection.

The first one was addressed to the Institute for Advanced Study.

"X asked me to write you a letter of recommendation for him; he says he plans to be on leave next year at the Institute.

"What can I say? He is a sound, unspectacular, non-prolific contributor to [his subject]. He won't do any harm, he might learn a lot, and he'll almost certainly get some work done."

The next one is the prose part of an NSF form I was asked to fill in.

"*Y* is a very good man—he has ideas and he works—he must be encouraged—award of the grant is strongly recommended.

"That is my conclusion; I admit that the legal evidence for it is not convincing. He has nothing actually published, and the description of research that accompanies the proposal is horribly badly written—it is almost complete nonsense. I have heard him talk several times and I have had private conversations with him many times; he talks even worse than he writes.

"Under all the confusion there are ideas. He has proved this to me three distinct times; he solved problems, hard ones, that many of us were actively working on and stuck on. His solutions were awful—they could always be simplified and shortened in a ratio of 10 to 1—but he had ugly solutions when other people were elegantly stuck.

"Question: what can you do with a guy like that? Answer: give him the backing he needs, give him a hard problem, and then stand by to decode the solution."

My last sample consists of excerpts from the strongest letter I have ever written. Its date is 1949. It is in answer to a form letter from MIT, which asked for my evaluation of various qualities such as ability, enthusiasm, and personality. Its subject is I. M. Singer.

"*Singer* is very good.

"*Ability*. Throughout his career here he was one of our best students. By now he is more—he is a mathematician. In the course of the last year I could see the transformation in his attitude from that of a student who can get good grades in courses to that of a grown man who can understand concepts, who has digested a lot of mathematics, and who has the creative point of view.

"*Enthusiasm and industry*. I can rarely come to Eckhart Hall without finding Singer around, talking shop to his friends, working, reading, attending seminars, giving seminar reports, and in general behaving like someone who enjoys what he is doing.

"*Teaching ability*. As a lecturer Singer is clear, coherent, and patient—I have never seen him rattled. He has been a teaching assistant for a couple of years. Some of my senior colleagues (e.g. Lane), who were in charge of the courses Singer was teaching, think highly of his work. Because the department is dissatisfied with the expedient of teaching calculus from Courant and selecting problems from Granville, it was recently decided to take our most competent teaching assistant and assign him for one quarter to the task of preparing a suitable set of exercises for the first two quarters of our calculus sequence—the one that was selected was Singer.

"*Personality*. Singer is one of the nicest people I know. He is warm, pleasant, charming, a good basketball player, and a lousy poker player.

"I can, of course, turn out to be wrong—it is impossible to tell what kind of mathematician someone really is until after reading at least a half dozen of his papers—but I am betting on Singer. If there were ten guys like him around to bet on, I know I'd win in nine of the cases.

The next time Ted Martin saw me, he joshed me about that letter. "Come on, now", he said; "didn't you go too far overboard on that one? Nobody is *that* good!" Ted turned out to be wrong; Singer was that good, as history since then has proved.

How to advise

Much of professional correspondence involves giving advice, some solicited, some not, some taken, some not. Let me give a few examples.

Kelley's *General Topology* was published in 1955. Kelley and I had been friends for years, he let me see various versions of the manuscript as he produced them, and on two occasions the spirit moved me to write to him and offer him some unsolicited advice. Both times the subject was terminology, a subject close to my heart.

Here is what I wrote in March, 1953.

"In re semi-metric versus pseudo-metric, I much prefer the latter. I am opposed to notation and terminology that refers to non-existent structure. I am opposed, for instance, to the tendency of some classical analysts to talk about '(x, y, z)-space'. What they mean is that they have a space and that they'll usually denote a point of that space by (x, y, z). What they do is to give the space a complicated and ugly name and, incidentally, freeze three perfectly good letters, or, at the very least, one perfectly good combination of them. 'Semi-metric' is not quite that bad, but it's bad because 'semi', meaning 'half', hints at the number 2. I would say that 'semi' is justified only if there is some hint of duality in sight: e.g., for semicontinuous functions."

While Kelley was writing his topology, I was studying algebraic logic. The two subjects are not total strangers. Topology can be defined in terms of the closure operator, a concept whose characteristic algebraic properties have been abstracted and studied by some logicians. My monadic algebras, motivated by the quantifier operators $\exists$ and $\forall$, contained the closure algebras motivated by topology as special cases with certain extreme properties. The relation between the two structures is what prompted my second terminological letter to Kelley nine months after the first.

"To the author of THE book on topology, a terminological comment. A set has two extreme topologies. Everybody agrees that the largest topology should be called 'discrete', but there isn't universal agreement on a name for the smallest one. (Some indiscreet people use *indiscrete*.) ... Associated with each topological space is a closure algebra, i.e., the class of all subsets of the space equipped with Boolean operations and the closure operator. Closure operators are decent algebraic objects, with an obvious ideal theory. Among the closure algebras associated with the topologies of a fixed set there is exactly one that is *simple* in the customary algebraic sense of possessing no non-trivial ideals. That topology is the nameless small one, the one with only two open sets. Forced conclusion: that small topology ought to be called the simple topology of the set it lives in."

Kelley adopted "pseudo-metric" (he mentions "semi-metric" in the index only), but he insisted on sticking to "indiscrete" (and "trivial") instead of my cogently argued "simple". You win some, you lose some.

Almost twenty years later, when Heydar Radjavi and Peter Rosenthal were writing their book on invariant subspaces, I wrote Peter.

"This comes to you as author from me as editor.

"You're an established mathematician now, far from under my thumb as a student, and if I disagree with one of your terminological decisions, too bad for me. That's why I held my peace when you mentioned, and later wrote papers about, something you call "Hermitian" algebras. Now, however, that you are about to put them in your book (I didn't realize you would), I must speak.

"The word is simply awful, Peter, and if it were a question of articles alone, I'm sure it would soon die a peaceful death. Your book, however, will be a standard reference for many years to come and should, therefore, be as near to perfect as you can make it.

"You make a distinction between Hermitian algebras and self-adjoint ones. There are two concepts there, but who can remember the arbitrary choice of otherwise synonymous terms? What's Hermitian about a Hermitian algebra, except one measly little property which Hermitian operators happen to have (among many others)?

"Please change the word, throughout, and establish a sane terminology in this corner of the world. I have a proposal. What does an algebra (or, for that matter, a single operator) do when it has the property that every invariant subspace is reducing? Answer: it tends to make things reduce. What's the English inflection that accomplishes that sort of tendency? Answer: '-ive'. Solution: A REDUCTIVE algebra (or operator) is one such that every invariant subspace is reducing."

I won that one; Heydar and Peter agreed. "Reductive operators" and "reductive algebras" are now standard terms in operator theory. Pure algebraists use "reduce" and "reductive" differently, but that's their problem.

As a sample of some different kind of advice, the commercially solicited kind, here is a sample of a report I once sent to a publisher who asked me to evaluate a manuscript.

"The presentation is mathematically elegant; the style is clear, crisp. The author's mastery of expository English is unusually good (though not always perfect). The organization is the somewhat brutal theorem-proof-theorem-proof kind, with no respite.

"The book is obviously not written for undergraduates. The author assumes that the reader has mastered things like the Cantor set and complete metric spaces, and he assumes that the reader is quickly receptive to such abstract concepts as sigma-algebras of sets. Neither concepts nor results are ever preceded by motivation.

"There are very few examples, which is a pity. The exercises at the ends of chapters are good, but they are mostly of the hard kind, that make students think (a good thing, of course), and only rarely of the illustrative kind. Besides, the exercises are frequently worked out, or nearly so, so that, in effect, they are just additions to the text, treated even more briskly than usual.

"Neither the ideas nor the techniques are new, and the same is true of the organization and arrangement. This is perfectly all right; the author makes no claims of novelty along these lines.

"In sum: the book is a good compressed treatment of standard material. Its main virtue (and main fault) is its brevity. It will almost certainly not have a wide market, but it is quite likely to find support; as a guess I'd predict a sale of 5000 copies in 3 or 4 years.

"An interesting possibility to explore is this: compress the book even more, eliminate the overlaps, perhaps eliminate Chapter VII, and make a virtue of smallness. Advertise that the specialist in another branch of mathematics can buy this book, can put it in his pocket, and, riding the subway, can learn from it all the measure and integration he will ever need."

The last sample advice I present here is my answer to a graduate dean who sent me a long and detailed proposal to evaluate. The plan was to start a Ph.D. program in mathematics at his college; the proposal was accompanied by many documents, such as catalogues, lists of courses, and curricula vitae. The mathematics department of the college had no research reputation or tradition to base its ambitious plan on. The year was 1969; the lean 70's were looming on the horizon.

"The proposal as a whole is unusually well written. It is free from much of the customary administrative double talk; it says what it has to say clearly and briefly. Most of the details have been carefully thought through.

"Some statements I disagree with. The very first sentence is one: I doubt that there is 'no question' that the proposed program will be of service to the nation and that 'it will have no problem placing its graduates'. I don't pretend to know anything that the author of that sentence doesn't know, but perhaps I interpret the facts differently. It is a tenable thesis that the nation has reached the saturation rate in its production of mathematical Ph.D.'s; possibly neither the nation nor the individual products of the proposed program will be well served by its realization.

"The difficulties that the proposal soberly anticipates (strong competition for good faculty and lack of an adequate supply of good students) are real. It is a tenable thesis that they may be insuperable.

"The preceding two paragraphs indicate reasons why the establishment of a Ph.D. program in mathematics may not be a good idea. Let me proceed on the assumption that the program will be established anyway and comment briefly on some details.

"I strongly support the statement that a sound basic training in 'traditional' mathematics is the right way to begin. The proposal's expressed concern about the present faculty is relevant here, and worrisome. A library of 182 journals would be good, if back files were available for a respectable distance into the past (at least 20 years); somewhere between 200 and 300 journals is what the leading research libraries consider the minimum essential. The number of monographs (5000) is impractically small; something like four or five times that number is needed for effective graduate teaching and research.

"The proposed list of Ph.D. courses is more comprehensive than the University of Chicago requires for an M.S.—but not very much. What is missing is a set of research-level courses, for second- and third-year graduate students.

"The proposed procedures (courses, examinations, supervision) are non-surprising variations on an often-used theme. Perhaps some of the details will be found unsatisfactory or unfeasible, but that's nothing that a few hundred man-hours of committee time cannot cure. (One parenthetical phrase suggested an idea that was new to me, and I hasten to praise it: to have a graduate student be 'guest lecturer' in a graduate course is an excellent notion.)

"It is hard to come to a conclusion about the faculty on the basis of such statistical data as the proposal gives, but I had to come to one. Twenty names are listed; my conjecture is that at most four or five of them (chiefly the younger ones) could be used for the training of Ph.D.'s worthy of the name.

"What then is to be done? I am inclined to say: drop it. Let the mathematics department (if it is not satisfied with its present place in the sun) do what it has been doing, and strive to do it the best of anyone, rather than try to do something else. I do *not* predict a smooth and easy establishment of the Ph.D. program, rapid and spectacular growth, and long years of successful operation. Like small businesses, more Ph.D. programs are doomed to fail than will succeed. That's my view, now, on the basis of the facts available to me. I am truly sorry if that disappoints your eager and ambitious young mathematicians. I wish them well. If I am proved wrong, I'll be glad."

The doctoral program was not established.

I can skim documents and find their high points pretty fast, but even so it took me about four hours to work through the material the dean sent me, and another hour or two to compose the report. It was a thankless task—literally thankless. I didn't get paid for my time, I didn't get thanks for my work. I felt pretty sour about that for a while, and the experience contributed to my resolution to undertake such tasks in the future on a strict contractual basis only. The job, the time, and the pay should be spelled out in advance; the laborer is worthy of his hire.

Oh, well, you can't win them all.

Honolulu, here I come!

In 1968 I was offered the chairmanship of the mathematics department at the University of Hawaii. I don't mean that I was sitting at home minding my own business when a sudden telegram with the offer arrived. I knew about it months in advance; it was preceded by a lot of thinking and a lot of negotiation.

I was the right age and the right stature (neither of them too high, neither of them too low) to be in line for chairmanships in those years. I think I received something like a dozen assorted chairmanship nibbles and offers in my 50's. The ones I remember most clearly are the nibbles from the University of Iowa (1967), the University of Pittsburgh (1968 and 1969), the University of Maryland (1968), and the University of Calgary (1972), and the firm offers from the University of Hawaii and the University of Texas (1968), and the University of Massachusetts (1975).

I knew something about Hawaii. My ophthalmologist brother moved there in the 1930's and has lived there ever since. Ed Mookini, a Chicago student in one of my classes in the 1940's, was born in Hawaii to a Hawaiian father and a Japanese mother; much of what I learned came from him. (When you pronounce his name, sound both the o's, so as to make his name four syllables, or three and a half anyway: Mo-oh-ki-ni.) I visited Hawaii in 1967; I spent one term at the University, taught a course, and fell in love with paradise.

Back home in Michigan I started thinking. "Look here, Halmos", I said to myself, "you're over 50, your days of research are bound to come to an end soon if they haven't already done so, and some day you're going to have to retire. Why not move to Hawaii now, and when the time comes, you'll be there?" That wasn't all I said; I kept heaping one argument on another. I would make a good chairman, I told myself. I know what needs to be done, I said; I am well organized, and I am eager to work hard. These are times of big money and big expansion, I said (how was I to know that I was saying it at almost the end of the golden age?), and with that and Hawaii's natural attractions I am bound to be able to build a good mathematical center out there in the middle of the Pacific.

E. Mookini, 1965

Another ingenious proof about the wisdom of going to Hawaii went like this. Half a million people lived there, and if only one tenth of one per cent had some mathematical talent and interest, and if only one tenth of *them* would come to study mathematics at the University of Hawaii, my time and effort wouldn't be wasted. It just goes to show what wishful thinking can do when it doesn't examine all the facts. Sure there are bright people on the Islands, and quite possibly my estimate of the number of mathematically inclined ones was below the truth. What I didn't know, however, was that the reputation and attraction of the U. of H. were not high in Hawaii—far away California had the quality and the glamor. Any student worth his salt would go to Berkeley or Los Angeles if he could afford it, or borrow the money if he couldn't—and the really good ones got scholarships to help them. My "proof" was based on an incomplete system of axioms.

Life was complicated in the fall of 1967. Hawaii wanted me, Maryland wanted me (matters reached the stage where all details, including salary, were agreed on by the deans in question—conversations broke off because the president vetoed the salary), and Texas wanted me. Texas meant the University of Texas, in Austin, and the man I was negotiating with there was John Silber—who acquired national fame when he moved to Boston University and started running it with a firm hand.

Ed Mookini was the chairman at Hawaii when negotiations with that university started. His term was coming to an end, and both he and Todd Furniss (the dean) were eager to make the U. of H. mathematically visible. The letters Furniss received about me were apparently satisfactory. The negotiations took time because such things always do: we had to agree on the proposed new size of the department, on teaching loads, salaries, colloquium support, classrooms, offices, and on dozens of other details of that kind, and despite the large number of transoceanic telephone calls, we still had to write everything down and keep sending the written versions back and forth.

John Silber and I had one long conversation in his office tête-à-tête, a couple more on the phone, and we wrote to each other in careful administrative detail. My longest letter to him was five single-spaced pages discussing twenty-five numbered points. He kept his answer down to three pages by cheating: "... without commenting on your [first] twelve distinct points, let me say simply that...". How much can you learn about a human being through professional contact such as that? Not enough perhaps, but what I learned caused me to like and to admire John Silber. An ex-philosopher, he is intelligent, he is sensitive to shades of meaning, he has a sense of humor, he understands the academic world, and he is tough—he is a masterful administrator.

Todd Furniss is a very different kind of personality. Silber could have made a living in the hard-sell second-hand car business; Furniss, the

W. T. Furniss, 1965

J. R. Silber, 1968

smoothie, would have been more at home in diplomacy or in real estate. He was always honest with me and friendly, and he really wanted me to come. He hated to be the bearer of bad news, as sometimes he had to be. He didn't like telling me that Tom Hamilton, the president of the University whose signature appeared on my official invitation, had resigned and would be out before I got there. (I was Hamilton's last straw?) Much later, after we had worked together on the campus for some months, he didn't like telling me that *he* was leaving; he had accepted a position with the American Council on Education in Washington. Despite his good qualities, however, he had an untrustworthy-seeming smoothness that I was never completely at ease with; I preferred John Silber's blunt wholeheartedness.

A newspaper feature appeared late that year, reporting the highlights of a recent statistical study in the U.S. The most expensive place to live in was Honolulu, the story said, and the cheapest was Austin, Texas. I had two offers in my hands: one from Honolulu, one from Austin. Financially the Honolulu offer was respectable but not spectacular; the one from Austin was much fatter.

Why did I accept the Honolulu offer? I am not sure I knew the answer then, and I am sure I don't know it now. Here is what I wrote to Ambrose at the time.

"Climate—yes, the climate has a lot to do with it. I suffer in this inhuman cold weather, and I get depressed by the gloom and dark and dank. Hawaii is more than negatively not bad—it is positively gorgeous. The sun really does shine all the time, the sky is poetically blue, the sand is gold, and the flowers are spectacular—it does almost as much for my soul to stand and inhale that plumeria as to hear a Mozart quartet."

The climate was not the only reason. I actually wanted to be chairman—at least for a while. I liked the idea of being a petty Caesar and putting some of my theories of administration into practice—just to prove to my ego that it could be done. Then, too, being busy as a chairman would be an excuse for not producing the quality and quantity of mathematics that my conscience kept telling me I should produce. In any event, paradise won, and, gleefully, I said "Honolulu, here I come!"

When I told Silber of my decision, he said that he wasn't giving up—he would continue to expect me in Texas, a year later, or maybe two.

Monday evening, January 8, 1968, I sent a cable to Todd Furniss, accepting the Hawaii chairmanship. The outdoor thermometer by my living room window said −3. The next week I offered jobs to Allen Shields and Larry Wallen. It took Allen two months to think about it. By several phone calls to Hawaii I was able to keep improving the offer, but to not much avail: he accepted for a one-semester visit only. Larry took only one month to decide. Late on a cold and snowy February evening my doorbell rang. There stood Larry, glaring at me glumly. "All right", he said, "I'll come with you." We celebrated with a couple of

L. J. Wallen, 1968

beers. That was 16 years ago, and Larry Wallen is still in Hawaii, one of the mainstays of the department of mathematics.

I tore up my Ann Arbor roots, all but one, the official one. Both the mathematics department and I found it convenient to call my departure not a resignation but a leave of absence. The department's angle was to keep alive the budget line that had my name on it while they looked for a replacement; my angle was, well, why not..., just in case....

CHAPTER 15

Service, one way or another

Democracy ad absurdum

How should chairmen be selected? At many universities, perhaps most, the mathematics faculty of ten people, or a hundred, selects its chairman by election. At the same time university regulations or traditions say that the chairman is not elected but appointed by a dean. The contradiction is resolved by a legal fiction: the department tells the dean the result of its vote, and the dean, thinking deeply, appoints the winner. In theory he could appoint someone else, but in practice he wouldn't dare.

Elections mean democracy, and they are in the modern spirit of self-determination. I seem to be a member of a small minority opposed to the academic application of the democratic method. Most people get upset at the suggestion that democracy doesn't *always* work, and the overwhelming opposition worries me. It worries me very much—but it doesn't convince me. I know I could be wrong, but my experience and my judgment tell me that I am right and, much as I wish I could be with the majority, I am stuck with what I believe. Some people call my views reactionary, meaning that they have worked in the past; I believe that they have intrinsic merit, and could work in the future.

André Weil's logarithmic law (first-rate people choose first-rate people, but second-rate people elect third-rate ones) works the same way whether the vote concerns a minor addition to the teaching staff or the elevation of a colleague to leadership. Possibly, perhaps, Harvard and Chicago are good enough that they can't go too far wrong in an election, but most places are not. I am convinced that most of the time elections yield low quality chairmen. A chairman might be elected for his popularity ("he gets along with people—and that's important, you know"), or because of his subtle campaign promises ("don't you think it's unfair elitism to weight research higher than teaching in the determination of salaries?"). Knowledge of the profession, wisdom, and dedication are what people talk about, but that's not what they seem most frequently to vote for.

I favor the appointment of chairmen by the administration. I can hear the howls of my colleagues: no!, no!, the dean is incompetent; we cannot trust the judgment of people who know nothing about our profession; surely we can choose a better leader for ourselves than an outsider can! I am sympathetic to the howls. Deans are sometimes incompetent, and their choices could but improve if they knew more. It is true that many deans have made many bad choices. On balance, however, I am convinced that well-meaning impartiality works better than well-informed self-interest.

Elections at the departmental level tend to produce second-rate leadership. To be sure, elected chairmen are not likely to be the worst imaginable, certain to lead the department to demoralized chaos, but they are almost certain to be compromisers who can neither show the way to excellence nor even maintain it. If, on the other side of the coin, a department has degenerated to uniform mediocrity, then making any of its members chairman is a guarantee of continued low quality—there is then no remedy except the appointment of a vigorous outsider by administrative intervention.

I have worked with fifteen chairmen in my life, eleven appointed and four elected. Thirteen of them ranged from medium to excellent; two of them were egregiously bad. The two were elected, and they turned out to be dilatory, disorganized, and ineffective. By the time the ends of their terms came around they were also deservedly unpopular.

Well-meaning impartiality is not enough to choose good chairmen; action must be based on knowledge, which takes time to acquire. Deans know deans and chairmen at other universities, and a conscientious dean must consult that external network before making an important internal appointment. Another technique that conscientious deans have used is to solicit the views of every member of the department, preferably in a private face-to-face interview or possibly in writing. If the department has forty members, that technique burns up something like forty hours of the dean's time, but it's worth it. What it amounts to is a detailed voting procedure that gives much more information than numbers (22 for Smith, 14 for Jones, and 4 for Robinson). It amounts to a weighted voting procedure that enables the dean to use some wisdom. If the interviews convince the dean that 11 of the 22 who favor Smith are prejudiced, or self-serving, or second-rate people of bad judgment, or are totally apathetic and voted only because Smith asked them to, then he might choose Jones over Smith, and, two years later, earn everybody's respect and gratitude.

In the choosing of chairmen, and at every other step of administering an intellectual service such as a department of mathematics is intended to provide, it is absolutely essential to give every member of the group a voice that is listened to. I don't know how a machine shop of forty workers should be run (quite possibly the same way), but I know that in an important and perforce cooperative academic enterprise collegiality is indispensable. A chairman, once in office, shouldn't even try to do his job without continually asking the opinion and the advice of his colleagues. The final responsibility is his (paint it blue or paint it green, statistics yes but fuzzy sets no, hire her and fire him), but the wisdom is collective. It's bad to come to all decisions by counting ayes and nays, but it's even worse (and dangerous) to ignore the consensus altogether. Nobody can know everything, and two heads are proverbially better than one.

Are forty heads better than two? Are seven heads better than three? Complete democracy demands a vote on every issue. Many departments try for a compromise between one head and forty heads (something like a geometric mean): delegate the thinking and the responsibility to a committee (or two committees, or eleven committees). Napoleon said that the only thing worse than a bad general is two good ones, and I agree: a bad chairman will hurt a department less than a committee can, even if it consists of competent people full of good intentions.

At Michigan I was on the Executive Committee for a while. When Ted Martin began to think about leaving the chairmanship at MIT, Yitz Singer wrote and asked me how that Michigan committee worked. Here is what I wrote him.

Dear Yitz:
   Please burn this letter before, after, and during reading it.
   Michigan has an Executive Committee. Its members are the chairman (ex officio), the associate chairman (ditto), four seniorish people with three-year terms, and one juniorish person with a one-year term. They are elected as follows. The department nominates two or more candidates each year for each vacant slot. There's an election, and all but exactly two candidates for each slot are eliminated. The actual person to fill the slot is chosen by the chairman from those two—in all known cases he simply chose the one with the larger vote.
   The committee meets on the average once a week—Fridays from 3 to 6. In the late fall the meetings get much more frequent; at other seasons they can be rarer. In *theory* all power is with the chairman and the committee is advisory only; in practice all major decisions are made by the committee. Example: a departmental flunky (not a member of the committee) chooses classrooms and makes up schedules, but the committee has to approve. Example: the committee discusses all applications for research instructorships (after previously deciding on the terms and designing the announcement poster), and selects the winners. The committee does all the hiring (except graduate assistants), from instructors on up through full professors. The committee looks at everyone's salary and decides on the exact amount of the raise; it decides promotions; etc.

The chairman has a lot of power, of course, if he chooses to exercise it. He is in the chair at the meeting, he can guide the discussion, he can influence the vote, and he is the one who talks to the dean and transmits the department's recommendation. His term is five years, and he may be reappointed.

How does the system work? Awful. It makes for no imagination, no leadership, second-rate appointments, delay, indecision. I personally prefer a fascist. A good fascist (Marshall Stone) can work wonders—a bad fascist (who shall remain nameless) can wreak havoc—and I have worked under both—but a guy whose principal qualification is that he doesn't rock the committee boat can do nothing. I don't know how Martin was from the point of view of an MIT worm, but from the outside point of view he has done very well. MIT has many good people and deserves its good reputation. I cannot tell from here what Michigan's reputation is, but these years it doesn't deserve a very high one. The boat isn't rocking.

Are you still reading this? Haven't you burned it yet?

I believe that the work of the world is done by people, not by committees. Socrates, the teacher, was not a committee, nor was Archimedes, the inventor and research mathematician. The great strides forward (in administration and in finance, as well as in science and in the humanities) have always been made by people, not by committees; Lincoln, Rothschild, Newton, and Goethe bear witness.

Even at the pedestrian day-to-day working level, the decisions of committees are suspect. At some universities there is a "recruitment" committee; its members are supposed to do as good a job as possible of improving the quality of the department by hiring the right people. That's what I think the job is, but that's not always what they think. They often regard themselves as having constituencies; the topologists want another topologist and the analysts another analyst. Lip service is paid to quality and balance, but in practice empire building tends to come first. Scientific judgment rarely wins in the horse-swapping compromises of practical politics.

The ideal way to raise quality is to improve the good; the committee way is to fill the pits by levelling the peaks. One final manifestation of the levelling tendency deserves mention here. It is the current insistence that chairmen should not serve long terms: that is the best way, it is argued, to keep them from acquiring royal powers. Three years is a popular length, two years is not infrequent, and even one year is sometimes tolerated. I think that's dreadful; it is certainly not the way to raise quality.

It takes at least one year on the job to learn how to be a chairman. A three-year term implies, therefore, that the department is led by a fumbling beginner every third year. Since, moreover, a chairman is likely to be regarded as a lame duck in his last year, a three-year term implies only one year of effective action.

The three-year terms at Chicago became, by unwritten tradition, six-year terms for Stone, Mac Lane, and Albert (and some of their successors too), and that proved to be a good length. Six years wasn't enough to kill the chairmen mathematically, but it was enough to enable them to carry out more than just short-term plans.

G. C. Evans at Berkeley, T. H. Hildebrandt at Michigan, and W. T. Martin at MIT were three great chairmen. Some of their colleagues grumbled about them—sure—they were unfair at times, autocratic, and even just plain wrong. I don't remember just how long each served, but I do know that none of their terms was less than 15 years, and I believe they were all close to 20. They built three great departments.

How to be a chairman

A chairman is simultaneously a paper-shuffler, a protocol officer, and a policy maker.

A chairman should take seriously *all* the details of the operation of his department. Marshall Stone, in addition to proving his celebrated theorems about unitary groups and Boolean algebras, as well as the Stone–Weierstrass approximation theorem and the Stone–Čech compactification theorem, and rebuilding a collapsed department to one of the best in the world, also put chalk into the classrooms and swept the halls. (Such extreme steps were necessary only twice in my memory, when the janitors were kept away from Eckhart Hall, once by a solid week of temperatures in the minus 20's and once by a sympathy strike for some other union.) I saw Bochner, when he was chairman at Rice University, inspect the departmental bulletin board and remove out-of-date items from it.

Division of labor and delegation of authority are good things, but I still think that a chairman should personally see to it that all the jobs of the department are properly done. The jobs vary—an untrained receptionist answers the telephone and a famous professor arranges the colloquia—but the chairman must consider himself to be in charge of it all. There may be a dozen secretaries in the department with an office manager as their immediate supervisor, and there may be two dozen colloquia during the year whose smooth functioning requires the colloquium committee to conduct extensive correspondence and post many notices—no matter, if anything goes wrong, the chairman should know about it and should see to it that it's set right.

Some chairmen don't agree. "I have much more important things to worry about", one might say. "I'd rather be sure that we hire the right person to teach partial differential equations next fall than see to it that two seminars don't conflict—you must ask Jim about that. And, besides, I must keep doing my own mathematics—these minor things, even if they matter to some people, would take an unfair, a disproportionate amount of my time." I can't help being sympathetic—but I disapprove. Yes, good hiring is more important than bad seminar timing, and many people regard even a temporary abandonment of their own professional interests a major sacrifice. Just the same, a chairman who cannot take the time to do the whole job shouldn't take the job at all.

The chairman should keep his eyes and ears open in the corridor; he should know who is on speaking terms with whom, and whose office will need repainting next spring. The chairman calls the meetings of the faculty and is in the chair during them; he must call them in time to meet the deadlines that the central administration has set, and he must have enough charisma and knowledge of parliamentary procedure to run them pleasantly and efficiently. The chairman is the visible representative of the department in the university as well as outside, in the profession. He must therefore maintain many diplomatic contacts, for example by answering letters promptly and courteously.

No one can do mathematics on a desert island for more than a short time; it is a collaborative, synergetic activity. One of a chairman's vital functions, therefore, is to foster and encourage contact in the faculty, contact of all kinds. Contact can mean lunches every day and seminars every week; it includes the afternoon coffee hour, as well as the fall and spring picnics; it includes the get-acquainted party at the beginning of the first semester, the Christmas thank-God-it's-over party at its end, and the shared ride to the winter meeting of the AMS. A department that does these things is likely to have higher morale, and higher professional quality too, than one that does not.

I know from experience that if a department has a tradition of such activities, they do a lot of good. They are almost universally popular, but they don't have enough momentum to keep going without support— they need a chairman who continually champions and stimulates them. It's the chairman's job to provide the perpetual coffee pot around which the faculty can goof off, exchange gripes, discharge frustrations, and recharge. (The coffee itself is just a symbol; you don't have to drink it to do the rest.) Similarly it's the chairman's job, the chairman's duty, to eat lunch with his colleagues as near to every day as possible, and it's the chairman's job to encourage (to goad, to help) students to organize the mathematics club and the professional and personal parties usually associated with such organizations. Not even the evenings are the chairman's own; he should regard it his duty to entertain the faculty and its visitors on occasions sprinkled generously through the year.

I have left to the last the most obvious and most obviously important tasks that a chairman has, the ones having to do with personnel and scholarship. "Personnel" includes problems of hiring, firing, salaries, and promotions; "scholarship" includes decisions about the library as well as about which ones of the "mathematical sciences" should be in the department, which subjects should be in the curriculum, how often and when they should be taught, and by whom.

A chairman must be a mathematician of vision, thoroughly familiar with the mathematical community and how it works. Some chairmen sit and wait for job applications to come; others go out and solicit them. (Stone used to say things like this: "Let's keep an eye on that young fellow; two or three years from now he may be ready for Chicago.") Some chairmen go after the best available person; others, the "realists", give up before they start. "There's no sense making him an offer", they say; "he's too good for us. We'd just be losing time, and end up being stuck with everybody else's rejects. It's more realistic to lower our sights and invite people on our own level in the first place." That attitude is shortsighted, I think, and is certain to reduce quality in the long-run. I couldn't help being gleeful one time when the lower-our-sights argument backfired. "We have a chance with A and B", a chairman said, "but X is too good; let's go after A, B and C instead." Fortunately the department didn't go along with that; they voted for A, B, and X. Outcome: A and B declined their offers, but X accepted.

The library is the mathematician's laboratory. A teacher and his students, and a research mathematician and his students, must have a mathematics library as good as the university can afford, and they must have it down the hall a few steps away. Facts and proofs must be looked up now, other questions will very likely arise before lunch, and still others during the afternoon colloquium. A historian might be able to continue working without immediately verifying an important reference—that can wait till tomorrow. For a mathematician each step of the intellectual structure under study depends crucially on the previous steps, and a missing step could be an obstacle to any progress at all. To have to put on your coat and walk ten minutes to the central science library is discouraging and inefficient; it is no way to go at either pedagogy or discovery. Would you put a chemist's laboratory a ten-minute walk from his office? Would you force him to waste twenty minutes two or three times every day?

Once the library is in the mathematics building, the faculty, led by the chairman, must still decide what it should contain. The answer used to be easy: everything. Even in the days when the number of mathematics books was small and the library budget was relatively large, the word "everything" had to be defined with some care. It never meant all books on freshman algebra and trigonometry, and it always did mean all the volumes of the AMS blue colloquium series. Between those extremes judgment has to be exercised.

What is obvious is that a library with any hope of being scholarly must represent every part of mathematics whether it is currently fashionable or not and whether anyone in the department is currently interested in it or not. Arguments of expediency in favor of excluding certain subjects, of ordering only books explicitly requested by a member of the faculty whose work requires it, and of refusing to order them even then if they are on, say, combinatorics—arguments like that are anti-scholarly and anti-intellectual. (I am not fantasying—I have seen such things happen.) A good chairman wouldn't countenance arguments like that for a second.

Curriculum decisions sometimes overlap with personnel decisions and the union of the two is more complicated than either one. Somebody—and the usual first approximation is the chairman—must decide whether statistics should be in the mathematics department (probably not) and whether computer science should be in the mathematics department (certainly not). Somebody—the chairman?—must decide to what extent the department should go along with current fashions in teaching and research. Do we replace calculus by discrete mathematics? Do we prefer minimal surface theory to low-dimensional topology? Chairmen must at least raise many questions such as these, and in some cases answer them.

A chairman is an administrator. He serves as a buffer between his colleagues and the university administration; and the way he leans is important: is he a mathematician protecting his department, or does he play the tune called by the one who pays the piper? A chairman is both a clerk and a director; he is both a servant and a leader. It takes a special kind of talent and temperament to be a chairman. It is an important job but a secondary one, in the sense that makes the Annals a primary journal and Mathematical Reviews secondary; a chairman is there not to do something but to enable other people to do something. If you are thinking about becoming a chairman, think again. It's an easy job to do badly and a thankless one to do well. You make enemies no matter what you do. Avaricious faculty and autocratic deans push in opposite directions, and you are in the middle. Before you start be very sure that that's where you want to be—and if you do start, good luck to you. You'll need it.

How not to be a chairman

I should have been worried about becoming a chairman, but I wasn't; I thought I knew it all. I knew it all in the same sense as I knew all about teaching before I ever did any. I've been watching chairmen do things, some right, some wrong, for so many years that it would be easy, I thought, to adopt all their virtues and avoid all their faults. I was wrong; it wasn't easy at all.

I landed in Honolulu at the end of August in 1968; I was met by Jessie with a lei and a scheduling problem. "Aloha", she said; "Dr.

Halmos, what would you like to do about the over-registration in Math. 103?" (The number may be wrong, but the spirit is right.) I was in no shape to understand the problem, but fortunately when I asked Jessie what it meant, and whether it had ever happened before, and what was done about it then, she had the answer. (She always had the answer to everything.) I told her to do the same thing this year as she did last year, and that was exactly what she wanted me to say; she went away happy and did it. By next morning, when I showed up at the office for my first day of work, there was no more problem.

J. Nakata, 1968

Jessie Nakata is wonderful. She was born and raised in Hawaii of pure Japanese parentage. Her mother tongue is colloquial American. She knows a little of the culture and the language of her ancestors, but only very little. She is a perfectly trained secretary, and could, for instance, spell every word I ever threw at her, but she is almost completely uneducated. (When I made a casual reference to Cain and Abel, she told me that she had never heard of them.)

She had been the chief secretary of the mathematics department at Hawaii with the preceding two chairmen; I was to be her third. In addition to her full-time job, she had a husband and three pre-school children. She must have been near to 30 when I met her. Both her personality and her shape (generously rounded) made her the ideal mother figure for the small flock of young typists and part-time secretaries she was in charge of, as well as for the students and the faculty of the department.

I liked her immediately and she liked me; it was obvious that we could work well together. My wife and I got to know her and her family slightly. A few Saturday afternoons Jessie and her children would visit us for soft drinks and the use of our swimming pool, and once or twice we attended large parties at her house. Mainly, however, our relationship, hers and mine, was professional, but very friendly; our conversation was efficient, but very informal.

Here is an example of Jessie's way of operating. The wall between her office and mine was only three quarter height—it didn't reach the ceiling. There was a buzzer on my desk for me to summon Jessie when I needed her. The first time I used it, she shouted over the partial wall: "What do YOU want?" That was the last time I used the buzzer.

She knew almost everything trained secretaries know, and she was extraordinarily intelligent and courageous. If I asked her to do something she had never done before, she would never tell me so; she would try anything and she would succeed. Here is a trivial sample: I asked her to arrange a mildly complicated visit to the mainland for me. I told her the days and the places and left the rest to her—and learned only later, after she arranged it all, that she had never spoken to a travel agent before and had only the vaguest ideas of how airline schedules were made up.

Once she did something, she remembered it from then on. When the dean's secretary called me before Christmas to remind me of my faculty-evaluation job, I asked Jessie to explain it to me. It was simple enough, in principle, as it turned out, but not, I thought, easy to make the appropriate decisions. What it meant was that I had to write the dean a large number of letters—one for each member of the department. Each letter was to say one of three things: minus, zero, or plus. (The ad hoc terminology is my own.) "Minus" meant that the person's work was not satisfactory, and no action should be taken about salary or rank (other than the routine "step" advancement that a rigidly formalized scheme called for). "Zero" meant that the work was satisfactory but not great; funds available were to be divided among the "zero" people, but promotion was certainly not recommended. "Plus" was the good grade, of course: it implied a recommended merit raise and, when possible, promotion. I sweated out the writing of the letters: I composed three forms, and left them on Jessie's desk Friday noon; I was going to spend the weekend deciding which letter applied to which person. I needn't have wasted my time. All the letters, with all the names, were on my desk, ready for signature on Monday morning. Jessie knew the department and she knew me; she knew exactly what I'd say about everybody. She "guessed" every grade right.

Other than Jessie almost everything at the University of Hawaii was a source of worry to me. The worry and the pressure were great enough to produce, for the first time in my life, a psychological symptom with a name that I knew. Of course I knew words such as "anxiety" and "depression", but they were remote abstractions like "pellagra" and "phthisis"—I didn't really understand what it meant to suffer from them. On my third night in Hawaii I woke up at 3:00 a.m., terrified and trembling, with what I later diagnosed as a genuine, just-like-in-the-books anxiety attack. There wasn't anything in the house to be frightened of— but I was frightened, perhaps literally "out of my wits". The problems that were lurking, ready to attack me (I couldn't name a single one, but they were monstrous) frightened me, I couldn't cope with them, I couldn't cope with anything. I woke my wife and I talked—I talked— I talked. We had a snack and a beer—and I talked. After an hour I felt better and we went back to sleep—the attack was over and from then on I coped.

What brought the attack on? The past and the future I suppose. For over six months I had been facing and solving chairman problems, I went through the usual traumas of a long-distance move, I had just returned from a long European conference-and-lecture tour, my diurnal rhythm was several hours out of phase—and I was facing a gigantic and probably impossible job. I wrote in my diary: "It'll never fly, but I'll give it a whirl."

And I did. I dived into it with energy and, almost, enthusiasm. Big job number 1: kill the T.V. course and replace it by people. O.K., let's go; how do we do that?

The T.V. course was somebody's inspiration a few years back. Why not solve the two big problems, quantity and quality, at the same time? We (Hawaii) don't have enough people to teach freshman mathematics on the scale demanded, and many of the ones we have are not good enough. Let our best and most popular teacher, Jimmy Siu, tape the course, and then let's distribute the tapes all over the campus for replay all around the clock. Students would have complete freedom—they could hear the lectures when and where they pleased, and they could hear them as often as they pleased. The faculty would still hold office hours, and, obviously, students would just love the perfect combination of easily available public performances and private attention.

Students hated it. They wanted a live teacher, not a canned one, and when they had questions, they wanted to talk to their teacher, not to a stranger. Morale went down, attendance went down, grades went down. The decision had to be made: let's cut our losses. The T.V. experiment was an expensive folly. Bodies to put in front of classes are more easily available now than three years ago, and a new chairman who knows the market is coming in. Let's kill T.V. at the end of this year and hire the 20 bodies it'll take to replace it.

Yes—20. That's what my orders said: hire 20 assistant professors at the going rate, which was about $10,000 for the academic year. Full professors were earning about $20,000 in those days, or a little more. When I told Alex Heller that I am offering 20 jobs at $10,000 each, he thought for a moment and then said: "I'll take three".

The logistics of hiring 20 people can be quite strenuous. It takes about ten exchanges of letters to make the initial approach, answer questions, get letters of references, arrange an interview, make the formal offer, and then, if the offer is accepted, be helpful about courses, travel, and housing—and that's ten for each name on the list. At one time (in November) I had 47 offers out at various stages: would you be interested...?—will you please fill in these forms...?—I am authorized to offer you.... If they had all accepted, I would have been in a very peculiar position—but, of course, they didn't. I kept a complicated chart on my office wall—a scorecard—to help me from overextending my gamble with Hawaii's money.

The winter meeting of the AMS was in San Francisco in 1969. My friends in Berkeley were kind to me: they lent me an office for a couple of days before the meeting and scheduled twenty of their graduate students for interviews with me. Once that was over, I moved my headquarters to San Francisco and, by moving fast from one restaurant table to another hotel lounge, interviewed twenty more candidates. The final score was not perfect, but it was pretty good: by April I acquired a total of $17\frac{1}{2}$ assistant professors (including half a wife) and three graduate assistants. The T.V. is dead, long live the people!

Permanent faculty were not easy to attract to Hawaii (quality of the university, isolation, high cost of living, long travel times), but I could have all the visitors I wanted, for a week, for a semester, for a year. Hasse spent the year there (the Hasse, the great algebraist, number theorist), as did Leo Moser (one of the most energetic combinatorial problemists of his day), and Don Fraser (leading Canadian statistician, an addicted surf-boarder). Leon Cohen (retired from both the NSF and the Maryland chairmanship) and Allen Shields (my Michigan colleague) came for a semester. Almost everybody on his way to or from the other side of the world (Australia, Japan, etc.) stopped off for a few days: I remember Michael Atiyah, Edward Collingwood (Sir Edward—one of the small number of knighted mathematicians, the one I bought a drink for in Dundee), Richard Courant (frail by then and ailing, and watched over by Peter Lax), and Joe Doob—and that's by no means the whole list.

Then there were the official one-week visitors. I wangled from Todd Furniss funding approximately equal to the pay of one junior faculty member to be spent in seven or eight pieces on distinguished senior people. They were to come at the rate of one a month and be temporary members of the family: they would give three or four lectures (at all levels from expository to fancy), and they would participate in our lunches, parties, and bull-sessions. The idea was to counteract the isolation, and it worked quite well. The list of one-week visitors included Ambrose, Bing, Paul Cohen, Lester Dubins, Erdös, and Ed Hewitt. It included Kaplansky too, in a different capacity. For him we were able to arrange a CBMS regional conference (the kind where the principal speaker lectures twice a day for five days in a row); it was a very well-attended success.

Chairmen are hired to worry about many routine administrative problems far less glamorous than visiting mathematicians: course and class assignments, college and departmental committees, the coffee budget, decisions about xerox, postage, and telephones, the welfare and promotion of secretaries, and long, boring, and ineffective meetings with other chairmen and deans. I worried appropriately about all those things, some well and some badly. Some of my colleagues thought I was doing fine, and others, it's probably needless to say, disapproved of and disliked almost everything I did. In an attempt to keep people from feeling out of things I sent out a lot of information memos; the only perceptible response was a sarcastic anonymous letter asking for more. (I didn't have much trouble detecting who wrote it.) One colleague sent me a memo with the rude (?) salutation "Chairman Halmos"; I was forced to tell him that we didn't have the money to buy the desk calculator he wanted. (It didn't really break my heart.) University regulations made it impossible to re-hire Hasse (he was too old), and, like the Persian kings of old he held the message-bearer (me) responsible for the bad news.

The main service a chairman provides is psychotherapy to the faculty, and that's not something I know how to do. Intuitively I knew, I guessed, long ago that chairman is the last thing in the world I wanted to be—it was specious reasoning based on the fear of approaching old age and the glamor of Hawaii that changed my mind. Even back at Chicago, when I was acting chairman for a short time, and Walter Bartky made it plain that I was in line for the chairmanship, the prospect did not fill me with glee. Just the opposite: it was one of the reasons that contributed to my leaving Chicago. In Hawaii I worked hard at every aspect of the job, including psychotherapy, but I didn't like it at all.

I worked hard at it for the full nine months, but by the end of the first three my mind was made up: I was going to quit. I told Jessie first, then Todd Furniss, then Mookini, and finally, a couple of weeks later, the entire department. Everyone reacted nicely: they tried to persuade me to keep the job, or, failing that, to let someone else have it but stay on in the department. I was pleased by the reaction, but I was eager to get back to the real world, my real world, the world where time and geography made it possible to make brief trips for colloquium talks and conferences, and I could always know what and where the action was.

The last big job Todd Furniss gave me was to find my successor. That took a little doing, but it too got done. H. S. Bear, known to one and all as Jake, agreed to join the faculty and become the chairman of the army of assistant professors I hired for him.

My last six months in Honolulu were infinitely more pleasant than the first three. I continued doing everything I had done before, but my attitude was relaxed, and my complaints disappeared. With a relaxed attitude I was able to do more, and, for instance, I could work harder at the Diamond Head Seminar. The seminar consisted of Larry Wallen, Allen Shields, and me; it met once or twice a week, at my house on Diamond Head Circle, at 7 in the morning. (Honolulu is an early city, and I picked up its ways. I got used to waking at 5:30 and having breakfast after a dip in the pool. The morning rush-hour traffic downtown was over by 7:30.) We would drink coffee, swim, and talk mathematics. Once I knew that I didn't have to serve a life sentence being chairman, I could think about theorems between memos, and found that I could pull my weight in the seminar.

There was one more problem to be solved: where was I going from here? In theory I could go back to Michigan, and once the word got out that I was on the loose, some job possibilities started sprouting. One day in November (I happened to be in my motel room in Austin, where some AMS committee business had called me), the telephone rang. "Paul, this is Arlen Brown. Would you be interested in a job here —in Bloomington—at Indiana University?" How could I say no?

M. A. Zorn, 1983

George Springer was chairman at Indiana then, and he invited me to come for a colloquium talk and an interview with a room full of deans. I came, I lectured, and I shook hands. I met Max Zorn, who was later to become a valued and good friend. He came to tea with a slip of paper bearing a list of questions—questions he had been saving to ask me! Max is an inveterate and superb question asker; in private conversations or in public colloquia he can always put his finger on the exact spot where the deep ideas are hidden. How could I resist? Two weeks later, still before Christmas, Springer made me an offer, which I accepted; in 1969 I became a Hoosier.

Life in Bloomington

The year before I moved to Bloomington the American Council on Education rated Indiana as number 26 among the universities of its class; in the revised rating the following year Indiana became 27. Some of my friends pretended to see the two events (my arrival and the change in rating) as cause and effect. They were nice to me anyway. The university as a whole was nice to me, both personally and officially, and I settled down to enjoy my "golden years" by working harder than ever.

The first step was to organize a functional analysis seminar, for which there appeared to be a good basis. Six of us were operator theorists, to begin with. Arlen Brown was one, and John Conway (not the games and groups one in Cambridge) was another. Peter Fillmore was still there (but before long he returned to Canada), as were two of his collaborators, Joe Stampfli and Jim Williams. Jim was a Ph.D. student of Arlen's, so that he was my grandson; by inclination and by training, the interests of all six of us were closely connected. There were others too who could either talk or at least listen to functional analysis: Billy Rhoades and Graham Bennett, who knew about summability, and Pesi Masani and Slim (= Seymour) Sherman, who knew about probability. (Masani went to Pittsburgh before long, and first Sherman and then, later, Williams died before their time.)

The functional analysis seminar got off to a running start, and, as these lines are being written, is still going on. Its first year was unusually successful because of one of George Springer's administrative inspirations. The university was about to celebrate its 150th birthday, and Springer established and told me to take charge of a Sesquicentennial Seminar. The idea was to invite one-week visitors, not, as Hawaii did, six or eight times in the course of the year, but each and every week. The way schedules usually misbehave, some weeks we had two visitors at once, but I don't remember any week without one.

Functional analysts at Indiana University, 1968. *Left to right*: P. A. Fillmore, J. G. Stampfli, J. P. Williams, J. B. Conway, H. A. Brown.

It sounds good, perhaps, but there are problems. The visitor would be with us from Sunday night till Friday afternoon. He would give a lecture, and come to lunch and to tea, but in between he would look bored and lost. The result was that we, the home team, felt we had to give him cocktail parties and show him the sights and listen to his latest mathematics—all of which is fine but too much of which is too much. We had over 25 one-week visitors between September and April, and I hope they had a good time; we, the hosts, were weary of it all and glad when it was over. Still and all, a seminar that is addressed by notables such as Jim Glimm, Shizuo Kakutani, and George Mackey (I picked three names out of the 25 almost at random) deserves to be called a success. The subjects the visitors talked about were fresh mathematics at the time (Sarason, for example, told us about his work on weak-star approximation by polynomials), and when I looked over the titles recently I was pleased to note that talks on those subjects would still sound interesting today.

Bloomington had many advantages for me. It was a small town for one thing, about the size of Chambana in the days of my youth. The law school was next door but one from Swain Hall, the mathematics building, and I took advantage of it: I audited the first-year courses on torts and criminal law. The music school, one of the best in the world, was four buildings from Swain Hall in the other direction, and it had a tremendous effect on my life. I met Janos Starker and Menahem Pressler, I heard them play, and I reaped the benefit of their presence: their students, their colleagues, and their visitors offered nearly a hundred high quality performances every month.

The functional analysts at Indiana were young enough and numerous enough to constitute a respectable group at first, but for various personal reasons the group kept declining in quality and in number. Aside from functional analysis the department was not strong. Algebra was almost invisible, as was topology (which has improved a great deal since then); there was a highly audible group of applied mathematicians who were mainly weak analysts (and that group too has changed a lot for the better). Bill Gustafson was with us for a few years and he served outstandingly well as my algebraic therapist. He knew not only algebra but he knew mathematics; he had a broad view over many subjects. He is also a cultured man with many interests; he is, for instance, a reliable source of information about detective stories and about the movie stars of the 1930's. Overlapping with Bill, and continuing still, is my topological informant John Ewing; since he also knows an indecently large amount of number theory and is willing to talk about mathematics, or anything, any hour of day or night, he is a pleasure to have as a colleague.

Gustafson and Ewing are examples of dedicated mathematicians, real pros, and the department had and still has others like them—but very few. For many their job is a job, and they do it with good grace and competence and even with an occasional spark of interest—but it is not a calling, not a vocation. For some the job is a chore, a frustration, and they do enough of it to keep it—no more. The effect on everyone is less than maximal morale.

It is a pleasure, a rewarding satisfaction, to belong to a community most members of which contribute to it to the best of their ability and have the self respect that comes from doing so. I am not talking about producing top quality research, not necessarily. A first-rate undergraduate teaching institution can be a community of that kind—think of Bryn Mawr and Wabash College. Both Wabash and Chicago take pride in trying to do as well as possible—so far as I could see the trouble with Indiana was that it wasn't trying.

Such a judgment is subjective, of course, and it is influenced by impressions that have nothing to do with mathematics. Mathematically I thought that Michigan was very much like Chicago, but not quite so good, whereas Indiana was very much like Michigan but nowhere near so good. More important than that, however, was that Ann Arbor is in the Middle West by most geometric criteria, but it has many of the attitudes of the East, whereas Bloomington is in the Middle West by most geometric criteria, but to me it always seemed a part of the South. The University of Michigan has a deservedly high reputation, based on a deeply rooted tradition of scholarship that the state recognizes and encourages. Indiana University is near the bottom of the "great" universities, and in the last twenty years or so has had neither academic leadership nor legislative support worthy of the name. In Ann Arbor I was, of course, "gown", but I sensed no enmity or even coolness from the natives; in Bloomington I felt like an unwelcome stranger much of the time and I could be convinced that the derisive adjective "redneck" describes many of the members of "town". I must hasten to say, however, that I am talking about averages, not individuals. The gang that I played poker with for a while contained car dealers, plumbers, and truck drivers, and they seemed to accept me as one of the boys even if I did talk funny.

Indiana students

The best way for a prospective Ph.D. candidate and his prospective supervisor to meet and feel each other out before any serious commitment is made is in graduate courses. (It's not the only way, but it's the safest.) Most of my students came to me that way. I would teach a first-year graduate course on "real variables", followed by a second-year course on functional analysis, and sometimes even followed by a third-year research level course on operator theory. I went through three or four such sequences in my life, and each one produced several Ph.D. students, anywhere between two and five.

I taught the basic real variable course in 1969, my first Indiana year. Three of the twenty students in the class were Joe Bastian, Don Rogers, and Russ Smucker. They didn't especially choose me—they were ready for the course, and when they came to class the first day there I was. The next year I taught functional analysis, and about half the first-year class, including those three, chose to come along. The second-year class also picked up about half of the other section of the real variable course that had been parallel with mine; the newcomers included José Barría and Bob Moore. I produced a total of eight Ph.D.'s in Indiana, and they included the five I just named.

They got along well, those five. They all worked on operator theory, they could all understand the problems the others were working on, and they had sympathy for each other's troubles. I saw them often, separately and together, in class, in public seminars, and in our private seminar (which met sometimes in my living room with beer and pretzels). When I spent one semester of a sabbatical year in Edinburgh I was able to wangle funding to have them come with me, and that too worked out well. They enjoyed the glamor of travel and of a strange country that wasn't too strange, and it was a source of comfort and support for each of them to have the others along. I enjoyed that gang, and I liked every one of them. They are all established by now, three of them (Barría, Moore, and Smucker) in the academic world, and the other two outside of it, and I am pleased and proud that I had a hand in making them the useful citizens that they have become.

I mentioned before that the shared thesis discussions of some of my students horrified my Indiana colleague Anna Hatcher; it was, in particular, this gang of five that precipitated her expression of uncomprehending disapproval. Despite her shortsightedness on this small point, she was a great woman and I like to remember meeting her and knowing her.

I met her soon after our names and pictures, hers and mine, appeared one day in the local newspaper, the Bloomington Herald-Telephone. I had been slow on the up-take a few months before that spring day. George Springer asked me some unusual questions (what honors and awards had I received?, what positions in the AMS had I held?, what prestigious lectures had I been invited to deliver?). I was already there, I already had a job; what good purpose could all that information serve? He mumbled about incomplete files, and I put the matter out of my mind. It came as a complete surprise to see my picture on the front page of the second section: two of the four members of the faculty whom the Board of Trustees promoted to the rank of Distinguished Professor were Anna Hatcher and I. It was a pleasing honor, of course, accompanied by a small but pleasant raise in pay. I retroactively caught on, and phoned George Springer to thank him for nominating me. He was pleased by my pleasure—and by the fact!—he didn't know that his effort was successful, and since I saw the paper before he did, I was the first to tell him the news. The comic relief to all this was provided by a silly printing error; the names under the pictures of Anna and me were reversed.

I phoned her, introduced myself, congratulated both of us, and suggested that under the circumstances we should at least have lunch together. I liked her immediately. She was about ten years older than I. Her departmental affiliation was double, but sounded quadruple: she had appointments in both the Department of French and Italian and the Department of Spanish and Portuguese. We shared an interest in words, professional on her part and dilettante on mine, and many of our conversations in the years that followed were about linguistics. One of her books had the (to me) attractive title *Modern English Word Formation and Neo-Latin*. We had her to dinner quite often, and, not wanting to be a free-loader, she asked us to dinner in her small apartment in return. The sociability was always pleasant, but she was an atrocious cook; the chicken was nearly raw but the rice was overcooked and soggy. She was hard of hearing and she smoked heavily. She retired when she reached 70 (mandatory), and she died a couple of years later. I miss her.

Let's get back to students. Ken Harrison was my student in all but name. He is an Australian and as an undergraduate he studied with John Miller at Monash (in Melbourne). Miller didn't force any particular subject on him, and when it turned out that Ken was attracted to the same kind of operator theory as I, Miller wrote and asked whether I'd be willing to have Ken come to study with me for a while. Why not?, said I, and the "for a while" began in Hawaii. Ken was very young then (20?, 21?), blond, eager, and bright, and had one of the thickest Aussie accents I had ever heard. (Bill Gustafson said later that Ken's dialect seemed to have the five vowels aye, aye, aye, aye, and u.) We got along well in Hawaii, and when I moved to Indiana, Ken came too. He wrote a thesis about a part of invariant subspace theory that had always

P. R. Halmos, 1970

interested me, and I was his thesis supervisor in every legitimate sense of the word—but technicality and his future career made it desirable that officially his Ph.D. be awarded by Monash, and so it was.

Next came Don Hadwin. In the spring of 1970 I was the principal lecturer at a CBMS regional conference at the Texas Christian University in Fort Worth. I worked very hard preparing my ten lectures for that conference, and the result was the best research-level expository paper I ever wrote. I felt deeply embedded in the subject, I enjoyed every part of it, and I tried to put into both the oral delivery and the subsequent paper not only new expository techniques but as much new mathematics as I could. Hadwin came to the conference, seemed to learn something and like it, and a year or two later showed up in Bloomington and said he wanted to be my Ph.D. student. That was not the canonical way to do things, but I gambled on him and won: he too wrote an invariant subspace thesis, a good one, and he too went on from there to become a full-fledged member of the operator community.

There is a small story behind the published version of my TCU lectures. The way the NSF funded regional conferences in those days (and the organization is still very much the same) was to pay the speaker N dollars for the lectures and later, when the speaker submitted an acceptable manuscript, N more dollars for the paper. (As I remember it now, N was 1500.) The idea was, and is, to publish the paper in a special series—and I was very unhappy about that idea. Special series do not have wide circulations. A few dozen individuals might buy each number, and a few hundred libraries might subscribe to the series, but most people never even know of most series, and when by chance they hear about something interesting in one of them, they cannot find it in their library. A randomly chosen non-major university, in the U.S., or in Japan, or in Poland, is unlikely to be sufficiently knowledgeable and sufficiently rich to subscribe to many special series, and a member of such a university will never see the most recent issue on the new-journal shelf. The same argument applies to the published "Proceedings" of conferences generally; I think they are a bad concept.

I was eager for my inspired words to be seen by everyone—you don't have to like them, but the decision to reject them should be your own. I said so, and the result, after a bit of official correspondence, was an agreement. Very well: I would get N dollars (but not $2N$), and I could do what I liked with the manuscript. Great!, I said, and I forthwith submitted it to the Bulletin of the AMS. I lucked out. A while before that, in an attempt to increase the number and quality of expository papers in the Bulletin, the AMS announced that as an experiment they would pay for ten such papers, the first ten acceptable ones from among however many would be submitted. (The payment was $1000.00, I think.) The experiment was not a great success: apparently people didn't want to write exposition, or couldn't, not even for money. It took a long time to get nine papers, and no tenth one was on the horizon when my manuscript arrived. The Bulletin accepted it, and I had my cake and ate it too—or two thirds of it anyway. I was proud enough of the result that I headed it with a dedication: "to my teacher and friend Joseph Leo Doob with admiration and affection". Joe sent me a postcard when the paper appeared: "It was a pleasant surprise to read that dedication", he wrote, "but you make me feel like a monument!"

That was the second time Joe wrote me a thank-you note; the first time was a few years earlier, when I saw a martingale in a mail-order catalogue and had it sent to him. Gamblers used to speak of a legendary Colonel Martingale who wanted to double the stakes after each loss. Later the word acquired a technical meaning in probability theory. For Doob a martingale was a specific kind of stochastic process, a kind to whose theory he made many deep contributions. The first meaning of "martingale", however, is "a part of a harness designed to prevent a horse from throwing back its head", and that's the kind I found in a catalogue. Joe was pleased to get one. "I am very grateful to you" he wrote, "for sending me just what my office needs. It is now hanging up on one of my bookshelves, next to my collected reprints, as is fitting."

My last Indiana student, and, in fact, my last Ph.D. student, was Sunder. His full name is Viakalathur Shankar Sunder, but *his* name, the name that he is always to be called by, is just Sunder. It's not like John Leroy Kelley, whom some ignoramuses call John, and whom some of his family called Leroy, but whom all his friends call Kelley—that was idiosyncratic. In Sunder's case the Viakalathur is a place name, and the Shankar is his father's name; the only name that's properly *his*, that his mother, his friends, and strangers all call him by, is Sunder.

He appeared in my first-year real variable class in 1973, straight from Madras, and immediately showed himself a bright and hard working student. That he was also polite, and charming, and a good conversationalist, with many interests, including some talent with Indian musical instruments not widely known in the west—all that was extra. He stayed with me for the second-year course in that cycle, and then asked if he could work on his Ph.D. with me. I was pleased to say yes. Since I was working at the time on a book on integral operators, I had him read some of the literature of that subject, as well as a draft of my book, and he took to it all with ease and apparent pleasure.

Then, suddenly, the sun bug hit me again. John Ernest, an old friend, phoned one day from Santa Barbara and asked if I would like to go there. I was very much interested, and told him so, but, at the same time, I protested. "I'm too old, John", I said, "and too expensive". It was the autumn of 1975, I was approaching 60, and my salary was $41,000. He wasn't fazed, and the discussion continued. Result: I moved to Santa Barbara.

It was a foolish move, but, fortunately, it didn't have much effect on me, either psychologically or professionally. There is nothing intrinsically wrong with Santa Barbara, but for me there was less to like and more to grumble about than at Indiana. The city is beautiful and the climate is great. Cultural facilities (e.g., music) are available but skimpy. The university was second rate, and the extraordinarily heavy-handed administrative bureaucracy of the University of California system made it hard to love. There were some friendly and talented people there, and, of course, some deadwood, but there had also been a Hatfield–McCoy schism in the department that made for bitter memories and still palpable friction. I taught a few courses, thought about some mathematics, finished the book on integral operators—and, after two years, returned to Indiana. Won't I ever learn that sunshine is not enough?

Sunder came to Santa Barbara with me and finished his thesis there. He got some beautiful results and it was exciting to watch him getting them. He put compactness back into the theory of integral operators; when the breakthrough arrived, around Christmas 1976, he would keep coming to me, pleased and breathless, morning or evening, to let me in on the latest secret. He answered so many questions that my book would have been incomplete without his results, and, at the same time, it would have been unfair to put them in and have them first see the light of day under my name. A possible solution was to ask him to collaborate with me on the book, and while that solution seemed fair to me, I worried about it. The student-teacher relation being what it is, it would have been almost impossible for him to decline my invitation, and I worried about his accepting it under duress. Worry or no, I asked him, and I am convinced that his yes was sincerely enthusiastic.

The story ends well: the collaboration was effective, we both worked hard, and we wrote a pretty good book. The major cloud in the silver lining appeared after the book was published: we overlooked a part of the literature. (We? That's passing the buck. I am the one who missed: as thesis supervisor I had the responsibility.) One of the theorems that Sunder proved had been proved before, and one of the problems the book proposed had been solved before. The literature in question is Russian; the relevant names are Korotkov and Bukhvalov. Too bad, but such things happen. I shouldn't have let it happen, but I did and it did.

Only once did we have a major disagreement. After the mathematics and its exposition were done, I wrote a draft of a paragraph describing the prerequisites needed for reading the book. One sentence of the paragraph looked like this: "... the reader should know x, and he should either know, be willing to take on faith, or be willing to look up y". Sunder put his foot down: he couldn't approve that sentence. Reason: it's a sexist sentence, it implies that the reader ("he") is necessarily male, and Sunder's convictions wouldn't permit him to sign his name to that implication. Then it was my turn to get mad. Stuff and nonsense, I said; there is nothing sexist about the correct use of the classical English neuter pronoun. It's a mere historical accident that it has the same form as the masculine pronoun, and it has no implications whatever about sex. The use of the ugly "he or she' would call attention to sexism where none was intended and none would have been perceived (except by pathologically sensitive and grammatically uneducated readers)—we leave that sentence the way it was, or else. Or else what? Sunder was firm, and I was firm. He refused to use "he" and I absolutely rejected "he or she". The result was, of course, a compromise, but I thought that Sunder really won; he probably thinks I did.

Committees of one: Wabash

Much of what is called "service" at a university is committee work, the most time-wasting and inefficient way of attending to trivialities ever invented. The departmental committees on graduate policy that I have served on, for instance, did nothing but establish and maintain mediocre standards, and the college committees on tenure and promotion that I did not serve on did nothing except needlessly lengthen the time between departmental recommendations and administrative decisions.

There is all the difference in the world between the effective delegation of authority (a good thing) and its ridiculous dilution. The decisions that the graduate policy committee had to make could have been made just as well by one conscientious director of graduate affairs. No; they could have been made much better by one person instead of six—better and quicker.

Some departmental committees frequently have only one member, and so far as I can tell they function just as well as the big ones, or better. The library committee is a common example; it can consist of one dedicated member of the faculty who loves and knows books, and can discuss technicalities with the librarian. Another common example is the colloquium committee; one person can organize the colloquium, run it, and coordinate it with other activities four times better than two and nine times better than three.

The Bulletin of the AMS is run by an editorial committee of three, but they are, in effect, three little editorial committees of one person each. There is an editor for research-expository articles, an editor for research announcements, and an editor for book reviews. The "committee" meets by accident only. Its three members are completely autonomous; each one does his job without interference from the others. One of the three is managing editor, a job that rotates among them, but the only thing he usually has a chance to manage is routine correspondence with the central office of the AMS in Providence.

The American Mathematical Monthly looks as if it had a gigantic editorial board of about twenty people, but once again appearances are deceiving. A small subset of the twenty (anywhere between one and three) edits the short Notes, and even that subset splits its work among its members, so that in effect each of them is a plenipotentiary autocrat who makes editorial decisions without orders from above or votes from his peers. The other departments of the Monthly are equally autonomous. The function of the Editor (the one with a capital E) is to edit his department (the long articles) and to keep a watchful and worried eye on his associates.

What happens when a committee of one is no good? When a committee of two, or three, or seven people has a member who is dilatory, disorganized, incompetent, or even malevolent, the others can shoulder the whole load and muddle through. That is so, and it can be counted as an advantage of the committee system. I have seen criminally lazy library committees of one, who put all the work on the librarian. The ordering of books those years was in bad shape. I have seen a hopelessly disorganized organizer (appointed by a democratically elected chairman with hopelessly bad judgment) botch the colloquium program (talks without titles, speakers without hotel rooms, weeks without speakers). I know about editors who allowed unacknowledged manuscripts to stack up on their desks for months (and even years). That is so, and it must be counted as a serious disadvantage of administration by committees of one.

We have to choose. Do we choose the reliable mediocrity of a boat that doesn't rock, and never gets anywhere, or do we gamble on creative imagination and dedicated hard labor, facing the danger of total stoppage and possibly even going backward? Which choice you make depends on your temperament. Me, I'm a gambler; I firmly believe that the big gains made possible by good committees of one are worth the occasional temporary losses.

I have done my share of serving on large departmental and professional committees, but I'm convinced that the main services I have been able to render to the mathematical community were the ones that started by someone putting a broom in my hands and saying "Sweep!". One such broom was the suggestion by Earl Berkson. In the 1930's Aristotle Demetrius Michal conducted a famous peripatetic seminar. It circulated among USC, UCLA, and Caltech, following Aristotle (just as the original one did) from one place to another. In the late 1960's the NBFAS (North British Functional Analysis Seminar, pronounced en-bee-fass) started meetings in Edinburgh, Glasgow, and Newcastle. Inspired possibly by these examples, Berkson suggested that the functional analysts in Bloomington (Indiana University), Lafayette (Purdue), and

Urbana (University of Illinois) get together at regular intervals and swap professional know-how and gossip.

It worked. Earl, and Jerry (Meyer Jerison, departmental chairman at Purdue in those days), and I ran the show. We were lucky in finding a host for our meetings: Wabash College (in Crawfordsville, Indiana) is as near as we could come to the center of gravity of Bloomington, Lafayette, and Urbana, and the mathematicians there, notably Bill Swift (chairman of his department at the time) went far out of their way to make the meetings pleasant and efficient.

Our "official" name is the Extramural Functional Analysis Seminar, but the usual identifying word is Wabash. ("Did you hear the last Wabash lecture on Gleason's theorem?") The lively exchange of ideas (formal lectures, informal conversations, challenging questions) is aided and abetted by tea and cookies before and between lectures, and drinks and dinner afterward. The festivities start at 2 o'clock Saturday afternoon (on about six to eight Saturdays during the academic year), and usually adjourn at about 7:30. The hours in between are sometimes confusing, often informative, and always profitable. We who attend learn what other people are doing (new techniques, new theorems, new jobs, new colleagues), and we go home tired but stimulated. We are glad it's over, but we look forward to the next time, and sometimes we can hardly wait for a chance to sit down and think peacefully more about what we have just heard.

At the beginning Wabash operated on enthusiasm alone—there was no funding, no one got travel expenses or anything else, and everyone chipped in to pay for the cookies. In three years we became sufficiently well-established and prestigious that we were able to arrange an international meeting. That one did get NSF support, and did pay several people from abroad to come for a week. There were a gross (12^2) of registered participants, including some from Hawaii, Ireland, Israel, and Rumania. It was pronounced a mathematical and culinary success.

After that big jamboree we went back to our original modest scale; the average monthly attendance is around 25 or 30. The participants are not restricted to members of the sponsoring groups. We have had and welcomed visitors from many neighboring colleges and universities, including for instance Kentucky, and once or twice even Michigan (Leon Brown flew his plane). After twelve years of operation, the seminar applied for and received a small NSF grant, which can be used to pay travel expenses for an occasional visiting lecturer from far away. Worried about the bureaucratization that might ensue, I disapproved—but so far my worries have been groundless.

The committee of three that runs the seminar changes every now and then; none of the founding fathers is still a member. A possible reason for the success of Wabash is that the committee doesn't have too much to do, and another is that it is really three committees of one. One of them is in charge of mailing out the reminder notices before each meeting, and each of them is responsible for twisting enough arms at home to produce his proportion of speakers during the year—and that's all there is to it.

One of my proudest accomplishments in Santa Barbara was the establishment of the SCFAS. You can probably guess that the name stands for Southern California Functional Analysis Seminar (pronounced ess-cee-fass). It seems to have become a permanent part of Southern California life, and only a few irreverent Hoosiers dare refer to it as Wabash West.

Committees of one: Bulletin

I became book review editor of the Bulletin in 1974 and served, in accordance with long standing tradition, two three-year terms. For the twelve years preceding, book reviews had fallen to a sad state; the editors in those years seemed to interpret their position as an honor, not a job. While in the 1930's the Bulletin published an average of about 70 reviews each year, the average in the years 1961–1973 was between 10 and 15. I started the job with enthusiasm and ambition: I intended to publish a lot of reviews, good ones, that people would read, learn from, and enjoy.

I subscribe to the New York Times Book Review, and my idea was to do for mathematics what it does for literature. I read the Times book reviews because I don't have the time to read all the books that come out, and reading reviews keeps me closer in touch with modern culture than not reading anything at all. The sheet of instructions that I sent to prospective reviewers contained the following paragraphs.

> An annotated table of contents is a bad book review. Nobody wants to know that theorem so-and-so appears in chapter such-and-such.
>
> A good book review is a chatty expository essay on a currently interesting subject. The purpose of the review is to provide a first approximation in three or four pages to what the book does (or could have done) in three or four hundred.

I told reviewers that they may, but need not, express value judgments ("it deserves to be on the bookshelf of every student and scholar"); in fact I went so far as to tell them that except in a courtesy paragraph somewhere near the end they need hardly even mention the book. Their aim was to keep readers in touch with modern mathematics and its growth, not to tell facts about a book.

In many subjects, such as history, or economics, or philosophy, it is a tradition that a review is a scholarly contribution that is taken seriously by both its author and its readers. I wanted to establish (re-establish?) that tradition in mathematics. My purpose was not to help readers decide whether they wanted to spend $29.50, but to put out an interesting magazine.

I made ground rules for myself as well as for reviewers. I decided, for instance, not to review research (e.g., Memoirs of the AMS, or most Springer lecture notes), not to review proceedings of symposia (with fifteen authors), not to review textbooks, or new printings, or translations, or books more than two years old. I decided also to stick to mathematics, and stay away from computer science, economics, and biology. I think I managed to break every one of those rules at least once, but they were pretty good rules to work by most of the time.

What does a book review editor do? He opens packages of books; he enters their authors, titles, publishers, and other bibliographic properties in a master log; he decides which ones to send out for review; he tries to dispose of the others in some intelligent way; he selects a reviewer; he reminds the reviewer (several times) that the review is overdue; he reads, edits, and proofreads the reviews; and he tries to retain enough interest in the job to be able to do it again next week. All this is impossible without an intelligent and diligent secretary. I estimate that the job took me about 10 hours each week and needed perhaps twice that much secretarial time.

So far as I could tell approximately 1500 mathematics books appeared each year during the years I was editor (don't ask me for an exact definition); slightly over 400 of those were sent to me for review. I found it easy to decide which ones (approximately 20%) to try to get reviewed (average time: 20 seconds), and I had no trouble disposing of the rest. (I donated most of them, let's say 79%, to the university library, and kept and treasured the remaining 1% as perquisites of the job.) The hardest job by far was to find a reviewer.

A review sheet was made out for each book, containing all the pertinent information, and, with that before me, I'd sit and glare glumly at the book and try to think of names. I'd look at the book's bibliography and if it suggested any possibilities, I'd scribble them on the review sheet. I'd try to think of famous names or obscure friends or someone at the college where I lectured last week and hope that one of them would be willing to write the review I needed. I'd hunt through Math. Reviews, looking at the names of authors and reviewers, and keep adding to the scribbles. I had to stay alert: the reviewer had better not be at the same university as the author, and the reviewer had better not be the author's Ph.D. student, or the other way around, and while differences in point of view can be interesting, feuds can be bloody, and the reviewer had better not be a mortal enemy of the subject, determined to pan the book no matter what it contained. Another obvious boundary condition: if I had already used the perfect reviewer twice that year, I had better not over-exploit his good nature the third time; I might get more of his golden eggs in the future if I gave him some rest now.

Once the review sheet had six or seven names on it, or better yet ten or twelve, I could easily enough choose one of them and shoot off a form letter. "The book titled... by... has been sent to the Bulletin of the AMS for review, and I am writing to ask if you would be willing to review it in something like the spirit described by the enclosed sheet." If the acceptance postcard came back to me in five or six weeks, all was well (for the time being); if not, I'd send the form letter to another name on the list. The average number of requests for one review was two, but "average" meaning what it does, many of them needed three or four; the maximum I remember is seven. (That book never got reviewed.) The average time it took actually to receive the manuscript of a review was six months; the maximum I remember is eighteen months.

I loved some of my reviewers and got boiling mad at others. Just plain nonfeasance was not enough to raise my blood pressure—it happened too often. Typical scenario: the acceptance card comes back promptly; the reminder four months later is ignored, as is the reminder two months after that, and the reminder three months after *that*; I shrug my shoulders and give up. Cost: a dollar's worth of postage, a few man-hours of work, one book, a lost review, and one reviewer's name struck off the list. Some reviewers who, in effect, did the same thing did it more gracefully. One conscience-stricken reviewer, for instance, explicitly admitted defeat. "I have decided that it will not be possible for me to write the review... I would like to keep the book (I have made some notes in it)... Please tell me how much it costs and I will send a check at once." My answer: "Don't worry about it, and, of course, do *not* send a check. That was a free review copy, and a review editor knows that he can't win them all."

Some reviewers selected themselves, and I cannot remember a time when the outcome was a success. Once or twice I had unsolicited reviews refereed, and was advised to reject them. More often I knew enough to find them unacceptable, but hid behind my own bureaucratic regulations: "Sorry, but reviews are obtained by invitation only." One volunteer didn't get discouraged when I told him that a review of the book he wanted to write about was already in press; he urged me to publish a second review, his. Since the detailed plan he described sounded interesting, I cautiously encouraged him to go ahead and write it; I refused to buy his piglet in a poke, but if it grew up, I was willing to heft it. Result: an enthusiastically appreciative letter accepting my terms, and then nothing more, ever; no pig.

I had known Benoit Mandelbrot for some time, and when I first learned, from him, about his book on fractals, I was eager to get it reviewed. It was a stroke of good luck, I thought, that just at the right time I received a note from a highly respected famous French mathematician volunteering to write the review. I answered immediately: "I was delighted to get your offer... the answer is enthusiastically yes; I'm looking forward to reading your essay." That was in November, 1977; the promised delivery time was Christmas, 1977—surprisingly and pleasingly quick. In February, 1978, I sent my usual reminder:

"How goes the review of Mandelbrot?

"The time limit you set yourself is in the past, but that doesn't worry me; the writing of high quality reviews cannot be rushed and forced.

"The purpose of this letter is only to remind you, and to ask you to give me a new time limit that is both finite and realistic. Scribble it on the enclosed card, and I'll leave you in peace for a while longer. O.K.?"

The card returned fast, with "April 1st, 1978" written on it. My subsequent inquiries, in May and in July, went unanswered, and I never got my chance to publish a Bulletin review of Mandelbrot.

The most spectacular renege in my experience was started by a volume of Lecture Notes by Ronald Lipsman. The answer to my invitation arrived promptly.

"I would be happy to review Lipsman's *Group Representations* for the Bulletin. Interest in the field has increased markedly in the past few years, and a number of books have recently appeared which treat various aspects of the subject. I feel that it would be useful to have reviews of all these books appear in the Bulletin. I propose the following program.

"I will review seven books... according to the following schedule.

"R. Lipsman, *Group Representations*; S. Gaal, *Linear Analysis...*;... by January 1, 1975.

"A. Borel, *Representations de groupes...*; N. Wallach, *Harmonic Analysis...*; G. Warner, *Harmonic Analysis...*; by April 1, 1975.

"I. M. Gelfand et al., *Representation Theory...*; S. Lang, $SL_2(\mathbb{R})$; by midsummer 1975."

My immediate answer: "You're the answer to a book review editor's prayers." I accepted five sevenths of the proposal: Gelfand's book was too old and Lang's too new (not out yet), and I wrote for, obtained, and sent to the "reviewer" review copies of the others. (Their total list price in 1974 was somewhat over $150.00.) The rest is a story of follow-up letters, frustration, and ultimate total defeat. When I wrote early in October 1975 and said I give up, the answer came six weeks later. "Never give up!... I'm almost finished with the Lipsman review. After that's done, we can hopefully set up a time schedule for the rest." That's all: no more letters from that source and never any reviews.

My files contain the names of all the sinners; it would take only a small bribe to make me reveal them.

These are only some of the tribulations of being an editor. I find them as fascinating as they are exasperating, and I could go on telling about them well beyond the boredom point for most people. The job had rewards too. During my first year, 1974, I corresponded about 75 reviews, received and accepted 22 (the rest didn't arrive till 1975), but since it took at least six months after I sent one to Providence to have it published, only two actually appeared. Once this transient delay was over, I published something like 75 reviews each year, but they were different from the contents-and-judgments reviews of fifty years ago; they were expository essays of about four printed pages on the average, and they constituted the most popular part of the Bulletin. Not everybody liked them, that's for sure, but people read them. I got hate mail as well as fan mail, but on balance I felt that I had accomplished something. You can't lose them all.

The Monthly

Some people sneer at the American Mathematical Monthly. "I never read it", they proudly say; "it's all junk". The meaning of "junk" in this context is "trivial"; not new, not deep, not research. And what's more, they say, it's full of all that education stuff.

Others scold the Monthly and write furious letters to the editor: the Monthly, they say, is supposed to be for teachers, and it should focus almost exclusively on mathematical education. All that pure math research junk (a different use of the word) is, they say, of no interest to us.

The Monthly is the mathematical journal with the largest circulation in the world, and, I suspect, the one most widely read. Its oldest features are its expository articles and its problem department, and the latter is, I suspect, the one people turn to first and love the most. When I was invited to succeed Ralph Boas as editor, I couldn't say yes immediately. Was I willing to stand in the middle and have brickbats thrown at me from both sides? I thought the Monthly was a fine magazine. Did I have a chance of keeping it good and possibly even making it better, or was I destined to go down in history as the editor under whom it sank?

When I finally decided to take a chance, about a year before my editorship was actually to begin, life became busy (busier) immediately, and the coefficient of busy-ness has been steadily increasing ever since. I had to make plans and I had to draw many people into them. I had detailed discussions with the officers of the Association (the Secretary, the Treasurer, the Executive Director, and others), I consulted my colleagues (was the department willing to support an activity that would, in effect, take about half my working time?), and I had to set up my team. Analogy: the editorial board is organizationally a little like the Cabinet of a governor or a president. The associate editors have full powers in their domain, and some of them even have a budget of their own, but sometimes they must coordinate their activities with those of the other editors, and they are in some sense under, or watched over, or monitored by the editor in chief. The editor's term is five years, and he picks his own associates; their terms are exactly coextensive with his.

I had long conversations with Alex Rosenberg (Boas's predecessor), and Harley Flanders (Rosenberg's predecessor), and I spent a wonderfully informative and exhausting day with Boas. They all gave me the advice I asked for, and, in particular, my decisions about the associate editors were based largely on what I learned from them. The main departments of the Monthly are (in current terminology) Articles, Notes, Teaching, Problems, and Reviews. The editor is in charge of articles, and small teams of two or three run each of the other departments. There are also a few associate editors "without portfolio"; the job of most of them is to stand by and advise the editor on subjects about which he is more than usually ignorant. Example: I asked Herb Wilf to be my computer science editor, and he has been useful beyond my expectations. He wrote some papers, solicited and edited others, and was always ready with solace and wisdom when I needed him.

Partly even before assembling my team, and partly during the process, in consultation with them, I had to decide what my emphasis on the Monthly would be; where did I want the magazine to go? Answer: mathematics. I wanted the articles and the notes to be expositions of mathematics, I wanted the teaching department to focus sharply on mathematics, and I wanted even the book reviews to be mathematical, not bibliographical. I wanted to de-emphasize "minor research" in favor of exposition; I wanted the discussions of teaching to be concerned with the ideas of the subject instead of the mechanics and the statistics of transmitting it; and I certainly wanted to stay away from sociology and politics.

Some people think that the Transactions is the journal for good papers, the Proceedings for weak ones, and the Monthly for the rest. If they wrote something they were not especially proud of, or, possibly, something the Proceedings rejected, they would submit it to the Monthly. Minor research papers on teratopology (the properties of monstrous topological spaces) and on the uninteresting parts of combinatorics appear in my mailbox with distressing frequency. Some past editors published such stuff; I tried not to. I rejected even good papers if they were research papers. The Monthly is not a research outlet, I kept telling authors. There is no law against a mathematical observation first seeing the light of day in the Monthly, but if the only thing a paper has going for it is that it's new, I would urge the author to send it elsewhere.

I changed the name of the department "Mathematical education" to

"Teaching of mathematics". One reason was emphasis: "mathematical education" is "education" with an adjective, not mathematics. Another reason was focus: if the educational wisdom that an author had to offer made sense when the word "mathematics" in it was replaced by "geography", say, throughout, then, I said, it should appear in a journal devoted to education, not mathematics.

My aim, to make everything in the Monthly mathematical, could not be attained, but I still think it was a good aim to keep in mind. Is the philosophy of mathematics mathematical? How about the history of mathematics? My answer is yes, but it must be a qualified yes, of course: depends on what the author says. Are the telegraphic book reviews introduced by Harley Flanders mathematical? My answer is no, but they are so popular, so much used by people trying to choose the ideal text to use next term, that it's all an editor's head is worth to try to eliminate them. Are the photographs of mathematicians that I introduced (reintroduced?) mathematical? No, but mathematicians like them, and nobody else would.

The editor of the Monthly has a lot of power over the magazine; what he says, goes. Various editors have tried to impress their personalities on their product in different ways. They changed the color of the cover, on the average about once every ten years, they put the table of contents on the front, and then they changed it to the back, and up front again. Boas went to a radical new cover design (which I abandoned in favor of the old contents in front) and introduced the department of Miscellanea (irreverently called "fillers" by some). Hating lists of all sorts and ephemeral ones worst of all, I introduced the discardable center section (over many howls), and letters to the editor. One of Boas's innovations was to change the cover color every year; the idea is that as you hunt for a specific issue on your bookshelf, you need help in keeping the volumes apart. I liked that, and continued it. As a result I must spend a few profound minutes each year deciding whether next year's Monthly should be purple or orange.

What else does an editor do? The answer that covers all his activities is: make decisions. As in most executive work, it doesn't really matter which decision is made; what matters is not to vacillate. Whose photograph shall I publish next March: a great English analyst, long dead, or a recent winner of the Fields medal? Decision time: about three minutes. Another ten minutes are needed to compose the captions (the teaser hint and the revelation) and to dictate the cover letter to the production editor, and I'm ready for the next job. To avoid wasting time you have to learn to make decisions firmly: choose and stick to your choice. You'd be surprised how many problems never even arise if you can manage to do that.

Where, I was often asked, do I get my pictures? Answer: I beg them, I borrow them, I steal them, I buy them, and, rarely, I make them.

Konrad Jacobs at Erlangen has a large collection (the basis of the impressive file in the Oberwolfach library), and George Bergman at Berkeley has one. Both of them have been helpful to me, as also has Jerry Alexanderson at Santa Clara, who is in charge of Pólya's lifelong accumulation. When I hear of a good picture, I write to ask for it (sometimes successfully), and I hoard the ones I get and keep counting them. At last count I had almost enough to last out my editorial term; après moi le déluge.

The editor of the Monthly sets the policy, chooses the colors and the photographs, tries to adjudicate complaints against (and sometimes between) associate editors, and (rarely) thinks about how to fire one of his associates. Technically speaking he can neither hire nor fire anyone. He nominates the associate editors, to be sure, but then they are "elected" by the Board of Governors, and once elected they are in, as firmly as the editor himself. If they do a bad job, or no job at all, all you can do is try to be diplomatic about persuading them to resign. The editor's most important job, however, is not policy, or photographs, or diplomacy; first and foremost the job has to do with the articles.

I emphasize: articles—not notes. What's the difference? Various distinctions have been proposed—tone, approach, unity—but the only one I really understand is length. A paper of eight manuscript pages or less is a note, and is processed by one of the Notes editors; the longer ones are articles. The editor's problem is to get enough articles, and to choose from among them; reject the bad ones, and edit the good ones.

How do you get articles? I tried soliciting them: "You know all about..., and you are a master expositor—could you be persuaded to share your secrets with the readers of the Monthly?" I kept trying, but my success ratio was discouraging; if I got as many as three papers out of a hundred solicitations, I could count myself lucky. The only effective way of getting articles seems to be to sit and wait for them.

A lot of what comes in is unusable. A hundred pages of explanations of some of the recent technicalities of algebraic geometry was one of the unusable manuscripts I received early on, and I tried to be as diplomatic as possible about explaining to the author that he misjudged the interests and the background information of the Monthly's readers. Crank papers are always with us, and most of them are not even very imaginative. They keep returning to the classic problems of antiquity: trisect the angle, square the circle, and duplicate the cube (in that order of popularity); and, from among the more modern problems, they almost always choose Fermat's last theorem. I feel that I must officially answer all official correspondents (but not necessarily more than once!). My usual answer to the ones who claim to have solved the unsolvable is to give a reference to a proof of unsolvability and insist that before I try to find the mistake in their work they find the mistake in that proof. What with hundred-page misjudgments, cranks, persistent contributors of minor research, and perfectly sensible writers with perfectly sensible aims who just happen to miss the target, the Monthly's acceptance rate has been around 20 percent.

The editor's hardest job is to get good advice. One trick I sometimes used was to ask authors to suggest referees. I'd ask the author to give me four or five names, and I'd tell the author that I don't promise to use any of them. Instead I would write to the author's suggestions, and ask *them* for further suggestions. The process converges remarkably fast; a pattern emerges, and the names of the most suitable referees seem to float to the top. Eureka!

I sent most papers to two referees at the same time. If the two agreed, it was relatively easy to come to a decision. If they didn't agree, I would try to achieve a reasonable compromise between them, or, if that couldn't work, to turn to a third referee, and, in any event, to come to a decision somehow—the buck could be passed only so far. If the decision was to reject, I would try to be firm but diplomatic—some referees are so venomously negative that it would be both pointless and grossly unkind to pass their words on to the authors undiluted. If the decision was to accept, then I could either rest on my oars—or go to work.

Just what does it mean to "edit" a paper once it has been accepted for publication? Every editor has a different answer to that question. To some it means an infinite amount of detailed fuss: change the words, change their order, change the punctuation. Sometimes it means rewrite the paper, or, in the case of papers written in Chinese or Hungarian or Pakistani, but using English words, to "translate" it into acceptable idiomatic usage. I did some of that kind of work, but as often as possible I gave the author his head. If his grammar was idiosyncratic, but clear, or if his writing retained a faint Japanese or Portuguese accent, I'd leave it be. Was I just being lazy? No, I told myself; I was just giving the author's personality a chance to show in the work.

That should give you some idea. If you're thinking of becoming editor of the Monthly, now you know what you're in for.

Here and there

Research, teaching, and service—yes, to be sure, but the mixture varies as the years go on. While I was teaching in Bloomington, and, in particular, turning out Ph.D. students, and while I was working at one or another form of service, at the same time I was still emotionally involved in research. I kept publishing papers, and some were not bad. I proved that every normal operator is a continuous function of a Hermitian one—a mildly surprising and definitely usable result. Later I studied the functor Lat that associates with each operator its lattice of invariant subspaces and proved that it is in a natural sense upper semi-continuous—a pleasant result, but one whose hoped for applications continue to be hoped for. All very sound and worth doing, but not spectacular. I passed 60 at about the middle of this work—forty years of mathematics—and I couldn't help but feel that both quality and quantity were on the wane. I am not ashamed—just rueful.

I kept in touch with the research of people in other parts of the world, and was stimulated by new ideas and got new energy, through quite a bit of travelling, lecturing, conferencing, spreading the gospel. Oberwolfach, Poços de Caldas (Brazil), Santiago, St. Andrews, Teheran, Perth (in Western Australia), and Budapest—almost every year in the late 1960's and through the 1970's I was some place I found glamorous, and I was not tempted to rest on my laurels.

To talk mathematics is the same whether you do it in southern Germany or central Scotland, and the mathematical jet set you meet isn't large, especially if you restrict your conference hopping, as most of us do, to one or two subjects. You keep bumping into the same people everywhere, talking about the same theorems. I'd meet Ron Douglas, my former Michigan colleague, in Oberwolfach, and Irving Kaplansky, my former Chicago colleague, in St. Andrews. The language in Iran is strange, the bazaars are noisy and colorful, the mosques are beautiful, and nowhere else are the invited speakers likely to be presented with a gaudy frame containing a gaudy color photograph of the Shah, but normal operators seem to have the same properties in Persian as in Portuguese. Polite children would speak to me in Teheran; what they said sounded like "so long", which puzzled me, but I answered with a friendly "hello" and a smile—and it wasn't till much later that I woke up and realized that they were pronouncing the Persian version of "salaam". As far as the Shah is concerned, he took his place behind a plastic sheet in an album and my wife replaced him in the frame by me.

Two of the places that I visited became especially dear to me, and I'd like to tell something about them; they are Oberwolfach and St. Andrews.

Oberwolfach is famous among mathematicians, and deservedly so. The Mathematical Research Institute sits on top of a hill, not very high but steep enough that at the end of the five minutes it takes to walk up you'll be puffing. To get there you fly to Frankfurt and then take a train, or, if you're feeling suicidal, accept a ride on the Autobahn, and go to Walke. It's in mountain country, full of tall, very tall trees, in the Black Forest, the lower corner of Germany, near France and Switzerland. The Institute sponsors one-week conferences, usually 50 of them each year, on various parts of mathematics. The first time I went there was in 1968 to attend a conference about "Abstract spaces and approximation". It was a success and it was repeated once every three years ever since then. I couldn't make it every time, but I did attend three different meetings with personal pleasure and mathematical profit. The organizers of the sequence were Paul Butzer and Béla Sz.-Nagy.

The 1980 session was more or less typical and I remember it with fondness. I arrived Saturday afternoon in the early part of August; the sessions started at 10 o'clock Sunday morning. Yes, Sunday morning; Butzer is a taskmaster. I was the first speaker—which was good (get it over with) and bad (the jet lag was not yet over with).

The Institute has a conference building, a guest building, and a couple of smaller buildings called bungalows. The conference building has a lecture room that can seat between 50 and 60 comfortably; more than that makes it undesirably crowded. That's on purpose; the idea is to keep the conferences small enough so that people can actually meet and talk. The conference building also has a lounge, a small but surprisingly effective mathematics library, a billiard table, a ping-pong table, easily available coffee, and a wine cellar; it is 20 steps from the guest building and the bungalows. The library has one of the most extensive collections of photographs of mathematicians; Konrad Jacobs, of Erlangen, a great picture taker and a superbly organized collector, established it and keeps an eye on it.

The sleeping rooms are clean, small, monastically simple, and quite adequate. There are a few double rooms, but most often people come without spouses and stay in singles. The furniture in a typical cell consists of a hard narrow cot, a couple of chairs, a table with a study lamp on it, and a flat bench big enough to put a large open suitcase on.

Meals are served in the ground-floor dining room of the guest building. They are not culinary thrills, but they too are adequate. Breakfast is bread and jam and ham; the main meal comes, European style, at about 12:30 p.m., followed by a substantial tea (cakes and tortes) around 3:30, and a light supper at 6:30. One ingenious feature of Oberwolfach meals is the system of stochastic napkins. Each guest is issued a cloth napkin bearing his name. When the tables, seating six, are set, the napkins are shuffled and placed by the settings at random. When you are ready to sit down you must wander around the room for a minute or so looking for your napkin, and the result is that you never know which five people you are going to share a table with. The purpose is to discourage the formation of cliques and encourage mixing; it works very well.

Next to the dining room is a help-yourself bar. Beer and wine and cognac and a few mixes are available, and you're on the honor system—put the money in the cigar box, or sign for what you consume and pay later. The bar is popular before supper, and, with many apparently infinitely energetic young people, long after supper too. Conversations that go on till 2:00 a.m. are rather common; the ones that go on till 6:00 a.m. are rare, but not unheard of.

Our meeting took six hard days, with a half day off for an excursion. There were eight scheduled lectures on each of the full days, and during meals and on walks and over drinks the unscheduled exchanges continued with full force. At Oberwolfach you work hard, and you have lots of chances to play hard. There are mountain walks, there is a chess board painted on the floor with one foot squares and chessmen about two feet tall, and decks of cards are available. The excursion gives you a choice: you can visit a cathedral, or a casino, or a French restaurant that Michelin gives three stars to.

In case the amenities miss one of your favorites, you are given a chance to express yourself in the Wunsch Buch, also known as the Kvetch Book. It's fun to leaf through it and see what's missing. People want drier wines, a pencil sharpener, better toilet paper, a Swedish actuarial journal, an electric train to bring them up the hill, T-shirts advertising the Institute, and the piano tuned.

The 1980 crew had representatives from many parts of the world: Iliev from Bulgaria, Zaanen from Holland, Szabados from Hungary, Lumer from Belgium, Ando from Japan, Milman from Israel, and, of course, large contingents from Germany and from the U.S. The one that was of special interest to me was Milman—yes, *the* Milman, David Milman, of Krein–Milman fame. He was in his late sixties then, and he loved to reminisce. He had abandoned his home in Odessa, his language, and his culture, and moved to Israel. There he learned Hebrew, and he learned English. He learned, to be more precise, a sort of one-third English—he knew about the first syllable or two of about one third of the words he needed—but he made out in it very well. He

D. Milman and P. R. Halmos, 1980

was completely bald, with light blue eyes, and many partly filled and partly false teeth made of an odd collection of assorted metals. He had an ebullient and instantly likable personality. He called most of his Russian ex-colleagues crazy. Gelfand, he assured me, is crazy—great mathematician but crazy; Kolmogorov is crazy—great mathematician but crazy; Shilov is strange, Raikov is strange. Krein?—no, Krein is not strange. (I did not voice my disagreement.)

The official language of the Oberwolfach sessions is usually English —it's hard to find another language in which a Japanese and a Bulgarian can communicate. German is used, of course, but surprisingly little. I tried to re-learn as much German as I could, and I was grateful to Berens. He had spent a considerable amount of time in the U.S., and, in particular in Santa Barbara, and he could speak fluent, correct, colloquial, idiomatic American. He elected, nevertheless, to give his lecture in German—he told me that he wasn't willing to relinquish 100 percent linguistic control, and I sympathized with and applauded his decision. He made an effort to enunciate even more clearly than usual, and I found that the very small extra effort that I had to make to listen in German made it easier, not harder, to understand the mathematics.

In a conversation about languages I learned that "Zorn" is the German word for "wrath" or "anger". The Roman historian Tacitus said once that he remembered an event "sine ira et studio"—roughly "without anger and bias". A possible German translation of his claim is "Blick zurück ohne Zorn", which comes to "look back without anger", and that phrase has come to be used by literate German mathematicians sometimes when the axiom of choice—a logical equivalent of Zorn's lemma—turns out not to be needed in a proof.

In those six and a half Oberwolfach days I needed every language that I knew more than two words in—English, French (even French!), German, Hungarian, Russian, and Spanish—and I could have used several others—Greek, Hebrew, and Japanese, for instance—if I had known them. Basic English would have been enough by itself, just to get along, but every extra little bit helps.

Thursday night there was a problem session, followed by a wine party —the last official social function of the week. Toasts, speeches of thanks, and speeches of farewell. Friday was still a full working day, but things were obviously winding down; most people leave early Saturday morning to catch trains and planes, and late Friday sociability wouldn't be popular. The final goodbyes on Saturday were said amid chairs piled up on dining room tables, and pails and mops everywhere—the next bunch were expected that afternoon.

The St. Andrews colloquium is a completely different sort of thing. It's not so famous, not so international, and not so sharply focused on one subject; it's somewhat larger, it lasts longer, and its purpose is more to spread the word than to enlarge humanity's knowledge.

St. Andrews is a small town (by now its population must be about 12000) 35 miles northeast of Edinburgh, looking out on the North Sea. Its university is the oldest in Scotland and the third oldest in Britain; it began in the early 1400's.

I always liked Scotland. I like its people and the way they talk; I like the countryside, the views, the beer, the sheep, and even the climate. To be sure it's dark from December till March, and it rains a lot—but the spring is breathtakingly beautiful, and the rain is of the soft, gentle kind that hardly dampens your spirit or your clothes. When I was there in 1972, the hottest day of the summer came in mid July; the thermometer reached 73°.

Once every four years—in leap years to be exact—the Edinburgh Mathematical Society holds a colloquium in St. Andrews. An official letter from Arthur Erdélyi reached me in November 1970, on behalf of the Planning Committee for the 1972 colloquium: would I deliver a course of six or seven lectures? Honor, yes; honorarium, no; travel expenses—we'll try our best to arrange something with the Science Research Council; but we'll definitely provide room and board for the ten days involved. Would I? Try and stop me! I accepted by return mail, and, in due course, arrived, lectured, and enjoyed every minute of it.

I made some new friends at that colloquium and met several old ones. Tom Blyth was the junior faculty member at St. Andrews who was saddled with the chore of local arrangements, and he discharged his duties very well. He had the three principal speakers to contend with (F. Harary, S. Vajda, and me), and about 150 others. Harary talked about graph theory, Vajda about mathematical programming, and I about the connection between linear algebra and operator theory—what else? John Howie, the chairman of the mathematics department at St. Andrews, was pleasantly and efficiently hospitable, and Paul Cohn and Jim Eells were in charge of the afternoon seminars (on algebra and global analysis, respectively) that, after the morning lectures, filled out the day.

The people I spent the most time with were Frank and Nora Smithies. Frank could be called the father (grandfather?) of functional analysis in Britain. He spent the greater part of his life at Cambridge, and his students and his students' students grew into many influential positions in the country. He is a wonderful man and always a pleasure to be with. He knows poetry and history and languages and, of course, mathematics; he has insight and wisdom and wit; he talks softly and precisely; he likes a glass or two of sherry before his dinner, and he smokes all the time.

Obviously not all of us at the colloquium were Britons, but most of the members were, and, in fact, as I remember, most of them either were Scots or lived in Scotland. There were many research-level conversations, but the main purpose of the colloquium seemed to be expository:

F. Smithies, 1968

open a window to let the light in. Most of the members were hard working teachers, who didn't have too much opportunity to see one another during the academic year, and didn't have enough time to stay in touch with the development of mathematics. The St. Andrews colloquium put them in contact with each other and with some of the rest of the world—and did it in an attractive unpressured way. They brought their golf clubs, or their violins, or both, and they made full use of both the professional and the recreational facilities.

The daily schedule was conspicuously different from the one in Oberwolfach. The three lectures (after a large breakfast) were at 9:15, 10:30, and (after a half hour coffee break) 12. After lunch you could nap, or work, or play golf till tea at 4. The seminars ran from 4:30 to about 6, and they were arranged (by their conductors) impromptu: "Would you speak at the algebra seminar tomorrow?".

As for recreation: of course chess and cards; and, very much of course, golf; tennis and squash courts and ping-pong tables were easily accessible. In the evening (after the dinner that started at 7) nothing professional was scheduled. You could chat with friends, or go to a movie—or you could participate in the scheduled entertainment. One evening was music (self made), on one a block of tickets to the theater in town was made available; there was a bridge evening, an evening with Scottish country dancing, and another with an organ recital. As at

Oberwolfach, one long half day was saved for an organized excursion, and, unlike at Oberwolfach, Sunday was completely free.

The St. Andrews colloquium is a serious mathematical gathering—but, at the same time, it is a holiday. It is a mathematical holiday. I loved it the first time, returned to every session since then, and am planning to keep doing so for as many more leap years as I can.

How to write mathematics

I didn't plan it that way, but in the 1960's, especially toward the end of the decade, and in the 1970's, much of my effort went to the writing of surveys and expository papers. There is a difference, isn't there? A survey is more likely to be on the research level, an organized presentation telling the expert or near expert about things he might have overlooked. An exposition expects no expertise from the reader, and very little effort; it is intended to show why one might want to, and how one might begin to, undertake a serious study. In other words, exposition is intended to attract and describe more than to explain and instruct. My early (1944) paper on the foundations of probability is expository; the CBMS lectures (1970) on ten problems in Hilbert space constitute a survey.

In 1956, at the height of my love affair with polyadic algebras, I published an exposition titled "The basic concepts of algebraic logic" because the spirit moved me to do so; I was bent on being a missionary. In 1963, back home in operator theory, I published a survey titled "A glimpse into Hilbert space" because Tom Saaty (with the Office of Naval Research at the time) commissioned a sequence of talks that were to be written up and put between hard covers as a sort of view from the top for those down below who were ready for a stiff climb. He definitely wanted a survey, on the research level, and he was upset when, early in our correspondence, I told him the title of my piece. A "glimpse" sounded too easy to him, too soft, too low; I suspect that he suspected me of trying to circumvent the difficulties of taking the high road.

Surveys are hard to write, but good expositions, the low road, are harder still—the lower the harder. The watershed that really established me as a semipro exposition-and-survey producer was the 1967 invitation from my alma mater. The University of Illinois was approaching its hundredth birthday, and I was invited, as part of the celebration of the centennial year, to address a "general" audience. What could I possibly talk about? I don't know anything except mathematics; what could I possibly tell a "general" audience about mathematics? My answer to that question then was different from what it would be now. Now I want the audience to be active, to challenge them, to make them think about the problems I "assign". All I could think of then was a flowery sales talk, or, perhaps a better metaphor, a sermon. I didn't quite threaten laymen (which meant everybody who wasn't a mathematician) with brimstone and hellfire, but I tried to paint the mathematical heaven in glorious colors, described mathematics as an art, a creative art, and all but promised eternal salvation to those who saw the light. The talk went over pretty well—I still get an occasional request for a reprint— and with that heady success behind me I found it hard in the years that followed to say no. Does the Society want to produce a pamphlet from which mathematicians can learn how to write intelligibly? Does the Encyclopaedia Britannica want an article about von Neumann? Does the Association want a report on the development of mathematics in America since World War II? Sure—just ask me, and I'll go to work.

Norman Steenrod was an admirable man. He was an excellent topologist, a likable warm human being, and a man of the highest professional and ethical standards. Many of us complain about how badly written most books and papers are; Norman wanted to do something about it. At his behest the AMS appointed a committee (with him as chairman, of course) "for the purpose of preparing a pamphlet on expository writing of books and papers at the research level and at the

N. E. Steenrod, 1938

level of graduate texts". In addition to Steenrod, the committee consisted of Dieudonné, Menahem Schiffer, and me.

I was appointed in 1969. A little more than a year later, in June 1970, I wrote to Ambrose as follows.

> Steenrod is a man of good will, with cast iron principles, as elastic as poured concrete, and with a style as elegant as a porcupine. Dieudonné despite being, ex officio, an arrogant Frenchman, is a nice guy whose ideas on exposition I do not deplore but merely disagree with. At the same time he is some kind of dean, he writes two and a half books a week, and he is chief flagbearer for the forthcoming [Nice] Congress. As a result he is way too busy to fuss with writing letters to a stupid committee. Schiffer was probably put on the committee because he knows something called applied mathematics, but I don't think he knows English. As for me, I am an arrogant know-it-all who once looked up the meaning of "cooperation" in a dictionary and then threw the dictionary away. After a couple of months of membership on the committee I decided the whole thing was hopeless, I resigned, and I set about the task of writing what the committee was supposed to write. I did. It's a piece called HOW TO WRITE MATHEMATICS, except, as I point out in the preface, it's really more like a diary and "TO" should be changed to "I". Anyway it's done. Lots of people (maybe a dozen) read it and criticized it, and after reading all their suggestions I followed none. Reason: they told me to do both x and $-x$, as well as both y and the orthogonal complement of y, at the same time with z and the negation of z. So I said, the hell with it, and decided to publish it as it was.

That's a rather brutal summary, and it doesn't even summarize all the facts. Steenrod put a lot of his heart into the project, and he had a terrible committee to work with. My "resignation" didn't bother him much; he just took it in his stride, and, in effect, refused to accept it. To be sure, I made it easy for him to refuse; my letter contained the following plan.

> As a private citizen I'll continue working on my exposition of exposition. If, when I finish, the committee too has finished, and especially if the result of the committee's work is a set of disjoint essays, and if thereupon my essay is accepted as an addition to that work, then I'd be glad if it could be published all together. If, however, our timing is very different, and if your output is not a collection of personal essays but something more formal and united, like a handbook, or impersonal and hygienic, like the AMS Manual of Style, then I'll submit mine for separate publication.

Originally Steenrod had hoped that we would produce the bible, or at the very least the Strunk and White of mathematical writing. He wanted a unified book that tells it all, not four separate essays; he was sad that his committee just wouldn't do things his way. Late in 1970 he wrote me.

My last letter to the Committee on Expository Writing (November 1969) asked for a round of criticisms of our initial essays. You answered promptly that my request was useless. You were certainly correct in this evaluation; neither Dieudonné nor Schiffer replied. This was most disappointing to me; I had put considerable effort into the composing of the letter; I flattered myself that it contained just the right amount of sharp criticism to stimulate action, but not enough to offend. It fell flat... I've given up hope that our committee will produce anything more than some independent essays.

Steenrod died in 1971. His own essay was nearly complete; all that remained was to "finish up the last few pages, and clean up the phrasing, grammar, etc." The quotation is from the letter Carolyn (his widow) sent me. Norman asked her to do so; she said that the last work he did in the hospital was to dictate it to her.

The phrasing and the grammar were just fine; the changes I had to make were only of the trivial, editorial kind needed at the seam to ensure a smooth and consistent flow. I received Schiffer's essay, and adapted a letter Dieudonné wrote to Steenrod to make it look like an article. My own piece had already appeared (with everyone's agreement) in L'Enseignement Mathématique in 1970. I added it to the others, wrote a report to the Council of the AMS to serve as a preface, and the job was done. The pamphlet containing it all was published in 1978 under the title "How to write mathematics".

How to write about von Neumann

The Britannica project started while the Steenrod project was still going on, and it sounded simple: would I write a piece, 1000 words, about von Neumann? Sure, why not? The only problem was that I had no feeling for how much 1000 words was.

I went about the task as I usually do. The first step was to sit down and start writing in a thoroughly helter-skelter disorganized fashion. On one of my favorite $8\frac{1}{2} \times 11$ yellow sheets I'd write a sequence of disconnected words intended to remind me of topics to be covered; on the next a list of possible documents to consult (obituaries, private correspondence, earlier survey articles); and on the next a possible outline of the whole article. A couple of hours of work produced a couple dozen of such sheets; the work was now off the ground.

The next step is the literature search and reading. I'd consult the documents I already thought of, take notes, and look for other sources that they referred me to. I'd chase footnotes and MR references in the library (scribbling rough notes every step of the way), and I'd look for possible models to follow among earlier EB biographies. This part of the work took a couple of months; it had to be squeezed into the normal daily workload. As it was going on, I'd keep mentioning it at lunch and at tea to everyone I bumped into, and I'd avidly scribble the suggestions I received ("are you going to tell that story about his driving?") on the back of the lunch receipt and the tea napkin.

The most difficult technical problem of written communication (whether the intended result is a long novel, a short biography, a research paper, or a recipe for cherry pie) is that of linear order. The way we usually receive information about the universe is multi-dimensional. We learn about something (or somebody) through signals that come to us simultaneously through our many senses. Our sense of balance tells us one thing, and the way our muscles stretch another; we see it, hear it, feel it, smell it, and taste it; we find it warm or cold, wet or dry. A lecturer uses words, but, at the same time, he controls how fast he speaks and how loudly; his facial expression, his gestures, and his tone of voice are all part of the show. The most highly distilled form of verbal communication is writing. The only raw material a writer has is his vocabulary, and the presentation of his words in a total order is the only way he has to produce an effect.

I would face the ordering problem by playing a complicated game of solitaire. The first step was to transcribe all the material I had accumulated—yellow sheets, books, reprints, napkins—to 3×5 slips, in the form of short cue phrases of five or six words each. The number of such slips is, of course, directly proportional to the amount of material to be covered; typically it is somewhere between 25 and 100. (I am talking about the problem of ordering the material on a few printed pages at a time. The problem of globally ordering the many sections that make up a book is solved similarly.) With a stack of slips in my hand, sitting at a cleared desk, I'd start putting them down in perhaps five or six piles. I'd have no fixed idea of what the piles meant—I'd let the subject suggest its own organization. For the von Neumann piece there were, as I recall, five piles corresponding approximately to early life, career, mathematics, war, and death. The process is not automatic. Some slips just don't seem to belong anywhere (drastic cure: throw them away); different choices are equally defensible and (chronic worry) quite possibly much better; and the problems of ordering the slips within the piles and ordering the piles themselves still remain to be faced.

Once the slips are in order, the writing itself can begin. The order might change—new local organizational insights can arrive suddenly in the middle of a sentence—but the changes are not likely to be great, and the actual writing of the first draft is usually quite smooth for me. The American Heritage Dictionary and Roget's Thesaurus are on my desk, open, and I consult them often; except for that, however, I write straight through to the beginning. Yes, that's what I said: the beginning. Not only the work as a whole, but each part of it (for example, each section in this book) should have a proper beginning and a proper end, and, the way I do it, they come in the other order. I think hard about the coda as I go, and try to end on a snappy upbeat, but I think even harder about the overture. Once the section (or the whole paper, or the whole book) is finished, I read it over, and then write (re-write) the beginning. For a short section the overture might be just one paragraph, or even just one sentence, but it's an important one, and, to get it right, I have to have a feeling for the total structure that it introduces.

The dictionary and thesaurus interruptions are usually not about meaning in the gross sense (what's the correct use of "oppugn"?), but about precision, about finding just the right word. When I describe Lefschetz's handwriting (p. 86 above), should I call it "wiggly" or "wriggly"? What did the examples that von Neumann and I constructed do to the conjugacy conjecture for shifts (p. 236 above): did they contradict, contravene, gainsay, dispute, disaffirm, disallow, abnegate, or repudiate it? (Roget yielded a list of 25 candidates for the word I wanted.) Writing can stop for 10 or 15 minutes while I search and weigh.

Spelling doesn't worry me except for "double" words. Is "letdown" one word or two, or should it be hyphenated? What about "high school" and "grownup" and "book review" and "by-pass" and "payoff"? The problem is not important or deep, but inconsistency is bad. I don't mind going against the dictionary if I have a good reason for doing so, but in the absence of one I ask the dictionary to make my mind up for me.

Back to the von Neumann piece: it got written and counted and revised and re-counted; the total came to about 7000 words. I wasn't surprised or disconcerted; I was expecting that. Cut, compress, trim, shorten, and just brutally discard: I found it easy to convert what I had into an informative and digestible piece of exactly 998 words.

Now I had two articles, a long one and a short one, and I was loath to throw either one away. Solution (completely aboveboard, with the agreement of all the parties concerned): I submitted the long one to the Monthly and the short one to EB. That's the end of the story, and it's almost a happy ending. The Monthly printed the paper I submitted, and I still rather like it. The Encyclopaedia Britannica, however, "improved" my submission by some grammatically illiterate and mathematically ignorant editing. The improvement looked, to me, mass-produced, mechanical, flat, and dull.

Consider, for instance, my first paragraph:

> John von Neumann was a brilliant mathematician who made important contributions to quantum physics, to logic, to meteorology, to war, to the theory and applications of high-speed computing machines, and, via the mathematical theory of games of strategy, to economics.

The EB version looks like this.

> A mathematician, John von Neumann made important contributions to the sciences of quantum physics, logic, and meteorology that had practical applications in national defense and in the development of high-speed computing machines. His mathematical theory of games of strategy had a significant impact on economics.

Do you see the differences? I said "war"; they (those ubiquitous malevolent they) said "national defense", and, what's more, they said that the contributions to quantum physics, logic, and meteorology had practical applications in it. They do, do they? I said "theory and applications"; they said "development". Is that truer? Easier to understand? As for "significant", I am sorry that lots of people nowadays think that the word is a synonym of "important" or "valuable", but do I have to use it that way? I think it means "meaningful", and when I am told that someone made a significant speech or painted a significant picture, I want to ask what the speech or the picture signify. And as for "impact"—well, really!

I said: "The basic insight was that the geometry of the vectors in infinite-dimensional Euclidean space has the same formal properties as the structure of the states of a quantum-mechanical system." They said: "He had a basic insight concerning vectors: the geometry of vectors in infinite-dimensional Euclidean space has the same mathematical characteristics as the structure of the states of a quantum-mechanical system."

Speaking of von Neumann's papers on "rings of operators", I said: "This is quite probably the work for which he will be remembered the longest." They said: "Of all his work, quite probably these concepts will be remembered the longest." I said: "A surprising outgrowth of rings of operators was 'continuous geometry'"; they replaced "surprising" by "important". They also added a sentence: "Hitherto, mathematicians had represented a given space only in terms of whole numbers." Conclusion: people who knew nothing of how the language of a subject is used shouldn't write about it.

My last sentence was an epitaph: "His insights were illuminating and his statements were precise." They omitted the second "were". An improvement?

The EB piece is not shorter than mine, not clearer, not better; they changed (I believe they ruined) my clarity, my poetry. Acrimonious correspondence ensued. I offered to withdraw my 1000 words and refund the payment; the compromise was that they used their version after all, but, at my insistence, removed my name from it. They got their piece, and I saved my honor; I thought their taste was egregiously bad, and they thought I was egregiously fussy.

How to write history?

The 1976 winter meeting in San Antonio celebrated the 200th birthday of the United States, and, in that connection, the Association arranged a sequence of historical talks intended to cover the growth of American mathematics. Lenny Gillman was chairman of the program committee, and he invited me to prepare the most recent part, the last 35 years or so. I didn't say yes immediately, but I didn't dilly-dally for long. The way Gillman tells the story, I asked for a month to think it over, then phoned three days later to accept.

I knew I couldn't do the job alone; I spent the three days asking a group of colleagues whether they would collaborate with me. When five said yes, I swallowed my insecurity and we all went to work. The five were John Ewing, Bill Gustafson, Suresh Moolgavkar, Bill Wheeler, and Bill Ziemer; we hoped that between us we knew enough mathematics to say something about everything that might have happened in the preceding 35 years.

What should we say and how should we say it? We got together, sometimes all of us, but more often in subgroups, and we discussed the possibilities. Should we talk about the hard facts, such as the spectacular growth of Mathematical Reviews? Should we describe the lives of the mathematicians who made the U.S. a leading world power in the subject? Should we concentrate on bibliography and prepare a list of papers and books from which anyone who wanted to could learn the whole truth? These were some of our choices. The first paragraph of our report describes them, and others, and then goes on.

> We decided to do none of these things, but, instead, to tell as much as possible about mathematics, the live mathematics of today. To do so within prescribed boundaries of time and space, we present the subject in the traditional "battles and kings" style of history. We try to describe some major victories of American mathematics since 1940, and mention the names of the winners, with, we hope, enough explanation (but just) to show who the enemy was. The descriptions usually get as far as statements only. We omit all proofs, but we sometimes give a brief sketch of how a proof might go. A sketch can be one sentence, or two or three paragraphs; its purpose is more to illuminate than to convince.

"Battles and kings" is all very well, but which battles and which kings?, and, incidentally, what should we call our work—what title can we put at the head of our list? The last question was the easiest to answer. After some discussion we had a short list of three titles, and the one that won the popularity poll was "American mathematics from 1940 to the day before yesterday". After considerably more discussion, I drew up a list of 32 possible topics that our final list would be chosen from, each accompanied by some of the names associated with it. Examples:

Invariant subspaces (Aronszajn, Bernstein, Beurling, Lomonosov, Robinson, Smith);
Model theory (Ax, Kochen, Robinson, Tarski);
Fourier series (Carleson);
Superposition of functions (Kolmogorov);
Prime number theorem (Erdös, Selberg).

We voted, we assigned numerical weights, and we argued; we ended up with ten topics. They were: Continuum hypothesis, diophantine equations, simple groups, resolution of singularities, Weil conjectures, Lie groups, Poincaré conjecture, exotic spheres, differential equations, and index theorem. Each topic was assigned to one of us, who was to prepare a first draft. The drafts were circulated among all of us, criticized, and revised, and ultimately they all ended up on my desk. I was the chief scribe. I prepared a draft of the entire paper, and that was the hardest mathematical writing I have ever done or ever hope to do.

I knew something about the continuum hypothesis and Lie groups; I could understand most of the language and the ideas of several other topics; and I was totally and abysmally ignorant of the rest. The idea was to explain recent progress to a large audience—the members of the Association, the readers of the Monthly—many of whom were probably as ignorant as I. I dared my colleagues to explain the topics to me and make me understand them well enough to transmit what I learned to my fellow ignoramuses. It worked—somehow it worked. The Atiyah-Singer index theorem was the toughest hurdle for me, but, somehow, we conquered it too. (To be sure, after it appeared in print, Singer told me that it didn't come out quite right—the relation with the Riemann-Roch theorem was unclear or perhaps even misstated—but there it was, and I feel sure that my fellow ignoramuses and I learned something worth knowing that we hadn't known before.) My draft was circulated, criticized, and revised—and the next thing we knew the process had converged to its limit.

I selected a subset of our ten topics small enough for an hour's lecture, and, at the San Antonio meeting, I presented the subset to an audience of allegedly 1776 people. Then I heaved several sighs of relief and went out to consume several beers—a little too soon. What happened was this. The manuscript, the "final" version on which my talk was based, contained the following two sentences about the general Poincaré conjecture for S^n. "The proof was obtained by Stallings for $n \geq 7$ (1960) and Zeeman for $n = 5$ and $n = 6$ (1961). At the same time Smale (1961), using completely different techniques, gave a proof for all $n \geq 5$." The reason my sighs of relief were premature is the very strong (aggressive?, belligerent?, cantankerous?, grouchy?, pugnacious?)

letter I received from Steve Smale a few days later. It referred to the "grossly distorted picture" that my talk gave and to the "harm" I had done him.

I looked into the matter, and I found that, indeed, my historical emphasis was in error; the two offending sentences were replaced by these two. "The proof was obtained by Smale (1960). Shortly thereafter, having heard of Smale's success, Stallings gave another proof for $n \geq 7$ (1960) and Zeeman extended it to $n = 5$ and $n = 6$ (1961)." The mistake was mine, I made the change, and I wrote Smale and told him so, but I couldn't keep my letter from sounding plaintive. If he had written a kinder letter, I said, I would have made the same change, and felt a lot better about it. And then, finally, it was time for a justified sigh of relief.

We, the six collaborators, submitted our paper to the Monthly, and, almost immediately, the answer came back from Alex Rosenberg, the editor.

> This will acknowledge receipt of the paper by Ewing et al.... I have read the paper and really like it very much and hereby accept it for the Monthly.

The next day I wrote Alex.

> You *made* six days (one each for Ewing, Gustafson, Halmos, Moolgavkar, Wheeler, and Ziemer), and an especially beautiful one for John Ewing who enjoyed having the authors referred to as 'Ewing et al.'.

Coda

How to be a mathematician

It takes a long time to learn to live—by the time you learn your time is gone. I spent most of a lifetime trying to be a mathematician—and what did I learn? What does it take to be one? I think I know the answer: you have to be born right, you must continually strive to become perfect, you must love mathematics more than anything else, you must work at it hard and without stop, and you must never give up.

Born right? Yes. To be a scholar of mathematics you must be born with talent, insight, concentration, taste, luck, drive, and the ability to visualize and guess. For teaching you must in addition understand what kinds of obstacles learners are likely to place before themselves, and you must have sympathy for your audience, dedicated selflessness, verbal ability, clear style, and expository skill. To be able, finally, to pull your weight in the profession with the essential clerical and administrative jobs, you must be responsible, conscientious, careful, and organized—it helps if you also have some qualities of leadership and charisma.

You can't be perfect, but if you don't try, you won't be good enough.

To be a mathematician you must love mathematics more than family, religion, money, comfort, pleasure, glory. I do not mean that you must love it to the exclusion of family, religion, and the rest, and I do not mean that if you do love it, you'll never have any doubts, you'll never be discouraged, you'll never be ready to chuck it all and take up gardening instead. Doubts and discouragements are part of life. Great mathematicians have doubts and get discouraged, but usually they can't stop doing mathematics anyway, and, when they do, they miss it very deeply.

400

"Mathematician" is, to be sure, an undefined term, and it is possible that some people called mathematicians nowadays (or ever) do not (or did not) love mathematics all that much. A spouse unsympathetic to mathematics demands equal time, a guilty parental conscience causes you to play catch with your boy Saturday afternoon instead of beating your head against the brick wall of that elusive problem—family, and religion, and money, comfort, pleasure, glory, and other calls of life, deep or trivial, exist for all of us to varying degrees, and I am not saying that mathematicians always ignore all of them. I am not saying that the love of mathematics is more important than the love of other things. What I am saying is that to the extent that one's loves can be ordered, the greatest love of a mathematician (the way I would like to use the term) is mathematics. I have known many mathematicians, great and small, and I feel sure that what I am saying is true about them. To mention some famous names: I'd be very surprised if Marston Morse, and André Weil, and Hermann Weyl, and Oscar Zariski didn't agree with me.

Mind you, I am not recommending or insisting that you love mathematics. I am not issuing an order: "If you want to be a mathematician, start loving mathematics forthwith"—that would be absurd. What I am saying is that the love of mathematics is a hypothesis without which the conclusion doesn't follow. If you want to be a mathematician, look into your soul and ask yourself how much you want to be one. If the wish isn't very deep and very great, if it is not, in fact, maximal, if you have another desire that takes precedence, or even more than one, then you should not try to be a mathematician. The "should" is not a moral one; it is a pragmatic one. I think that you would probably not succeed in your attempt, and, in any event, you would probably feel frustrated and unhappy.

As for working hard, I got my first hint of what that means when Carmichael told me how long it took him to prepare a fifty-minute invited address. Fifty hours, he said; an hour of work for each minute of the final presentation. When, many years later, six of us wrote our "history" paper ("American mathematics from 1940..."), I calculated that my share of the work took about 150 hours; I shudder to think how many man-hours the whole group put in. A few of my hours went toward preparing the lecture (as opposed to the paper). I talked it, the whole thing, out loud, and then, I talked it again, the whole thing, into a dictaphone. Then I listened to it, from beginning to end, six times—three times for spots that needed polishing (and which I polished before the next time), and three more times to get the timing right (and, in particular, to get a feel for the timing of each part). Once all that was behind me, and I had prepared the transparencies, I talked the whole thing through one final rehearsal time (by myself—no audience). That's work.

Archimedes taught us that a small quantity added to itself often enough becomes a large quantity (or, in proverbial terms, that every little bit helps). When it comes to accomplishing the bulk of the world's work, and, in particular, the work of a mathematician, whether it is proving a theorem, writing a book, teaching a course, chairing a department, or editing a journal, I claim credit for the formulation of the converse: Archimedes's way is the only way to get something done. Do a small bit, steadily every day, with no exception, with no holiday. As an example, I mention the first edition of my *Hilbert Space Problem Book*, which had 199 problems. I wrote most of the first draft during my Miami year, and I forced myself, compulsively, to write a problem a day. That doesn't mean that it took 199 days to write the whole book —the total came to about three times that many.

As for "never give up", that doesn't need explanation, and I have been trying to illustrate it with anecdotal evidence all along, but here, just for fun, is a pertinent little story. Somewhere around 1980 I was invited to give a talk to a "general" audience, and, having given it, I wrote it up and submitted it for publication in The Mathematics Teacher. In due course I received reports from a couple of referees; they read, in part, as follows. "The author apparently feels she/he is illustrating the thrill and power of abstraction. While his/her examples contain the potential for doing that, I don't feel that the presentation creates that effect.... Much of the problem with the paper is that of a meandering style. The progression of ideas is not clear.... The mathematical topics focused on are of only moderate interest." The paper was firmly rejected. I didn't give up—I just shrugged my shoulders, and submitted the same article, word for word, to what was then called The Two-Year College Mathematics Journal. It was accepted and printed, and, a year after that, it received the Pólya award from the Association.

All these prescriptions and descriptions about how to be a mathematician arose, inevitably, from my own attempts to become one. Nobody can tell you what mathematicians should do, and I am not completely sure I know what in fact they do—all I can really say is what I did.

How close did I come? What is my total mathematical contribution? The first answers that occur to me are a small but pretty proof (the monotone class theorem), a few reasonable theorems (mainly in my ergodic papers "Approximation theories..." and "In general a measure-preserving theorem is mixing"), and a good idea in logic (polyadic algebras).

One thing that I am pretty good at is asking questions. Given a mathematical problem that I understand the statement of, one whose history I know something about and that I spent some time studying, one for which I am moderately up to date with the standard techniques —given those circumstances, I have some talent for identifying and formulating the central questions. If I look hard at a question for a month, try to answer it, and fail, then I am confident that it is not trivial. I am confident that a question like that is fit for mathematicians much better than I; when one of them solves it, he is bound to feel at least a small tingle of pride and pleasure. (Examples: the power inequality, and the Weyl–von Neumann theorem for normal operators.)

Related to the questions I have asked are the concepts I discovered and introduced, notably subnormal and quasitriangular operators, and, possibly, capacity in Banach algebras. Important theories have grown out of such concepts; I think it's fair to call them contributions.

I wrote a few good surveys and some pretty good books. Perhaps the best are *Finite-Dimensional Vector Spaces* and *A Hilbert Space Problem Book*—but then my vote on such matters is probably the one that weighs the least.

My most nearly immortal contributions are an abbreviation and a typographical symbol. I invented "iff", for "if and only if"—but I could never believe that I was really its first inventor. I am quite prepared to believe that it existed before me, but I don't *know* that it did, and my invention (re-invention?) of it is what spread it through the mathematical world. The symbol is definitely not my invention—it appeared in popular magazines (not mathematical ones) before I adopted it, but, once again, I seem to have introduced it into mathematics. It is the symbol that sometimes looks like ∎, and is used to indicate an end, usually the end of a proof. It is most frequently called the "tombstone", but at least one generous author referred to it as the "halmos".

That's it; that's a life, that's a career. I was, in I think decreasing order of quality, a writer, an editor, a teacher, and a research mathematician.

What's next? The writing of this book took a lot out of me. It took a year and a half of time and energy, during which I paid no attention to research. It was a deliberate gamble. I wanted to write this book, I wasn't at all sure that I could do it the way I dreamt, I wasn't sure I could tell the audience I had in mind what I wanted to say. If it turns out that I succeeded, I'll be happy; if not, I'll be sad. But, in either case, I'm not ready to crawl into a hole and pull it in after me. I'd like to write some more mathematics, to teach some more, and, who knows, even to prove a theorem. I'm going to try, that's for sure. I thought, I taught, I wrote, and I talked mathematics for fifty years, and I am glad I did. I wanted to be a mathematician. I still do.

Index of Photographs

Index